T0332158

Radio-Frequency Electronics

Circuits and Applications

This second, much updated edition of the best-selling *Radio-Frequency Electronics* introduces the basic concepts and key circuits of radio-frequency systems. It covers the fundamental principles applying to all radio devices, from wireless single-chip data transceivers to high-power broadcast transmitters.

New to this edition:

- Extensively revised and expanded throughout, including new chapters on radar, digital modulation, GPS navigation, and *S*-parameter circuit analysis.
- New worked examples and end-of-chapter problems aid and test understanding of the topics covered.
- Numerous extra figures provide a visual aid to learning, with over 400 illustrations throughout the book.

Key topics covered include filters, amplifiers, oscillators, modulators, low-noise amplifiers, phase lock loops, transformers, waveguides, and antennas. Assuming no prior knowledge of radio electronics, this is a perfect introduction to the subject. It is an ideal textbook for junior or senior courses in electrical engineering, as well as an invaluable reference for professional engineers in this area.

Praise for the first edition:

This book is wonderfully informative, and refreshingly different from the usual rehash of standard engineering topics. Hagen has put his unique insights, gleaned from a lifetime of engineering and radio science, into this volume and it shows. There's an insight per page, at least for me, that makes it truly enjoyable reading, even for those of us who think we know something about the field! *Paul Horowitz, Harvard University*

Jon B. Hagen was awarded his Ph.D. from Cornell University in 1972, where he went on to gain 30 years' experience as an electronic design engineer, as well as establishing and teaching a Cornell electrical engineering course on RF electronics. Now retired, he has held positions as Principal Engineer at Raytheon, Electronics Department Head at the Arecibo Observatory in Puerto Rico, and Director of the NAIC Support Laboratory at Cornell.

Radio-Frequency Electronics

Circuits and Applications

Second Edition

Jon B. Hagen

CAMBRIDGE
UNIVERSITY PRESS

CAMBRIDGE
UNIVERSITY PRESS

Shaftesbury Road, Cambridge CB2 8EA, United Kingdom

One Liberty Plaza, 20th Floor, New York, NY 10006, USA

477 Williamstown Road, Port Melbourne, VIC 3207, Australia

314–321, 3rd Floor, Plot 3, Splendor Forum, Jasola District Centre, New Delhi – 110025, India

103 Penang Road, #05–06/07, Visioncrest Commercial, Singapore 238467

Cambridge University Press is part of Cambridge University Press & Assessment,
a department of the University of Cambridge.

We share the University's mission to contribute to society through the pursuit of
education, learning and research at the highest international levels of excellence.

www.cambridge.org
Information on this title: www.cambridge.org/9780521889742

© Cambridge University Press & Assessment 1996, 2009

First published 1996
Second edition 2009

A catalogue record for this publication is available from the British Library

Library of Congress Cataloging-in-Publication data
Hagen, Jon B.
Radio-frequency electronics : circuits and applications / Jon B. Hagen. – 2nd ed.
    p.   cm.
ISBN 978-0-521-88974-2
1. Radio circuits.   I. Title.
TK6560.H34   2009
621.384′12–dc22

                                                                    2009007355

ISBN   978-0-521-88974-2   Hardback

Contents

Preface

This book was written to help the reader to understand, analyze, and design RF circuits. Developed as a textbook for an RF engineering course at Cornell University, it can also be used for self-study and as a reference for practising engineers. The scope of topics is wide and the level of analysis ranges from introductory to advanced. In each chapter, I have tried to convey an intuitive "how things work" understanding from which the mathematical analysis follows. The initial chapters present the amplifiers, filters, modulators, and demodulators, which are the basic building blocks of radio systems, from AM and FM to the latest digital radio systems. Later chapters alternate between systems, such as television, and radio astronomy, and theoretical topics, such as noise analysis and radio spectrometry. The book provides the RF vocabulary that carries over into microwave engineering, and one chapter is devoted to waveguides and other microwave components.

In this second edition, many chapters have been expanded. Others have been rearranged and consolidated . New chapters have been added to cover radar, the GPS navigation system, digital modulation, information transmission, and S-parameter circuit analysis.

The reader is assumed to have a working knowledge of basic engineering mathematics and electronic circuit theory, particularly linear circuit analysis. Many students will have had only one course in electronics, so I have included some fundamental material on amplifier topologies, transformers, and power supplies. The reader is encouraged to augment reading with problem-solving and lab work, making use of mathematical spreadsheet and circuit simulation programs, which are excellent learning aids and confidence builders. Some references are provided for further reading, but whole trails of reference can be found using the internet.

For helpful comments, suggestions, and proofreading, I am grateful to many students and colleagues, especially Wesley Swartz, Dana Whitlow, Bill Sisk, Suman Ganguly, Paul Horowitz, Michael Davis, and Mario Ierkic.

Jon B. Hagen
Brooklyn, NY
July 2008

1

Introduction

Consider the magic of radio. Portable, even hand-held, short-wave transmitters can reach thousands of miles beyond the horizon. Tiny microwave transmitters riding on spacecraft return data from across the solar system. And all at the speed of light. Yet, before the late 1800s, there was nothing to suggest that telegraphy through empty space would be possible even with mighty dynamos, much less with insignificantly small and inexpensive devices. The Victorians could extrapolate from experience to imagine flight aboard a steam-powered mechanical bird or space travel in a scaled-up Chinese skyrocket. But what experience would have even hinted at *wireless* communication? The key to radio came from theoretical physics. Maxwell consolidated the known laws of electricity and magnetism and added the famous displacement current term, $\partial D/\partial t$. By virtue of this term, a changing electric field produces a magnetic field, just as Faraday had discovered that a changing magnetic field produces an electric field. Maxwell's equations predicted that *electromagnetic waves* can break away from the electric currents that generate them and propagate independently through empty space with the electric and magnetic field components of the wave constantly regenerating each other.

Maxwell's equations predict the velocity of these waves to be $1/\sqrt{\varepsilon_0 \mu_0}$ where the constants, ε_0 and μ_0, can be determined by simple measurements of the forces between static electric charges and between current-carrying wires. The dramatic result is, of course, the experimentally-known speed of light, 3×10^8 m/sec. The electromagnetic nature of light is revealed. Hertz conducted a series of brilliant experiments in the 1880s in which he generated and detected electromagnetic waves with wavelengths very long compared to light. The utilization of Hertzian waves (electromagnetic waves) to transmit information developed hand-in-hand with the new science of electronics.

Where is radio today? AM radio, the pioneer broadcast service, still exists, along with FM, television and two-way communication. But radio now also includes digital broadcasting formats, radar, surveillance, navigation and broadcast satellites, cellular telephones, remote control devices, and wireless data communications. Applications of radio frequency (RF) technology outside

Figure 1.1. The radio spectrum.

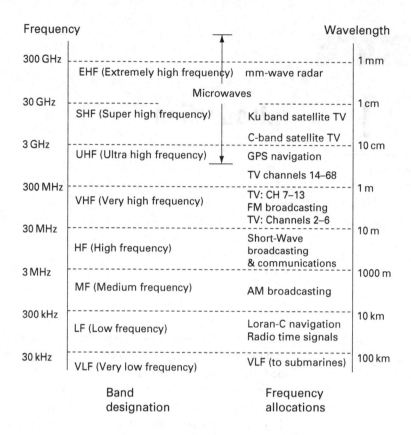

radio include microwave heaters, medical imaging systems, and cable television. Radio occupies about eight decades of the electromagnetic spectrum, as shown in Figure 1.1.

1.1 RF circuits

The circuits discussed in this book generate, amplify, modulate, filter, demodulate, detect, and measure ac voltages and currents at radio frequencies. They are the blocks from which RF systems are designed. They scale up and down in both power and frequency. A six-section bandpass filter with a given passband shape, for example, might be large and water-cooled in one application but subminiature in another. Depending on the frequency, this filter might be made of sheet metal boxes and pipes, of solenoidal coils and capacitors, or of piezo-electric mechanical resonators, yet the underlying circuit design remains the same. A class-C amplifier circuit might be a small section of an integrated circuit for a wireless data link or the largest part of a multi-megawatt broadcast transmitter. Again, the design principles are the same.

1.2 Narrowband nature of RF signals

Note that most frequency allocations have small fractional bandwidths, i.e., the bandwidths are small compared to the center frequencies. The fractional band-width of the signal from any given transmitter is less than 10 percent – usually much less. It follows that the RF voltages throughout a radio system are very nearly sinusoidal. An otherwise purely sinusoidal RF "carrier" voltage[1] must be *modulated* (varied in some way) to transmit information. Every type of modu-lation (audio, video, pulse, digital coding, etc.) works by varying the amplitude and/or the phase of the sinusoidal RF wave, called the "carrier" wave. An unmodulated carrier has only infinitesimal bandwidth; it is a pure spectral line. Modulation always broadens the line into a spectral band, but the energy clusters around the carrier frequency. Oscilloscope traces of the RF voltages in a transmitter or on a transmission line or antenna are therefore nearly sinusoidal. When modulation is present, the amplitude and/or phase of the sinusoid changes but only over many cycles. Because of this narrowband characteristic, elemen-tary sine wave ac circuit analysis serves for most RF work.

1.3 AC circuit analysis – a brief review

The standard method for ac circuit analysis that treats voltages and currents in linear networks is based on the linearity of the circuit elements: inductors, capacitors, resistors, etc. When a sinusoidal voltage or current generator drives a circuit made of linear elements, the resulting steady-state voltages and currents will all be perfectly sinusoidal and will have the same frequency as the gen-erator. Normally we find the response (voltage and current amplitudes and phases) of driven ac circuits by a mathematical artifice. We replace the given sinusoidal generator by a hypothetical generator whose time dependence is $e^{j\omega t}$ rather than $\cos(\omega t)$ or $\sin(\omega t)$. This source function has both a real and an imaginary part since $e^{j\omega t} = \cos(\omega t) + j\sin(\omega t)$. Such a nonphysical (because it is complex) source leads to a nonphysical (complex) solution. But the real and imaginary parts of the solution are separately good physical solutions that correspond respectively to the real and imaginary parts of the complex source. The value of this seemingly indirect method of solution is that the substitution of the complex source converts the set of linear *differential* equations into a set of easily solved linear *algebraic* equations. When the circuit has a simple topology, as is often the case, it can be reduced to a single loop by combining obvious series and parallel branches. Many computer programs are available to

---

[1] There is no low-frequency limit for radio waves but the wavelengths corresponding to audio frequencies, hundreds to thousands of kilometers, make it inefficient to connect an audio amplifier directly to an antenna of reasonable size. Instead, the information is impressed on a carrier wave whose wavelength is compatible with practical antennas.

find the currents and voltages in complicated ac circuits. Most versions of SPICE will do this steady-state ac analysis (which is much simpler than the transient analysis which is their primary function). Special linear ac analysis programs for RF and microwave work such as Agilent's ADS and MMICAD include circuit models for strip lines, waveguides, and other RF components. You can write your own program to analyze ladder networks (see Problem 1.3) and to analyze most filters and matching networks.

1.4 Impedance and admittance

The coefficients in the algebraic circuit equations are functions of the complex *impedances* (V/I), or *admittances* (I/V), of the RLC elements. The voltage across an inductor is LdI/dt. If the current is $I_0e^{j\omega t}$, then the voltage is $(j\omega L)I_0e^{j\omega t}$. The impedance and admittance of an inductor are therefore respectively $j\omega L$ and $1/(j\omega L)$. The current into a capacitor is CdV/dt, so its impedance and admittance are $1/(j\omega C)$ and $j\omega C$. The impedance and admittance of a resistor are just R and $1/R$. Elements in series have the same current so their total impedance is the sum of their separate impedances. Elements in parallel have the same voltage so their total admittance is the sum of their separate admittances. The real and imaginary parts of impedance are called *resistance* and *reactance* while the real and imaginary parts of admittance (the reciprocal of impedance) are called *conductance* and *susceptance*.

1.5 Series resonance

A capacitor and inductor in series have an impedance $Z_s = j\omega L+1/(j\omega C)$. This can be written as $Z_s = j(L/\omega)(\omega^2 - 1/[LC])$, so the impedance is zero when the (angular) frequency is $1/\sqrt{LC}$. At this *resonant frequency*, the series LC circuit is a perfect *short circuit* (Figure 1.2). Equal voltages are developed across the inductor and capacitor but they have opposite signs and the net voltage drop is zero.

At resonance and in the steady state there is no transfer of energy in or out of this combination. (Since the overall voltage is always zero, the power, IV, is always zero.) However, the circuit does contain *stored* energy which simply sloshes back and forth between the inductor and the capacitor. Note that this circuit, by itself, is a simple bandpass filter.

Figure 1.2. Series-resonant *LC* circuit.

At resonance   =   Short circuit

1.6 Parallel resonance

A capacitor and an inductor in parallel have an admittance $Y_p = j\omega C + 1/(j\omega L)$ which is zero when the (angular) frequency is $1/\sqrt{LC}$. At this resonant frequency, the parallel LC circuit is a perfect *open circuit* (Figure 1.3) – a simple bandstop filter.

Like the series LC circuit, the parallel LC circuit stores a fixed quantity of energy for a given applied voltage. These two simple combinations are important building blocks in RF engineering.

Figure 1.3. Parallel-resonant *LC* circuit.

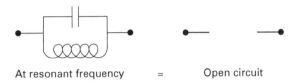

At resonant frequency   =   Open circuit

1.7 Nonlinear circuits

Many important RF circuits, including mixers, modulators, and detectors, are based on nonlinear circuit elements such as diodes and saturated transistors used as switches. Here we cannot use the linear $e^{j\omega t}$ analysis but must use time-domain analysis. Usually the nonlinear elements can be replaced by simple models to explain the circuit operation. Full computer modeling can be used for accurate circuit simulations.

Problems

Problem 1.1.   A generator has a source resistance r_S and an open circuit rms voltage V_0. Show that the maximum power available from the generator is given by $P_{max} = V_0^2/(4r_S)$ and that this maximum power will be delivered when the load resistance, R_L, is equal to the source resistance, r_S.

Problem 1.2.   A passive network, for example a circuit composed of resistors, inductors, and capacitors, is placed between a generator with source resistance r_S and a load resistor, R_L. The *power response* of the network (with respect to these resistances) is defined as the fraction of the generator's maximum available power that reaches the load. If the network is *lossless*, that is, contains no resistors or other dissipative elements, its power response function can be found in terms of the impedance, $Z_{in}(\omega) = R(\omega) + jX(\omega)$, seen looking into the network with the load connected. Show that the expression for the power response of the lossless network is given by

$$P(\omega) = \frac{4r_S R(\omega)}{(R(\omega) + r_S)^2 + X(\omega)^2}$$

where $R = \mathrm{Re}(Z_{in})$ and $X = \mathrm{Im}(Z_{in})$.

Problem 1.3. Most filters and matching networks take the form of the ladder network shown below.

Ladder network topology.

Series inductors, capacitors, or resistors

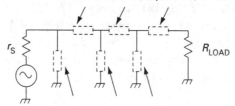

Parallel inductors, capacitors, or resistors

Write a program whose input data is the series and shunt circuit elements and whose output is the power response as defined in Problem 1.2.

Hints: One approach is to begin from the load resistor and calculate the input impedance as the elements are added, one by one. When all the elements are in place, the formula in Problem 1.2 gives the power response – as long as the load resistor is the only resistor. The process is repeated for every desired frequency.

A better approach, which is no more complicated and which allows resistors, is the following: Assume a current of $1 + j0$ ampere is flowing into the load resistor. The voltage at this point is therefore $R_L + j0$ volts. Move to the left one element. If this is a series element, the current is unchanged but the voltage is higher by IZ where Z is the impedance of the series element. If the element is a shunt element, the voltage remains the same but the input current is increased by VY where Y is the admittance of the shunt element. Continue adding elements, one at a time, updating the current and voltage. When all the elements are accounted for, you have the input voltage and current and could calculate the total input impedance of the network terminated by the load resistor. Instead, however, take one more step and treat the source resistance, r_S, as just another series impedance. This gives you the voltage of the source generator, from which you can calculate the maximum power available from the source. Since you already know the power delivered to the load, $(1)^2 R_L$, you can find the power response. Repeat this process for every desired frequency.

The ladder elements (and, optionally, the start frequency, stop frequency, frequency increment, and source and load resistances) can be treated as data, that is, they can be located together in a block of program statements or in a file so they can be changed easily. For now, the program only needs to deal with six element types: series and parallel inductors, capacitors, and resistors. Each element in the circuit file must therefore have an identifier such as "PL", "SL", "PC", "SC", "PR", and "SR" or 1, 2, 3, 4, 5, 6, or whatever, plus the value of the component in henrys, farads, or ohms. Organize the circuit file so that it begins with the element closest to R_L and ends with some identifier such as "EOF" (for "End Of File") or some distinctive number.

An example program, which produces both tabular and graphical output, is shown below, written in MATLAB, which produces particularly compact and readable code. The input data (included as program statements) is for the circuit shown below, of an LC network designed to connect a 50-ohm load to a 1000-ohm source. You will find this, or your own equivalent program, to be a useful tool when designing matching networks and filters. In the problems for Chapters 4, 10, 14, and 17, the program will be enhanced to plot phase response and to handle transmission lines, transformers, and transistors, making it a powerful design tool.

```
%MATLAB program to solve ladder networks
%Problem 1.3 in "Radio-Frequency Electronics"
%Save this file as "ladder.m" and run by typing "ladder" in command window
%INPUT DATA(circuit components from load end; 'SL' is series inductor,%etc.)
%—————————————————————
ckt={'SL',23.1e-6,'PC',463e-12,'EOF'};%'EOF' terminates list
Rload=50; Rsource=1000; startfreq=1e6; endfreq=2e6; freqstep=5e4;
%—————————————————————

f=(startfreq:freqstep:endfreq);%frequency loop
w=2*pi.*f;%w is angular frequency
I=ones(size(w));V=ones(size(w))*Rload;%set up arrays for inputI(f) and V(f)
ckt_index=0; morecompsflag=1;
while morecompsflag==1 %loop through string of components
ckt_index=ckt_index+1;%ckt_index prepared for next item in list
component=ckt{ckt_index};
morecompsflag=1-strcmp(component,'EOF');%zero after last component

if strcmp(component,'PC')==1
ckt_index=ckt_index+1; capacitance=ckt{ckt_index};
I=I+V.*(1j.*w.*capacitance);
elseif strcmp(component,'SC')==1
ckt_index=ckt_index+1; capacitance=ckt{ckt_index};
V=V+I./(1j.*w.*capacitance);
elseif strcmp(component,'PL')==1
ckt_index=ckt_index+1; inductance=ckt{ckt_index};
I=I+V./(1j.*w.*inductance);
elseif strcmp(component,'SL')==1
ckt_index=ckt_index+1; inductance=ckt{ckt_index};
V=V+I.*(1j.*w.*inductance);
elseif strcmp(component,'PR')==1
ckt_index=ckt_index+1; resistance=ckt{ckt_index};
I=I+V/resistance;
elseif strcmp(component,'SR')==1
ckt_index=ckt_index+1; resistance=ckt{ckt_index};
V=V+I*resistance;
end %components loop

end %frequency loop

Z=V./I; V=V+I.*Rsource;
frac=Rload./((abs(V).^2)/(4.*Rsource));
db=10/log(10)*log(frac);
heading = 'freq(MHz) frac dB' %print heading in command window
A=[(1E-6*f)' frac' db'] %print table of data in command window
plot(f,db);%graph the data
grid;xlabel('Frequency');ylabel('dB');title('Frequency response');
```

The circuit corresponding to the input data statements in the example program above is shown below, together with the analysis results produced by the program.

Example circuit.

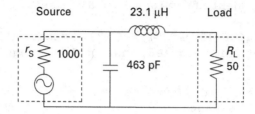

Analysis results.

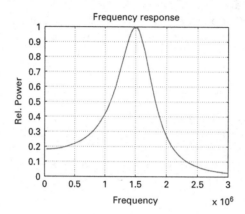

Problem 1.4. (AC circuit analysis review problem.) For the circuit in the figure, derive an expression for $I_R(t)$. Use a complex source voltage, $V_0 e^{j\omega t}$, the real part of which is $V_0 \cos(\omega t)$. The impedances of C, L, and R are $(j\omega C)^{-1}, j\omega L$, and R, respectively. Find the complex current through the resistor. $I_R(t)$ will be the real part of this complex current.

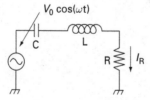

Problem 1.5. (More AC circuit analysis review.) For the circuit in the figure, derive an expression for $I_R(t)$. Note that the source voltage, $V_0\sin(\omega t + \theta)$, is equal to the *imaginary* part of $V_0 e^{j(\omega t+\theta)}$. Therefore, if we take the complex voltage to be $V_0 e^{j(\omega t+\theta)}$, $I_R(t)$ will be the imaginary part of the complex current through R. Alternatively, you can let the complex voltage source have the value $-jV_0 e^{j(\omega t+\theta)}$, the real part of which is $V_0\sin(\omega t)$. With this source, $I_R(t)$ is the real part of the complex current through R.

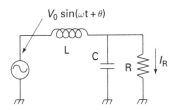

2   Impedance matching

Matching normally means the use of a lossless (nonresistive) network between a signal source and a load in order to maximize the power transferred to the load. This presupposes that the source is not capable of supplying infinite power, i.e., that the source is not just an ideal voltage generator or an ideal current generator. Rather, the source is assumed to be an ideal voltage generator in series with a source impedance, i.e., a Thévenin equivalent circuit, or an ideal current generator in parallel with a source admittance, a Norton equivalent circuit. Note that these equivalent circuits are themselves equivalent; each can be converted into the form of the other. An antenna that is feeding a receiver is an example of an ac signal source connected to a load. Figure 2.1 shows the simplest situation, a dc generator driving a resistive load. The generator is represented in Thévenin style (a) and in Norton style (b).

Figure 2.1. DC generator driving a resistive load, Equivalent circuits.

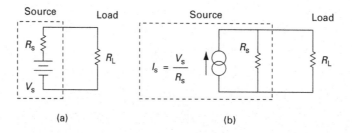

(a)   (b)

You can see the equivalence by inspection: the generators have the same open-circuit voltage and the same short-circuit current. Maximum power is transferred when the load resistance is made equal to the source resistance. You can show this by differentiating the expression for the power, $P_{load} = [V_S R_L/(R_L+R_S)]^2/R_L$. Figure 2.2 plots the relative transferred power (Pwr/MaxPwr) as a function of the normalized load resistance ($r = R_L/R_S$). In (a) the scales are linear and in (b) the scales are logarithmic so the relative power is expressed in dB. Note that R_L can differ by a factor of 10 from R_S and the power transferred is still 33% of the maximum value.

Figure 2.3. Transformer
converts R_L to R_S for maximum
power transfer.

2.1 Transformer matching

Figure 2.2. Relative power
transfer as a function of R_L/R_S, (=r)

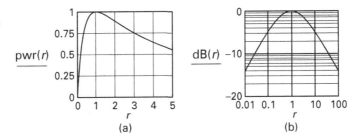

In the case of an ac source, a transformer can make the load resistance match
(equal) the source resistance (and vice versa) as shown in Figure 2.3. The
impedance is transformed by the square of the turns ratio.[1]

The ac situation often has a complication: the source and/or the load may be
reactive, i.e., have an unavoidable built-in reactance. An example of a reactive
load is an antenna; most antennas are purely resistive at only one frequency.
Above this resonant frequency they usually look like a resistance in series with an
inductor and below the resonant frequency they look like a resistance in series
with a capacitor. An obvious way to deal with this is first to cancel the reactance to
make the load and/or source impedance purely resistive and then use a transformer to match the resistances. In the circuit of Figure 2.4, an inductor cancels

Figure 2.4. A series reactor
makes the load a pure
resistance.

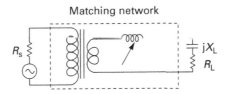

Matching network

---

[1] Let the secondary winding be the load side. Then $V_{sec} = V_{pri} N_{sec}/N_{pri}$. For energy to be conserved,
$V_{pri}I_{pri} = V_{sec}I_{sec}$. Therefore $I_{sec} = I_{pri} N_{pri}/N_{sec}$ and $V_{pri}/I_{pri} = (V_{sec}/I_{sec}) (N_{pri}/N_{sec})^2$ or
$Z_{pri} = Z_{sec}(N_{pri}/N_{sec})^2$.

the reactance of a capacitive (but not purely capacitive) load. If we are working at 60 Hz, we would say the inductor corrects the load's power factor.

From the standpoint of the load, the matching network converts the source impedance, $R_S + j0$, into the complex conjugate of the load impedance. When a matching network is used between two devices, each device will look into an impedance that is the complex conjugate of its own impedance. As a result, the reactances cancel and the resistances are equal. Whenever the source and/or load have a reactive component, the match will be frequency dependent, i.e., away from the design frequency the match will not be perfect. In fact, with reactive sources and/or reactive loads, *any* lossless matching circuit will be frequency dependent – a filter of some kind – whether we like it or not.

2.2 L-networks

More often than not, matching circuits use no transformers (i.e., no coupled inductors). Figure 2.5 shows a two-element L-network (in this figure, a rotated letter L) that will match a source to a load resistor whose resistance is smaller than the source resistance. The trick is to put a reactor, X_P, in *parallel* with the *larger* resistance. Consider a specific example: $R_S = 1000$ and $R_L = 50$.

The impedance of the left-hand side is given by

$$Z_{\text{left}} = R_{\text{left}} + jX_{\text{left}} = \frac{1000\,jX_P}{1000 + jX_P} = \frac{(1000jX_P)(1000 - jX_P)}{(1000 + jX_P)(1000 - jX_P)}$$

$$= \frac{1000^2 jX_P + 1000X_P^2}{1000^2 + X_P^2}. \tag{2.1}$$

We can pick the value of X_P so that the real part of Z_{left} will be 50 ohms, i.e., equal to the load resistance. Using Equation (2.1), we find that $X_P^2 = 52\,628$ so we can pick either $X_P = 229$ (an inductor) or $X_P = -229$ (a capacitor). The left-hand side now has the correct equivalent series resistance, 50 ohms, but it is accompanied by an equivalent series reactance, X_{left}, given by the imaginary part of Equation (2.1). We can cancel X_{left} by inserting a series reactor, X_S, equal to $-X_{\text{left}}$. Figure 2.6 shows the matching circuits that result when X_P is an inductor and when X_P is a capacitor.

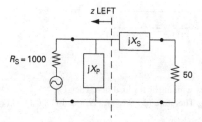

Figure 2.5. Two reactors in an L-network match R_L to R_S.

Figure 2.6. The two realizations for the L-network of Figure 2.5.

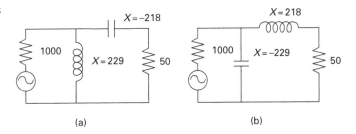

(a)   (b)

Figure 2.7. Frequency response (power vs. frequency) for the L-networks of Figure 2.6.

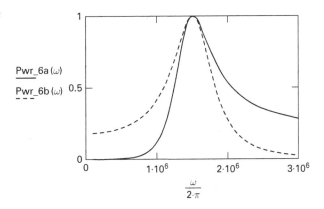

$\underline{\text{Pwr\_6a}}(\omega)$

$\text{Pwr\_6b}(\omega)$

The final step is to find the values of L and C that produce the specified reactances at the given frequency. For the circuit of Figure 2.6(b), $\omega L = 218$. Suppose the design frequency is 1.5 MHz ($\omega = 2\pi \cdot 1.5 \cdot 10^6$), near the top of the AM broadcast band. Then $L = 23.1\ \mu\text{H}$ and $C = 462\ \text{pF}$. Note that the values of the two reactors are completely determined by the source and load resistances. Except for the choice of which element is to be an inductor and which is to be a capacitor, there are no free parameters in this two-element matching circuit. The match is perfect at the design frequency but, away from that frequency, we must accept the resulting frequency response. The frequency responses (fractional power reaching the load vs. frequency) for the two circuits of Figure 2.6 are plotted in Figure 2.7. Note that around the design frequency, i.e., around the resonant peak, the curves are virtually identical. Otherwise, the complete cutoff at very low frequencies of Figure 2.6a and the complete cutoff at very high frequencies of Figure 2.6b can be predicted from inspection of the circuits.

Quick design procedure for L-networks

If you remember only that the parallel reactance goes across the larger resistance you will be able to repeat the steps used above and design L-networks. But if you are doing these things often it may be worth memorizing the following "Q factor" for L-network design:

$$Q_{\mathrm{EL}} = \sqrt{\frac{R_{\mathrm{high}}}{R_{\mathrm{low}}} - 1}. \tag{2.2}$$

You can verify (Problem 2.6) that the ratios $R_{\mathrm{high}}/X_{\mathrm{parallel}}$ and $X_{\mathrm{series}}/R_{\mathrm{low}}$ are both equal to this factor, Q_{EL}. Remember the definition of Q_{EL} and these ratios immediately give you the L-network reactance values. You can also verify that, when Q_{EL} is large, the two elements in an L-network have nearly equal and opposite reactances, i.e., together they resonate at the design frequency. In this case the magnitude of the reactances is given by the geometric mean of R_{high} and R_{low} (especially easy to remember).

When the ratio of the source resistance to the load resistance is much different from unity, an L-network produces a narrowband match, i.e., the match will be good only very close to the design frequency. Conversely, when the impedance ratio is close to unity, the match is wide. The width of any resonance phenomenon is described by a factor, *the effective Q* (or *circuit Q* or just Q), which is equal to the center frequency divided by the two-sided 3-dB bandwidth (the difference between the half-power points). Equivalently, Q_{eff} is the reciprocal of the fractional bandwidth. When an ideal voltage generator drives a simple RLC series circuit, Q_{eff} is given by X/R where X is either X_{L} or X_{C} at the center frequency (since they are equal). The L-network matching circuit is equivalent to a simple series RLC circuit, but Q_{EL} is twice Q_{eff} because the nonzero source resistance is also in series; the matching circuit makes the effective source resistance equal to the load resistance so the loop's total series resistance is twice the load resistance. As a result, the fractional bandwidth is given by $1/Q_{\mathrm{eff}} = 2/Q_{\mathrm{EL}}$. In many applications the bandwidth of the match is important and the match provided by the L-network (which is completely determined by the source and load resistances) may be too narrow or too wide. When matching an antenna to a receiver, for example, one wants a narrow bandwidth so that signals from strong nearby stations won't overload the receiver. In another situation the signal produced by a modulated transmitter might have more bandwidth than the L-network would pass. Networks described below solve these problems.

2.3 Higher Q – pi and T-networks

Higher Q can be obtained with back-to-back L-networks, each one transforming down to a center impedance that is lower than either the generator or the source resistance. The resulting pi-network is shown in Figure 2.8.

With the simple L-networks we had $Q_{\mathrm{EL}} = \sqrt{19} = 4.4$. In this pi-network both the 1000-ohm source and the 50-ohm load are matched down to a center impedance of 10 ohms (a free parameter). The bandwidth is equivalent to that of an L-network with $Q_{\mathrm{EL}} = 11.95$. When $R_{\mathrm{HIGH}} \gg R_{\mathrm{LOW}}$, the pi-network has a bandwidth equivalent to that of an L-network with $Q_{\mathrm{EL}} = \sqrt{R_{\mathrm{HIGH}}/R_{\mathrm{CENTER}}}$. Again, the fractional bandwidth is given by $1/Q_{\mathrm{eff}} = 2/Q_{\mathrm{EL}}$.

Figure 2.8. Pi-network (back-to-back L-networks) provides higher Q.

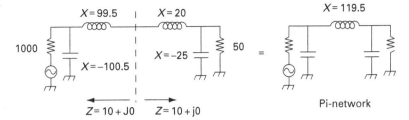

$Z = 10 + J0$　　$Z = 10 + j0$

Figure 2.9. Response of the pi-network of Figure 2.8 compared with the L-networks of Figure 2.6.

Pwr_8a(ω)

Pwr_8b(ω)

Pwr_6(ω)

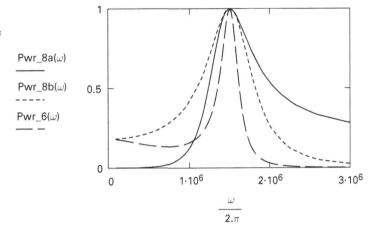

Figure 2.10. The T-network, like the pi-network, provides higher Q.

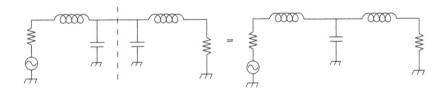

The response of this pi-network is shown in Figure 2.9 together with the responses of the L-networks of Figure 2.6.

You can guess that we could just as well have used "front-to-front" L-networks, each one transforming up to a center impedance that is higher than both the source and load impedances. This produces the T-network of Figure 2.10. Note that both the pi-network and the T-network have a free parameter (the center impedance) which gives us some control over the frequency response while still providing a perfect match at the center or design frequency.

2.4 Lower Q – the double L-network

In a double L-network (Figure 2.11) the first stage transforms to an impedance value between the source and load impedances. The second stage takes it the

Figure 2.11. Double L-network
for lower Q (wider bandwidth).

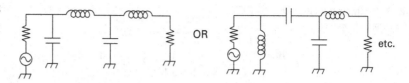

rest of the way. The process can, of course, be done in smaller steps with any
number of cascaded networks. A long chain of L-networks forms an artificial
transmission line that tapers in impedance to produce a frequency-independent
match. Real transmission lines (i.e., lines with distributed L and C) are sometimes physically tapered to provide this kind of impedance transformation.
A tapered transmission line is sometimes called a transformer, since, like the
transformer in Figure 2.3, it provides frequency-independent matching.

2.5 Equivalent series and parallel circuits

To design the L-network we used the fact that a two-element parallel XR circuit,
where $1/Z = 1/R_{\text{parallel}} + 1/jX_{\text{parallel}}$, has an equivalent series circuit, where
$Z = R_{\text{series}} + jX_{\text{series}}$. Conversion between equivalent series and parallel representations is used so often it is worth a few more words. If you are given, for
example, an antenna or a black box with two terminals and you make measurements at a *single frequency* you can only determine whether the box is "capacitive," i.e., equivalent to an RC combination, or is "inductive," i.e., equivalent to
an RL combination. Suppose it is capacitive. Then you can represent it equally
well as a series circuit where $Z = R_{\text{series}} + 1/j\omega C_{\text{series}}$ or as a parallel circuit where
$1/Z = 1/R_{\text{parallel}} + j\omega C_{\text{parallel}}$. As long as you're working only at (or never very far
from) the single frequency, either representation is equally valid, even if the box
contains a complicated circuit with discrete resistors, capacitors, inductors,
transmission lines, metallic and resistive structures, etc. If you measure the
impedance at more than one frequency you might determine that the box does
indeed contain a simple parallel RC or series RC circuit or that its impedance
variation at least resembles that of a simple parallel circuit more than it
resembles that of a simple series circuit.

2.6 Lossy components and efficiency of matching networks

So far we have considered networks made of ideal inductors and capacitors.
Real components, however, are lossy due to the finite conductivity of metals,
lossy dielectrics or magnetic materials, and even radiation. Power dissipated in
nonideal components is power that does not reach the load so, with lossy
components, we must consider a matching network's efficiency. As explained
above, a lossy reactor can be modeled as an ideal L or C together with either

a series or parallel resistor. Normally we can make the approximation that the values of L or C and the value of the associated resistor are constant throughout the band of interest. Let us consider the efficiency of the L-network that uses a series inductor and a parallel capacitor. We shall assume that the loss in the capacitor is negligible compared to the loss in the inductor. (This is very often the case with lumped components.) We shall model the lossy inductor as an ideal inductor in series with a resistor of value r_S. The ratio of the inductive reactance, X_L, to this resistance value is the *quality factor*, Q_U, where the subscript denotes *"unloaded Q"* or *component Q*. (Less series resistance certainly implies a higher quality component.) Note that this resistance, like the inductor, is in series with the load resistor so the same current, I, flows through both. The power delivered to the load is $I^2 R_L$ and the power dissipated in r_S is $I^2 r_S$. Using the relations $X_S = Q_{EL} R_{load}$ and $Q_U = X_S / r_S$, we find the efficiency of the match is given by

$$\eta = \text{Efficiency} = \frac{\text{Power Out}}{\text{Power In}} = \frac{I^2 R_L}{I^2 R_L + I^2 r_S} = \frac{1}{1 + Q_{EL}/Q_U}. \qquad (2.3)$$

Efficiency is maximized by maximizing the ratio Q_U/Q_{EL}, i.e., the ratio of unloaded Q to loaded Q. If we model the lossy inductor as a parallel LR circuit and define the unloaded Q as r_P/X we would get the same expression for efficiency (Problem 2.7). Likewise, if the loss occurs in the capacitor we will also get this expression, as long as we define the unloaded Q of the capacitor again as parallel resistance over parallel reactance or as series reactance over series resistance. When the load resistance is very different from the source resistance, the effective Q of an L-network will be high so, for high efficiency, the unloaded Q of the components must be very high. The double L-network, with its lower loaded Q's, can be used to provide higher efficiency.

Q factor summary

Loaded Q, the Q factor associated with *circuits*, can be either high or low depending on the application. Narrowband filters have high loaded Q. Wideband matching circuits have low loaded Q. Loaded Q is therefore not a measure of quality. Unloaded Q, however, which specifies the losses in *components*, is indeed a measure of quality since lowering component losses always increases circuit efficiency.

Problems

Problem 2.1.   A nominal 47-ohm, $\frac{1}{4}$-watt carbon resistor with 1.5 inch wire leads is measured at 100 MHz to have an impedance of 48+j39 ohms. Find the component values for (a) an equivalent series RL circuit, and (b) an equivalent parallel RL circuit.

Problem 2.2.   (a) Design an L-network to match a 50-ohm generator to a 100-ohm load at a frequency of 1.5 MHz. Let the parallel element be an inductor. Use your circuit analysis program (Problem 1.3) to find the frequency response of this circuit from 1 MHz to 2 MHz in steps of 50 kHz.

(b) Same as (a), but let the parallel element be a capacitor.

Problem 2.3.   Design a double L matching network for the generator, source, and frequency of Problem 2.2(a). For maximum bandwidth, let the intermediate impedance be the geometric mean of the source impedance and the load impedance, i.e., $\sqrt{50 \cdot 100}$. Use your circuit analysis program (Problem 1.3) to find the response as in Problem 2.2.

Problem 2.4.   Suppose the only inductors available for building the networks of Problems 2.2(a) and 2.3 have a Q_U (unloaded Q) of 100 at 1.5 MHz. Assume the capacitors have no loss. Calculate the efficiencies of the matching networks at 1.5 MHz. Check your results using your circuit analysis program.

Problem 2.5.   The diagram below shows a network that allows a 50-ohm generator to feed two loads (which might be antennas). The network divides the power such that the top load receives twice as much power as the bottom load. The generator is matched, i.e., it sees 50 ohms. Find the values of X_{L1}, X_{L2} and X_C. Hint: transform each load first with an L-section network and then combine the two networks into the circuit shown.

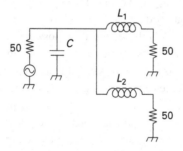

Problem 2.6.   Verify the prescription given for calculating the values of an L-network: $X_P = \pm R/Q$ and $X_S = \mp rQ$ where $R > r$ and $Q = \sqrt{R/r - 1}$.

Problem 2.7.   At a single frequency, a lossy inductor can be modeled as a lossless inductor in series with a resistance or as a lossless inductor in parallel with a resistance. Convert the series combination r_S, L_S to its equivalent parallel combination r_P, L_P and show that Q_U defined as X_S/r_S is equal to Q_U defined as r_P/X_P.

3 Linear power amplifiers

An amplifier is a circuit designed to impose a specified voltage waveform, $V(t)$, or, sometimes, a specified current waveform, $I(t)$, upon the terminals of a device known as the "load." The specified waveform is often supplied in the form of an analog "input signal" such as the millivolt-level signal from a dynamic microphone. In a public address system, an audio amplifier produces a scaled-up copy of the microphone voltage (the input signal) and this amplified voltage (the output signal) is connected to a loudspeaker (the load). An audio amplifier generally supplies more than a watt to the loudspeaker. The microphone cannot supply more than milliwatts, so the audio amplifier is a power amplifier as well as a voltage amplifier. The ability to amplify power is really the defining characteristic of an amplifier. Of course energy must be conserved; amplifiers contain or are connected to power supplies, usually batteries or power line-driven ac-to-dc converter circuits, originally known as "battery eliminators" but long since simply called "power supplies" (see Chapter 29). The amplifiers discussed in this chapter are the basic "resistance-controlled"[1] circuits in which transistors (or vacuum tubes) are used as electronically variable resistors to control the current through the load. Such circuits span the range from monolithic op-amps to the output amplifiers in high-power microwave transmitters.

3.1 Single-loop amplifier

Figure 3.1 shows the simplest resistance-controlled amplifier. This circuit is just a resistive voltage divider. The manually variable resistor (rheostat) in (a) represents the electronically variable resistor (transistor) in (b). Remember that the main current path through the transistor is between the emitter and the

---

[1] Resistance-controlled amplifier are also called *linear* amplifiers, to distinguish them from *switching* amplifiers, which are discussed in Chapters 9 and 29. Note, however, that *linear amplifier* is also used to denote amplifiers whose output waveform is a faithful (linearly proportional) copy of the input.

Figure 3.1. Basic single-loop amplifier.

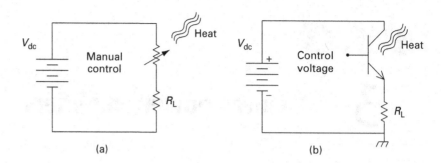

(a)   (b)

collector; the current through the control terminal, the base, is typically less than 1% of the emitter-collector current. The base voltage can vary the transistor's resistance from infinity to almost zero so any arbitrary current waveform (within the range of zero to V_{dc}/R_L) can be obtained by using an appropriate corresponding control voltage waveform. The load, represented as a resistor, R_L, could be, for example a heating element, a loudspeaker, a servomotor, or a transmitting antenna.

3.2 Drive circuitry: common-collector, common-emitter, and common-base

We will concentrate on the topologies of the output circuit or "business end" of amplifiers, i.e., the high-current paths between the load and the power supply(s). But let us briefly discuss the drive circuitry that controls the resistance of the transistor(s). The discussion is illustrated with circuits using bipolar transistors, but the basic concepts apply also to FETs and tubes.

In Figure 3.1(b), only one terminal is shown for the control voltage (drive signal) input. When the return connection for the drive signal is made at the bottom of the load resistor, we get the amplifier shown in Figure 3.2. This circuit is called a *common-collector* amplifier because the collector, in common with the drive signal return, is a ground point with respect to *ac* signals, even though it does have a dc voltage.

This circuit is also called an *emitter follower*. The transistor will adjust its current flow to make the instantaneous emitter voltage almost equal to the instantaneous base voltage. Here "almost identical" means a small dc offset (the emitter voltage will be about 0.7 volts less than the base voltage) along with a one or two percent reduction in signal amplitude. The reason the emitter voltage closely follows the base voltage is that the emitter current is a rapidly increasing (exponential) function of the base-to-emitter voltage.

Figure 3.3 shows the *common-emitter* drive arrangement, in which the return for the drive signal is connected to the transistor's emitter. It is common to rearrange the circuit as in (b), so that the return connection for the drive signal and the negative terminal of the supply are at the same point (ground).

Figure 3.2. The emitter follower.

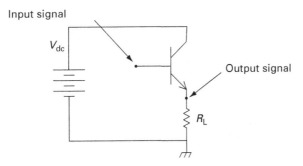

Figure 3.3. Common-emitter amplifier.

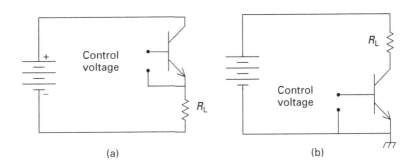

Figure 3.4. Common-base amplifier.

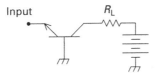

With the drive voltage placed directly across the base-emitter junction, the transistor current is a nonlinear (exponential) function of the drive voltage, and the output voltage (voltage across the load) will not be an accurate scaled version of the drive voltage. A special driver circuit can be used to generate an inverse exponential (logarithmic) drive signal to linearize the amplifier. This happens automatically if the base drive is a current waveform; the output current (and hence the voltage across the load) will have the same waveform shape as the base current. It is also common to use a negative feedback correction loop to force the output signal to follow the input signal. The common-emitter amplifier can supply voltage amplification as well as power amplification.

The third and final base drive arrangement, common-base, is shown in Figure 3.4. In this circuit the drive current flows through the main loop. Gain is obtained because, while the driver and load have essentially the same current (the base current might be only one percent as large as the collector-emitter current), the voltage swing at the collector, determined by the supply voltage, is much greater than the voltage swing at the emitter, determined by the near short-circuit base-emitter junction.

Note that, as in the common-emitter amplifier, the drive signal is applied across the transistor's base-to-emitter junction, so the signal developed across the load resistor will be a nonlinear function of the input voltage. But if the drive voltage is applied to the emitter through a series resistor, the drive current and, hence, the output current, will be essentially proportional to the drive voltage.

3.3 Shunt amplifier topology

In the amplifiers discussed above, the load current is controlled by a transistor in series with the load and the power supply. Another way to vary the current in the load is to divert current around it, as in the shunt circuit amplifier shown in Figure 3.5, where the supply and a series resistor form a (nonideal) current source.

Figure 3.5. Shunt amplifier.

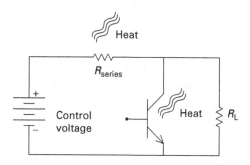

This shunt circuit appears inferior to the series circuit; power will be wasted in the series resistor and the full supply voltage is not available to the load. We will see later, however, that shunt circuits can be used to advantage in *ac* amplifiers which, unlike the general-purpose amplifiers above, are amplifiers designed for signals whose average dc value is zero, e.g., audio and RF signals.

3.4 Dual-polarity amplifiers

If the amplifier must supply output voltages of either polarity and also must handle arbitrary waveforms (as opposed to ac waveforms, whose average dc value is zero) it will require a circuit with two power supplies (or a single "floating" power supply, as we shall see later). The two-loop circuit shown in Figure 3.6 still uses only one transistor.

Here R_B pulls the output toward the negative supply as much as the transistor allows. The voltage on the load is determined by a tug of war between R_B and the transistor. Note that this circuit is a combination of the series and shunt amplifier arrangements. If we make $V_{NEG} = -2\,V_{POS}$ and $R_B = R_L$, you can see

Figure 3.6. A dual-supply, single-transistor amplifier (drawn in two ways).

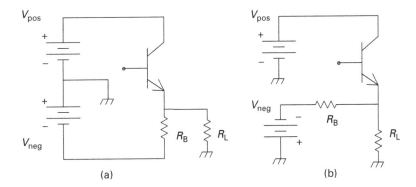

(a)   (b)

that the maximum negative swing will be equal to the maximum positive swing. A constant *bias* current is maintained in the transistor to set the output voltage at zero when the input signal is zero. The maximum efficiency is calculated in the next section, and is only $\frac{1}{12}$. Biased amplifiers (this one and everything discussed so far), which draw current from the supply(s) even when the input signal is zero, are known as *class-A* amplifiers. They are commonly used where their low efficiency is not a problem.

3.5 Push–pull amplifiers

The two-transistor *push–pull* configuration shown in Figure 3.7 provides output voltage of both polarities and has high efficiency compared to the single-transistor amplifiers. (Note: non-push–pull amplifiers are often called "single-ended" amplifiers.)

Figure 3.7. Totem pole push–pull amplifier.

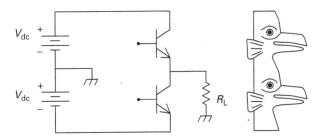

The top transistor allows the top supply to "push" current into the load. The lower transistor lets the lower supply "pull" current from the load. The push–pull circuit is the circuit of choice for arbitrary waveforms. The efficiency, calculated in the next section, is $\pi/4$ (78%) for a sine wave of maximum amplitude. Since there are no series resistors, both positive and negative load currents are limited only by the size of the transistors and power supplies. By contrast, the single-transistor circuit of Figure 3.6 can deliver high positive

Figure 3.8. Complementary (PNP/NPN) push–pull amplifier.

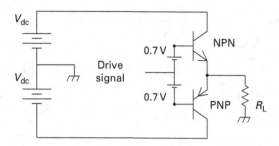

current but the maximum negative current is limited by R_B.[2] Push–pull amplifiers are normally set up to run as *class-B* amplifiers, which means that, when the input voltage is zero, both transistors are just turned off and there is no power drawn from the supply(s). For low distortion, it is important that the crossover at $I = 0$ be continuous, so sometimes push–pull amplifiers are run class-AB which means that each transistor is given some bias current. Note that the amplifier of Figure 3.7 uses two NPN transistors, placed one above the other like faces on a totem pole. The top transistor acts as an emitter follower; when it is conducting, the output voltage will be almost equal to that transistor's base voltage. The bottom transistor, however, is driven in the common-emitter mode. The two transistors need drive signals of opposite polarities and present different drive impedances, requiring separate and different drive circuits. This unappealing asymmetry is eliminated in the complementary push–pull amplifier shown in Figure 3.8.

The complementary push–pull amplifier uses an NPN and a complementary (identical, except for polarity) PNP transistor. Both operate as emitter followers. Except for the 0.7 V offsets, their bases could be tied together. This can be taken care of, as shown in the figure, by using a pair of 0.7 V batteries. (In practice, this is done with some diode circuitry, rather than batteries.) There is no vacuum tube analog to this circuit, because there are no PNP tubes.[3] For completeness, we note that a third type of push–pull amplifier is obtained by interchanging the transistors in the complementary push–pull amplifier. This circuit, in which both transistors are driven in the common-emitter mode, is found in some switching (one transistor full-on while the other is full-off) servo amplifiers.

So-called bridge amplifiers are shown in Figure 3.9(a). They use a single power supply, but can supply the load with either polarity.

---

[2] As long as the load is a pure resistor, the pull-down resistor, R_B, is not a problem; any waveform not exceeding the power supply voltage limits can be faithfully amplified. But if the load contains an unavoidable capacitance C_L in parallel with R_L, the amplifier must be able to deliver current $V_{out}/R_L + C_L \, d/dt \, (V_{out})$. See Problem 3.6.

[3] The charge carriers in semiconductors are both electrons (negative) and "holes" (positive). Vacuum tubes use only electrons as charge carriers, although unwanted positive ions are sometimes produced by electron impact.

Figure 3.9. Bridge amplifiers: (a) full bridge (b) half bridge.

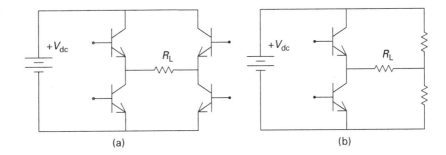

(a)   (b)

These circuits have no direct connection between the power supply and the load; either the supply or the load must "float," i.e., have no ground connection. In the circuit of Figure 3.9(a), the top pair or bottom pair of transistors can operate as on–off switches (fully conducting or fully turned off). This circuit is a true push–pull amplifier, while the half bridge of Figure 3.9(b) is equivalent to a push-pull amplifier that has resistors in series with the power supplies. This reduces the maximum voltage swing as well as the efficiency.

3.6 Efficiency calculations

As we are assuming that the drive power is small compared to the output power, we will calculate efficiency as the ratio of the average output power to the dc input power. Depending on the devices used, this ratio is known as the *collector efficiency*, *drain efficiency*, or *plate efficiency*. Most often, we want to compute the efficiency for the situation in which the amplifier is producing a sinusoidal output at full power (the condition for which the efficiency is usually a maximum).

When calculating efficiency, it is important to remember that average power is the time average of instantaneous power, i.e., the time average of voltage × current. Consider, for example, a power supply of voltage V_{dc} that is furnishing a current $I = I_0 + I_1\cos(\omega t)$. The average power is $\langle V_{dc}(I_0 + I_1\cos(\omega t))\rangle$, where the brackets $\langle\ldots\rangle$ indicate averaging. Since the average of a sum is equal to the sum of the averages, we can expand this expression.

$$\langle V_{dc}(I_0 + I_1\cos(\omega t))\rangle = \langle V_{dc}I_0\rangle + \langle V_{dc}I_1\cos(\omega t)\rangle = V_{dc}I_0, \quad (3.1)$$

since the average value of $\cos(\omega t)$ is zero. When a sine wave $V_0\sin(\omega t)$ is applied to a resistor, the instantaneous power is $VI=[V_0\sin(\omega t)]^2/R$ and the average power is $V_0^2\langle\sin^2(\omega t)\rangle/R=V_0^2/(2R)$.[4] It is also useful to remember that $\langle\sin(\theta)\cos(\theta)\rangle=0$ and that the average value of $\sin(\theta)$ or $\cos(\theta)$ over one positive loop is equal to $2/\pi$.

[4] To see that $\langle\sin^2(\theta)\rangle = \frac{1}{2}$, note that $\langle\sin^2(\theta)\rangle = \langle\cos^2(\theta)\rangle$, since the waveforms are identical, and use the identity $\sin^2(\theta) + \cos^2(\theta) = 1$.

We will first calculate the efficiency of the amplifier of Figure 3.6, under the conditions that $R_B = R_L$, $V_{pos} = V_{dc}$, and $V_{neg} = -2V_{dc}$. Let I_L denote the current downward into the load; I_B, the bias current leftward into R_B; and I_E, the current downward from the emitter. Note that $I_E = I_L + I_B$. Assume the maximum signal condition: $V_L = V_{dc}\cos(\omega t)$. This lets us write $I_B = (V_{dc}\cos(\omega t) + 2V_{dc})/R_B$. The power from the negative supply is therefore $P_{neg} = \langle I_B\, 2V_{dc}\rangle = 4V_{dc}^2/R_B$. The current into the load is just $I_L = V_{dc}\cos(\omega t)/R_L$. Adding I_L and I_B, we have $I_E = V_{dc}\cos(\omega t)/R_L + (V_{dc}\cos(\omega t) + 2V_{dc})/R_B$. Since this is the same as the current supplied by the positive supply (ignoring the transistor's small base current), we find that the power from the positive supply is given by $P_{pos} = 2V_{dc}^2/R_B$. The total dc power, P_{dc}, is the sum of P_{pos} and P_{neg}: $P_{dc} = 6V_{dc}^2/R_B$. The power into the load is $V_0^2/(2R_L)$, so the efficiency is given by

$$\eta = \frac{V_{dc}^2/(2R_L)}{6V_{dc}^2/R_B} \tag{3.2}$$

which reduces to $\eta = \frac{1}{12}$ when $R_B = R_L$.

Next we will find the efficiency of the push–pull amplifiers of Figures 3.7 and 3.8. Again, we assume the maximum signal condition, $V_L = V_{dc}\cos(\omega t)$. Consider the top transistor, which conducts during the positive half of the cycle. Assuming negligible base current, the current through this transistor will be the same as the current in the load, $I = (V_{dc}\cos(\omega t))/R_L$. During the positive half-cycle, the positive supply furnishes an average power given by $\langle V_{dc} \times (V_{dc}/R_L)\cos(\omega t)\rangle = (V_{dc}^2/R_L)\,2/\pi$, since here the average is just over the positive loop. The negative supply, during its half-cycle, furnishes the same average power, since the circuit is symmetric. Thus, the efficiency is given by

$$\eta = \frac{V_{dc}^2/(2R_L)}{(V_{dc}^2/R_L)2/\pi} = \frac{\pi}{4} = 78\%. \tag{3.3}$$

3.7 AC amplifiers

All the amplifiers discussed above are known as *dc amplifiers* because they can handle signals of arbitrarily low frequency. (They might well be called universal amplifiers since they have no high-frequency limitations except those set by the transistors.) Audio and RF signals, however, are pure ac signals: their average value, i.e., their dc component, is zero. For these signals, special *ac amplifier* circuits provide simplicity and efficiency.

The circuit in Figure 3.10(a) is an ac version of the class-A amplifier of Figure 3.6. The drive signal at the base is given a positive offset (bias) which will create the same bias voltage at the emitter and a bias current through the pull-down resistor, R_E. The coupling capacitor (*dc blocking capacitor*) eliminates the dc bias from the load, and the output signal swings both positive and

Figure 3.10. Common-collector
single-ended ac amplifiers.

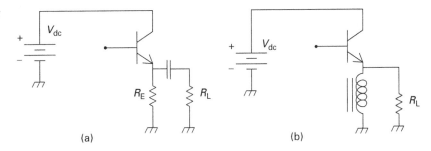

(a)   (b)

negative. The capacitance is chosen to be high enough that the full ac signal at
the emitter will appear at the load. Only one power supply is required for this ac
version. If $R_E = R_L$, the maximum peak-to-peak output swing is 2/3 V_{dc} and
efficiency is again only $\frac{1}{12}$.

A major improvement is to replace the power-dissipating pull-down resistor
with an inductor (ac *choke*) as shown in Figure 3.10(b). The inductor allows the
output to go negative as well as positive[5] and makes possible a maximum output
swing from $-V_{dc}$ to $+V_{dc}$. The inductance is chosen to be high enough to
eliminate currents at the signal frequencies. No capacitor is needed; assuming
the choke has negligible dc resistance, the average dc on the load will be zero.
There must be sufficient bias current through the inductor to keep the transistor
always on for the continuous control needed in linear operation. You can
calculate (Problem 3.1) that the maximum efficiency of this circuit is 50%;
the inductor improves the efficiency by a factor of 6 and the output swing by a
factor of 3.

It might seem that the maximum efficiency of the class-B amplifier (78%) is
only slightly better than the maximum efficiency of this class-A amplifier
(50%). But these maximum efficiencies apply only when the amplifier is
delivering a sine wave of maximum amplitude. For speech and music, the
average power is much less than the maximum power. The class-B amplifier
has little dissipation when the signal is low but a class-A amplifier, with its
constant bias current, draws constant power equal to twice the maximum output
power. A class-A audio amplifier rated for 25 watts output would consume a
continuous 50 watts from its supply while a class-B amplifier of equal power
rating would consume, on average, only a few watts, since the average power of
audio signals is much lower than the peak power.

Common-emitter versions of this class-A amplifier are shown in Figure 3.11.
The circuit of Figure 13.11(b) uses the shunt amplifier topology of Figure 3.5.
Here, the inductor provides a wideband constant current source. (If the signal
has a narrow bandwidth (RF), a parallel-resonant LC circuit will serve the same

---

[5] The bias current flowing downward in the inductor is essentially constant since the inductance is
large. At the part(s) of the cycle when the current through the transistor becomes less than the
inductor current, the inductor maintains its constant current by "sucking" current out of the load
resistor and thus producing the negative output voltage.

Figure 3.11. Common-emitter single-ended ac amplifiers.

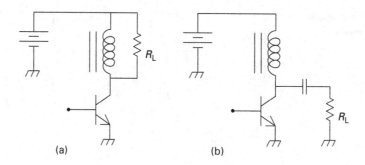

Figure 3.12. Transformer-coupled single-ended ac amplifier.

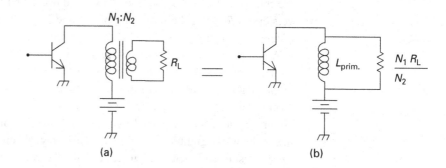

purpose.) A blocking capacitor allows one end of the load to be grounded, which is often a convenience. As we have seen, current drive is called for to achieve linearity if the emitter is tied directly to ground. At the expense of some efficiency, however, the emitter can be tied to ground through a resistor to allow the emitter voltage to follow the drive (base) voltage. Then the emitter current, collector current, and output voltage will all be linearly proportional to the input voltage. This technique of linearizing a common-emitter amplifier is known as "emitter degeneration" or "series feedback." To its credit, the common-emitter arrangement requires only a small and always positive drive signal, whereas the circuit of Figure 3.10(b) requires a drive voltage identical to the output signal, swinging both positive and negative.

Often the resistance of a given load is unsuitable for obtaining the desired power with the given power supply voltage. In this case, the choke and blocking capacitor can be replaced by a transformer as shown below in Figure 3.12.

This circuit is equivalent to that of Figure 13.11(a). The equivalence is shown in (b). Note how the load resistor is R_L, multiplied by the square of the transformer turns ratio. The transformer's primary winding provides the inductor. Again, these choke-coupled and transformer-coupled class-A amplifiers can provide a peak-to-peak collector swing of twice V_{dc}. Note that if we used only the simple "ideal transformer" model to replace the load resistor by its transformed value, we would have no inductance and would predict a maximum

Figure 3.13. Symmetric transformer-coupled push–pull ac amplifier.

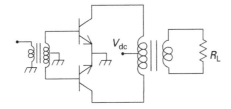

Figure 3.14. Half-bridge ac amplifier (a) and an equivalent version (b).

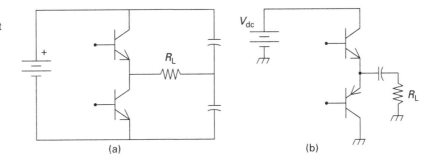

peak-to-peak swing of only V_{dc} (see Chapter 14). A center-tapped transformer can be used to make the symmetric push–pull amplifier shown in Figure 3.13.

This push–pull circuit, like the transformerless push–pull circuits, can be operated class B for high efficiency. Some high-power tube-type audio and RF amplifiers use this symmetric transformer circuit. Note the use of a center-tapped driver transformer – one way to supply the bases with the required opposite polarity signals.

Transistor audio amplifiers are usually push–pull and transformerless. They can be built with a single power supply by using capacitor coupling, as in the half-bridge circuit shown below in Figure 3.14(a). The circuit of Figure 3.14(b) is equivalent and uses only a single capacitor. The transistors can be in either the totem pole arrangement (a), or in the complementary NPN/PNP arrangement (b).

Since an audio or RF waveform has no dc component, the capacitors each charge to half the supply voltage and are equivalent to batteries. The capacitors thus form an artificial center tap for the power supply so this is an efficient push–pull amplifier, unlike the resistive half-bridge circuit of Figure 3.9(b).

3.8 RF amplifiers

RF amplifiers form a subset of ac amplifiers. RF signals are narrowband ac signals. Besides having no dc component, they have an almost constant waveform; while the amplitude and phase can vary, the shape remains sinusoidal.

Figure 3.15. Single-ended class-B RF amplifier.

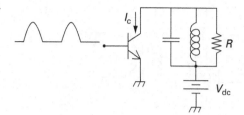

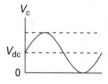

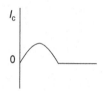

Figure 3.16. Collector voltage and current in the single-ended class-B RF amplifier.

This makes it possible to build a class-B RF amplifier with only a single transistor. The circuit (which looks no different from a class-A amplifier, except for the input waveform) is shown in Figure 3.15.

Here the transistor "pulls" current from the load, but there is no pusher transistor. Instead, a parallel resonant LC circuit provides a flywheel (energy storage) effect which maintains the sinusoidal waveform during the half-cycle in which the transistor is not conducting. The drive waveform consists only of positive loops. Between these loops, the transistor is nonconducting. This amplifier has the same 78% maximum efficiency of a push–pull class-B amplifier and also the class-B virtue of drawing no power when the signal level is zero. Let us analyze this circuit. Assume that, during the active half-cycle, the transistor current is $I_0 \sin(\theta)$, where $\theta = \omega t$ and I_0 is to be determined. Assume the resonant circuit provides enough energy storage (high enough Q) that the output voltage (the voltage across the load resistor) can be written as $A \sin(\theta)$. The output power is therefore $A^2/(2R)$. These voltage and current waveforms are shown in Figure 3.16.

During the active half-cycle, the power (IV product) into the RCL parallel circuit is $I_0 \sin(\theta) A \sin(\theta)$. Over a cycle, this averages to $I_0 A/4$, where one factor of $\frac{1}{2}$ is the time average of $\sin^2(\theta)$ and the other factor of $\frac{1}{2}$ comes from the transistor being turned off during half of every cycle. The power delivered to the parallel circuit must be equal to the power delivered to the load:

$$\frac{I_0 A}{4} = \frac{V_{dc}^2}{2R}. \tag{3.4}$$

From this we find $I_0 = 2A/R$. We can calculate the power delivered by the supply by noting that over the half-cycle, the instantaneous power is $V_{dc} I_0 \sin(\theta)$. The average over the half-cycle is $V_{dc} I_0 2/\pi$ and the average over the entire cycle is again half of this or $V_{dc} I_0/\pi = V_{dc} 2A/(\pi R)$. The efficiency, power out divided by power supplied by the supply, is therefore

$$\eta = \frac{A^2/(2R)}{V_{dc} 2A/(\pi R)} = \frac{A}{V_{dc}} \frac{\pi}{4}. \tag{3.5}$$

At the maximum output, where $A = V_{dc}$, the efficiency of this "single-ended" class-B amplifier is $\pi/4$, the same as the maximum efficiency for a push–pull class-B amplifier.

The circuit of Figure 3.15, if made to conduct throughout the complete cycle, would be a class-A amplifier. The LC would not be needed as a flywheel, but it does act as a bandpass filter. Moreover, the inductance, or part of it, serves to cancel out the unavoidable collector-to-emitter parasitic capacitance inherent in the transistor.

3.9 Matching a power amplifier to its load

In Chapter 2 we saw that, for maximum power transfer, a load should have the same impedance as the source that drives it. However, the power amplifiers discussed in this chapter all have essentially zero output impedances. Their Thévenin equivalent circuits are almost perfect voltage generators. Do we therefore try to make the load resistances as low as possible? No. The amplifiers are designed to deliver a specified power to a specified load. This determines the power supply voltages and current capacities and the required current-handling capacity of the transistor(s). Therefore, power amplifiers are deliberately mismatched to their loads. But a power amplifier does still have some very small output impedance. Won't it therefore supply the most power to a load of that impedance? The answer is yes, as long as the amplitude is kept very low. If the amplitude is turned up, such a load will simply "short out" the amplifier.

The output amplifier in a transmitter usually includes an impedance transforming network, often called an *antenna tuner*. The purpose of this network is *not* to make the antenna impedance equal to the amplifier's very low output impedance. Rather, the network transforms the antenna impedance to the impedance needed for the amplifier to produce its rated power.

Problems

Problem 3.1.   Calculate efficiency of the class-A amplifier of Figure 3.10(b). Assume the output is a sine wave whose peak-to-peak amplitude is $2V_{dc}$, symmetric about $V=0$.

Assume that the dc bias current in the inductor is just enough to allow the amplifier to produce the maximum signal.

Problem 3.2.   The class-A amplifier shown below is operating at maximum power, applying a 24-volt peak-to-peak sine wave to the load resistor.

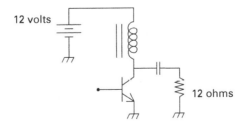

12 volts

12 ohms

Assume the choke has zero dc resistance and enough inductance to block any ac current and that the capacitor has enough susceptance to prevent any ac voltage drop.

(a) Draw the waveform of the collector voltage. Hint: Remember that there can be no dc voltage drop across the choke.

(b) Draw the waveform of the collector current. Hint: Remember that there is no ac current through the choke.

(c) What power is drawn from the supply under the maximum signal sine wave condition?

(d) Show that the efficiency under this maximum signal sine wave condition is 50%.

(e) What power is drawn from the supply if the signal is zero?

Problem 3.3.   An ideal push–pull amplifier does not have the current limitation of the emitter follower, so it can drive capacitive loads at high frequencies. But what about inductive loads? Suppose a load has an unavoidable series inductance but large voltages at high frequencies must be produced across the resistive part of the load. How does this impact the amplifier design?

Problem 3.4.   Justify the statements made about the voltage gain (about 99%) and the offset (about 0.7 volts) of the emitter follower. Use the relation between the emitter current and base-to-emitter voltage of a (bipolar) transistor, $I \approx I_{sat}\exp([V_b - V_e]/0.026)$. To get a value for I_{sat}, assume that $I = 10$ ma when $V_b - V_e = 0.7$ volts. Remember that in the emitter follower, $V_e = I_e R$. Assume a reasonable value for R such as 1000 ohms and find V_e for several values of V_b.

Problem 3.5.   Find the power gain of an emitter follower, i.e., the ratio of output signal power to input signal power. Use the fact that the input current (base current) is less than the emitter current by a factor $1/(\beta+1)$ where β is the transistor's current gain (typically on the order of 100). Remember that the output voltage is essentially the same (follows) the input voltage.

Problem 3.6.   The emitter follower amplifier shown below has a load which includes an unavoidable parallel capacitance.

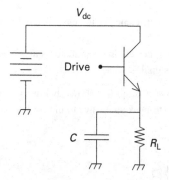

(a) What is the maximum peak-to-peak voltage that can be delivered to the load at low frequencies (where the capacitor can be neglected)?

(b) At what frequency will a sine-wave output signal of half the maximum amplitude become distorted? Hint: Express the emitter current as the sum of the resistor current and capacitor current and note that distortion will occur if this current should ever be negative (the transistor can only supply positive current). Answer: $\omega = \sqrt{3(RC)}$.

Problem 3.7. Consider the push–pull amplifier of Figure 3.8 when it is being driven by a sine-wave signal, $V(t) = V_0\sin(\omega t)$, and is connected to a load that is an inductor, L, rather than a resistor. Draw a graph showing the current flowing into the inductor and the individual emitter currents flowing in the direction of the load. Based on your graph, would you agree with the statement: The top transistor applies positive voltage to the load and the bottom transistor applies negative voltage to the load?

Problem 3.8. The maximum efficiency (the efficiency when the signal is a maximum-amplitude sine wave) of the amplifier in Figure 3.6 is $\frac{1}{12}$ when $R_B = R_L$ and $V_{NEG} = -2V_{POS}$.
(a) Calculate the maximum efficiency when $R_B = \sqrt{2}R_L$ and $V_{NEG} - (1 + \sqrt{2})\,V_{POS}$.
(b) Show that this combination of R_B and V_{NEG} yields the greatest maximum efficiency.

Problem 3.9. Draw a circuit for a "double push–pull" amplifier with four transistors. Two of the transistors connect the load to supply voltages $V_{dc}/2$ and $-V_{dc}/2$. The other two transistors connect the load to a second pair of supplies with voltages V_{dc} and $-V_{dc}$. The transistors connecting the smaller power supplies are turned off (nonconducting) when $|V_{out}| > V_{dc}/2$ and the transistors connecting the larger power supplies are turned off when $|V_{out}| < V_{dc}/2$. Calculate the efficiency when the output is a sine wave swinging from $-V_{dc}$ to $+V_{dc}$. This circuit is sometimes called a class-K amplifier.

4 Basic filters

Bandpass filters are key elements in radio circuits, for example, in radio receivers, to select the desired station. Here we will discuss lumped-element filters made of inductors and capacitors. We will first look at lowpass filters, and then see how they serve as prototypes for conversion to bandpass filters. We begin with the well-established lowpass filter prototypes – Butterworth, Chebyshev, Bessel, etc. These lowpass prototypes are simple *LC* ladder networks with series inductors and shunt capacitors, as shown in Figure 4.1.

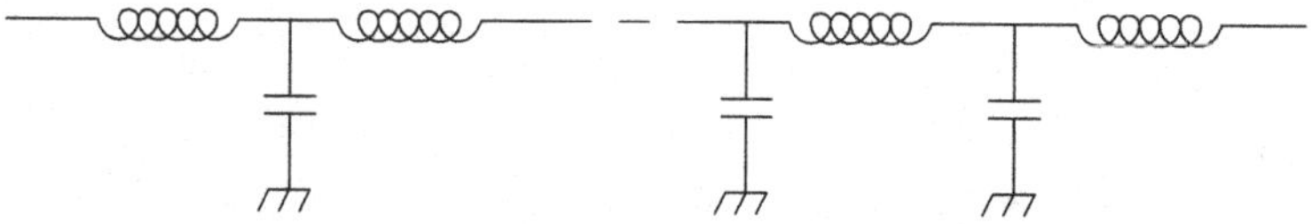

Figure 4.1. Lowpass ladder network.

An *n*-section lowpass filter has *n* components (capacitors plus inductors). The end components can be either series inductors, as shown above, or shunt capacitors, or one of each. Since they contain no (intentional) resistance, these filters are *reflective filters*; outside the passband, it is mismatch that keeps power from reaching the load. The ladder network can be redrawn as a cascade of voltage dividers as in Figure 4.2.

Figure 4.2. Ladder network as a cascade of voltage dividers.

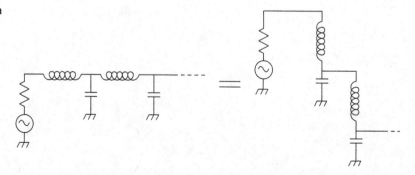

At high frequencies the division ratio increases so the load is increasingly isolated from the source. For frequencies well above cutoff, each circuit element contributes 6 dB of attenuation per octave (20 dB per decade). Within the passband, an ideal lowpass filter provides a perfect match between the load and the source. Filters with many sections approach this ideal. When the source and load impedances have no reactance (either built-in or parasitic) it is theoretically possible to have a perfect match across a wide band.

4.1 Prototype lowpass filter designs

The Butterworth filter is maximally flat, that is, it is designed so that at zero frequency the first $2n - 1$ derivatives with respect to frequency of the power transfer function are zero. The final condition (needed to determine the values of n elements) is the specification of the cutoff frequency, f_0, often specified as the 3-dB or half-power frequency. The frequency response of the Butterworth filter turns out to be

$$\left|\frac{V_{out}}{V_{in}}\right|^2 = \frac{1}{1 + (f/f_0)^{2n}}. \tag{4.1}$$

While it is the flattest filter, the Butterworth filter does not have skirts as sharp as those of the Chebyshev filter. The trade-off is that the Chebyshev filters have some passband ripple. The design criterion for the Chebyshev filter is that these ripples all have equal depth. The response is given by

$$\left|\frac{V_{out}}{V_{in}}\right|^2 = \frac{1}{1 + (V_r^{-2} - 1)\cosh^2\left(n\cosh^{-1}(f/f_0)\right)}, \tag{4.2}$$

where V_r is the height above zero of the ripple valley (in voltage) relative to the height of the peaks.

You will find tables of filter element values in many handbooks and textbooks. Two tables from Matthaei, Young and Jones [2] are given in Appendix 4.1 at the end of this chapter. These tables are for normalized filters, i.e., the cutoff frequency[1] is 1 radian/sec ($1/2\pi$ Hz). The value of the n-th component is g_n farads or henrys, depending on whether the filter begins with a capacitor or with an inductor. The proper source impedance is $1 + j0$ ohms. This is also the proper load impedance except for the even-order Chebyshev filters, where it is $1/g_{n+1} + j0$ ohms. Figure 4.3 shows plotted power responses of a Butterworth filter and several Chebyshev filters.

---

[1] The cutoff frequency for the Butterworth filters is the half-power (3 db) point. For an n-dB Chebyshev filter it is the highest frequency for which the response is down by n dB (see Figure 4.3).

Figure 4.3. Butterworth and
Chebyshev responses.

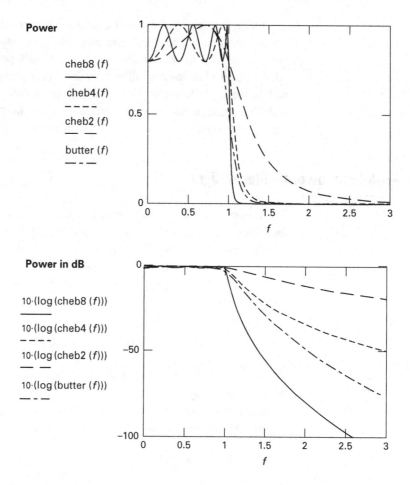

Power

cheb8 (*f*)
⎯⎯⎯

cheb4 (*f*)
⎯ ⎯ ⎯ ⎯

cheb2 (*f*)
⎯⎯ ⎯ ⎯⎯

butter (*f*)
⎯⎯ ⎯ ⎯⎯

Power in dB

$10 \cdot (\log (\text{cheb8} (f)))$
⎯⎯⎯

$10 \cdot (\log (\text{cheb4} (f)))$
⎯ ⎯ ⎯ ⎯

$10 \cdot (\log (\text{cheb2} (f)))$
⎯⎯ ⎯ ⎯⎯

$10 \cdot (\log (\text{butter} (f)))$
⎯⎯ ⎯ ⎯⎯

4.2 A lowpass filter example

As an example, we will look at the three-section Butterworth lowpass filter.
From the table, the filter has values of 1 H, 2 F, and 1 H (Figure 4.4a) or 1F, 2H,
and 1F (Figure 4.4b). The (identical) responses for these two filters are given in
Table 4.1 and plotted in Figure 4.5. Note that they work as advertised; the 3-dB
point is at 0.159 Hz.

Suppose we need a three-section Butterworth that is 5 kHz wide and works
between a 50-ohm generator and a 50-ohm load. We can easily find the element
values by scaling the prototype. The values of the inductors are just multiplied
by 50 (we need 50 times the reactance) and divided by $2\pi \cdot 5000$ (we need to
reach that reactance at 5 kHz, not 1 radian/sec). Similarly, the capacitor values
are divided by 50 and divided by $2\pi \cdot 5000$. Figure 4.6 shows the circuit resulting
from scaling the values of Figure 4.4b.

The response of the scaled filter is shown below in Table 4.2 and Figure 4.7.

Basic filters

Figure 4.4. Equivalent three-section Butterworth lowpass filters.

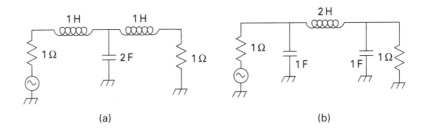

(a)   (b)

Table 4.1 Frequency response for filters of Figure 4.4.

Frequency (Hz)	Power	Response (dB)
0.00	1.000	− 0.0
0.0321	0.000	− 0.0
0.0640	0.996	− 0.02
0.095	0.955	− 0.20
0.1270	0.792	− 1.01
0.1590	0.500	− 3.01
0.1910	0.251	− 6.00
0.2230	0.117	− 9.31
0.2540	0.056	− 12.5
0.2860	0.029	− 15.4
0.3180	0.015	− 18.1

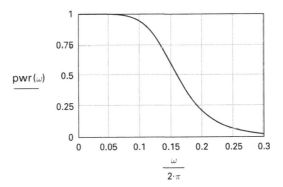

$\mathrm{pwr}(\omega)$

$\dfrac{\omega}{2 \cdot \pi}$

Figure 4.5. Plotted response of filters of Figure 4.4.

Figure 4.6. Filter of Figure 4.4(b), after conversion to 50 ohms and 5 kHz cutoff frequency.

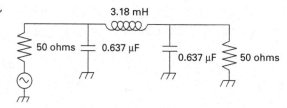

Table 4.2 Response of the scaled lowpass filter of Figure 4.6.

Frequency (Hz)	Power	Response (dB)
0	1.000	−0.0
1000	1.000	−0.0
2000	0.996	−0.02
3000	0.956	−0.20
4000	0.793	−1.01
5000	0.500	−3.01
8000	0.056	−12.5
6000	0.251	−6.00
7000	0.117	−9.31
9000	0.029	−15.4
10000	0.015	−18.1

Figure 4.7. Plotted response of the scaled lowpass filter of Figure 4.6.

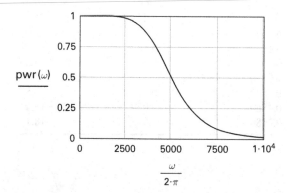

4.3 Lowpass-to-bandpass conversion

Here we will see how to convert lowpass filters into bandpass filters. Remember how the lowpass filters work: as frequency increases, the series arms (inductors), which are short circuits at dc, begin to pick up reactance. Likewise, the shunt arms (capacitors), which are open circuits at dc, begin to pick up susceptance. Both effects impede the signal transmission, as we have seen. To convert these lowpass filters in the most direct way to bandpass filters, we can replace the inductors by series LC combinations and the capacitors by parallel LC combinations. The series combinations are made to resonate (have zero impedance) at the center frequency of the desired bandpass filter, just as the inductors had zero impedance at dc, the "center frequency" of the prototype lowpass filter. It is important to note that as we move away from resonance, a series LC arm picks up reactance at twice the rate of the inductor alone. This is easy to see: The reactance of the series arm is given by

$$X_{\text{series}} = \omega L - \frac{1}{\omega C}. \tag{4.3}$$

Differentiating with respect to ω, we find

$$\frac{\mathrm{d}X}{\mathrm{d}\omega} = L + \frac{1}{\omega^2 C}. \tag{4.4}$$

At $\omega = \omega_0$,

$$\frac{\mathrm{d}X}{\mathrm{d}\omega} = L + \frac{1}{\omega_0{}^2 C} = 2L. \tag{4.5}$$

As we move off resonance, the inductor and the capacitor provide equal contributions to the reactance. Likewise, the parallel LC circuits, which replace the capacitors in the prototype lowpass filter, pick up susceptance at twice the rate of their capacitors. With this in mind, let us convert our 5-kHz lowpass filter into a bandpass filter. Suppose we want the center frequency to be 500 kHz and the bandwidth to be 10 kHz. As we move up from the center frequency, the series arms must pick up a reactance at the same rate the inductors picked up a reactance in the prototype lowpass filter. Similarly, the shunt arms must pick up susceptance at the same rate the capacitors picked up susceptance in the proto- type. This will cause the bandpass filter to have the same shape above the center frequency as the prototype had above dc. If the 3-dB point of the prototype filter was 5 kHz, the upper 3-dB point of the bandpass filter will be at 5 kHz above the center frequency. The bandpass filter, however, will have a mirror-image response as we go below the center frequency. (Below center frequency the reactances and susceptances change sign but the response remains the same.)

Let us calculate the component values. As we leave center frequency, the series circuits will get equal amounts of reactance from the L and the C, as explained above. Therefore the series inductor values should be exactly half what they were in the low pass prototype. Note: no matter how high we make the center frequency, the values of the inductors are reduced only by a factor of 2 from the those of the scaled lowpass filter. The series capacitors are chosen to resonate at the center frequency with the new (half-value) series inductors. The values of the parallel arms are determined similarly; the parallel capacitors must have half the value they had in the prototype lowpass filter. Finally, the parallel inductors are chosen to resonate with the new (half-value) parallel capacitors. These simple conversions yield the bandpass filter shown in Figure 4.8.

The response of this bandpass filter is given below in Table 4.3 and Figure 4.9.

While this theoretical filter works perfectly (since its components are lossless), the component values are impractical; typical real components with these values would be too lossy to achieve the calculated filter shape. When a bandpass filter is

Figure 4.8. Bandpass filter.

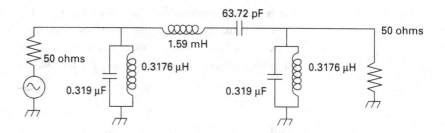

Table 4.3 Response of the bandpass filter of Figure 4.8.

Frequency (kHz)	Power	Response (dB)
490	0.014	−18.1
492	0.053	−12.8
494	0.241	−6.19
496	0.785	−1.05
498	0.996	−0.18
500	1.000	−0.00
502	0.996	−1.16
504	0.801	−0.966
506	0.260	−5.84
508	0.059	−12.9
510	0.016	−17.9

Figure 4.9. Plotted response for Table 4.3.

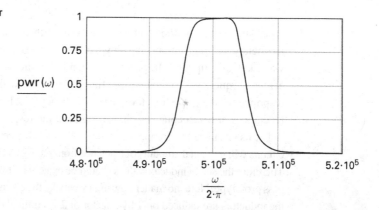

to have a large fractional bandwidth (bandwidth divided by center frequency) this direct conversion from lowpass to bandpass can be altogether satisfactory. It is when the fractional bandwidth is small, as in this example, that the direct conversion gets into trouble.[2] We will see later that the problem is solved by

---

[2] The component problem with the straightforward lowpass-to-bandpass conversion is that the values of the series inductors are very different from the values of the parallel inductors. (The same is true of the capacitors, but high-Q capacitors can usually be found.) In the above example, the inductors differ by a factor of about 5000 and it is normally impossible to find high-Q components over this range. (Low-Q inductors, of course, make the filter lossy and, if not accounted for, distort the bandpass

transforming the prototype lowpass filters into somewhat more complicated bandpass circuits known as *coupled resonator filters*. Those filters retain the desired shape (Butterworth, Chebyshev, etc.) and can serve, in turn, as prototypes for filters made from quartz or ceramic resonators and for filters made with resonant irises (thin aperture plates that partially block a waveguide).

Appendix 4.1. Component values for normalized lowpass filters[3]

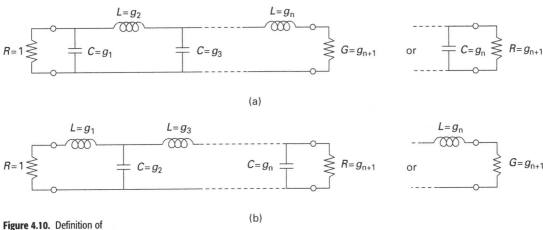

Figure 4.10. Definition of prototype filter parameters, g_1, g_2, ..., g_n, g_{n+1}. The prototype circuit (a) and its dual (b) give the same response.

Table A4.1 Element values for Butterworth (maximally flat) lowpass filters (the 3-dB point is at $\omega = 1$ radian/sec).

Value of n	g_1	g_2	g_3	g_4	g_5	g_6	g_7	g_8	g_9	g_{10}	g_{11}
1	2.000	1.000									
2	1.414	1.414	1.000								
3	1.000	2.000	1.000	1.000							
4	0.7654	1.848	1.848	0.7654	1.000						
5	0.6180	1.618	2.000	1.618	0.6180	1.000					
6	0.5176	1.414	1.932	1.932	1.414	0.5176	1.000				
7	0.4450	1.247	1.802	2.000	1.802	1.247	0.4450	1.000			
8	0.3902	1.111	1.663	1.962	1.962	1.663	1.111	0.3902	1.000		
9	0.3473	1.000	1.532	1.879	2.000	1.879	1.532	1.000	0.3473	1.000	
10	0.3129	0.9080	1.414	1.782	1.975	1.975	1.782	1.414	0.9080	0.3129	1.000

shape.) The inductors in coupled-resonator filters are all of about the same value. If a high-Q inductor can be found, the coupled resonator filter is designed for whatever impedance calls for that value of inductor and then transformers or matching sections are used at each end to convert to the desired impedance.

[3] From Matthaei, Young, and Jones [2].

Table A4.2 Element values for Chebyshev lowpass filters (for a filter with N-dB ripple, the last N-dB point is at $\omega = 1$ radian/sec).

Value of n	g_1	g_2	g_3	g_4	g_5	g_6	g_7	g_8	g_9	g_{10}	g_{11}
0.01-dB ripple											
1	0.0960	1.0000									
2	0.4488	0.4077	1.1007								
3	0.6291	0.9702	0.6291	1.0000							
4	0.7128	1.2003	1.3212	0.6476	1.1007						
5	0.7563	1.3049	1.5773	1.3049	0.7563	1.0000					
6	0.7813	1.3600	1.6896	1.5350	1.4970	0.7098	1.1007				
7	0.7969	1.3924	1.7481	1.6331	1.7481	1.3924	0.7969	1.0000			
8	0.8072	1.4130	1.7824	1.6833	1.8529	1.6193	1.5554	0.7333	1.1007		
9	0.8144	1.4270	1.8043	1.7125	1.9057	1.7125	1.8043	1.4270	0.8144	1.0000	
10	0.8196	1.4369	1.8192	1.7311	1.9362	1.7590	1.9055	1.6527	1.5817	0.7446	1.1007
0.1-dB ripple											
1	0.3052	1.0000									
2	0.8430	0.6220	1.3554								
3	1.0315	1.1474	1.0315	1.0000							
4	1.1088	1.3061	1.7703	0.8180	1.3554						
5	1.1468	1.3712	1.9750	1.3712	1.1468	1.0000					
6	1.1681	1.4039	2.0562	1.5170	1.9029	0.8618	1.3554				
7	1.1811	1.4228	2.0966	1.5733	2.0966	1.4228	1.1811	1.0000			
8	1.1897	1.4346	2.1199	1.6010	2.1699	1.5640	1.9444	0.8778	1.3554		
9	1.1956	1.4425	2.1345	1.6167	2.2053	1.6167	2.1345	1.4425	1.1956	1.0000	
10	1.1999	1.4481	2.1444	1.6265	2.2253	1.6418	2.2046	1.5821	1.9628	0.8853	1.3554
0.2-dB ripple											
1	0.4342	1.0000									
2	1.0378	0.6745	1.5386								
3	1.2275	1.1525	1.2275	1.0000							
4	1.3028	1.2844	1.9761	0.8468	1.5386						
5	1.3394	1.3370	2.1660	1.3370	1.3394	1.0000					
6	1.3598	1.3632	2.2394	1.4555	2.0974	0.8838	1.5386				
7	1.3722	1.3781	2.2756	1.5001	2.2756	1.3781	1.3722	1.0000			
8	1.3804	1.3875	2.2963	1.5217	2.3413	1.4925	2.1349	0.8972	1.5386		
9	1.3860	1.3938	2.3093	1.5340	2.3728	1.5340	2.3093	1.3938	1.3860	1.0000	
10	1.3901	1.3983	2.3181	1.5417	2.3904	1.5536	2.3720	1.5066	2.1514	0.9034	1.5386
0.5-dB ripple											
1	0.6986	1.0000									
2	1.4029	0.7071	1.9841								
3	1.5963	1.0967	1.5963	1.0000							
4	1.6703	1.1926	2.3661	0.8419	1.9841						
5	1.7058	1.2296	2.5408	1.2296	1.7058	1.0000					
6	1.7254	1.2479	2.6064	1.3137	2.4758	0.8696	1.9841				
7	1.7372	1.2583	2.6381	1.3444	2.6381	1.2583	1.7372	1.0000			
8	1.7451	1.2647	2.6564	1.3590	2.6964	1.3389	2.5093	0.8796	1.9841		
9	1.7504	1.2690	2.6678	1.3673	2.7239	1.3673	2.6678	1.2690	1.7504	1.0000	
10	1.7543	1.2721	2.6754	1.3725	2.7392	1.3806	2.7231	1.3485	2.5239	0.8842	1.9841
1.0-dB ripple											
1	1.0177	1.0000									
2	1.8219	0.6850	2.6599								
3	2.0236	0.9941	2.0236	1.0000							

Table A4.2 (cont.)

Value of n	g_1	g_2	g_3	g_4	g_5	g_6	g_7	g_8	g_9	g_{10}	g_{11}
4	2.0991	1.0644	2.8311	0.7892	2.6599						
5	2.1349	1.0911	3.0009	1.0911	2.1349	1.0000					
6	2.1546	1.1041	3.0634	1.1518	2.9367	0.8101	2.6599				
7	2.1664	1.1116	3.0934	1.1736	3.0934	1.1116	2.1664	1.0000			
8	2.1744	1.1161	3.1107	1.1839	3.1488	1.1696	2.9685	0.8175	2.6599		
9	2.1797	1.1192	3.1215	1.1897	3.1747	1.1897	3.1215	1.1192	2.1797	1.0000	
10	2.1836	1.1213	3.1286	1.1933	3.1890	1.1990	3.1738	1.1763	2.9824	0.8210	2.6599
2.0-dB ripple											
1	1.5296	1.0000									
2	2.4881	0.6075	4.0957								
3	2.7107	0.8327	2.7107	1.0000							
4	2.7925	0.8806	3.6063	0.6819	4.0957						
5	2.8310	0.8985	3.7827	0.8985	2.8310	1.0000					
6	2.8521	0.9071	3.8467	0.9393	3.7151	0.6964	4.0957				
7	2.8655	0.9119	3.8780	0.9535	3.8780	0.9119	2.8655	1.0000			
8	2.8733	0.9151	3.8948	0.9605	3.9335	0.9510	3.7477	0.7016	4.0957		
9	2.8790	0.9171	3.9056	0.9643	3.9598	0.9643	3.9056	0.9171	2.8790	1.0000	
10	2.8831	0.9186	3.9128	0.9667	3.9743	0.9704	3.9589	0.9554	3.7619	0.7040	4.0957
3.0-dB ripple											
1	1.9953	1.0000									
2	3.1013	0.5339	5.8095								
3	3.3487	0.7117	3.3487	1.0000							
4	3.4389	0.7483	4.3471	0.5920	5.8095						
5	3.4817	0.7618	4.5381	0.7618	3.4817	1.0000					
6	3.5045	0.7685	4.6061	0.7929	4.4641	0.6033	5.8095				
7	3.5182	0.7723	4.6386	0.8039	4.6386	0.7723	3.5182	1.0000			
8	3.5277	0.7745	4.6575	0.8089	4.6990	0.8018	4.4990	0.6073	5.8095		
9	3.5340	0.7760	4.6692	0.8118	4.7272	0.8118	4.6692	0.7760	3.5340	1.0000	
10	3.5384	0.7771	4.6768	0.8136	4.7425	0.8164	4.7260	0.8051	4.5142	0.6091	5.8095

Problems

Problem 4.1. Design a five-element lowpass filter with a Chebyshev 0.5-dB ripple shape. Let the input and output impedances be 100 ohms. Use parallel capacitors at the ends. The bandwidth (from dc to the last 0.5-dB point) is to be 100 kHz. Use Table A4.2 to find the values of the prototype 1 ohm, 1 rad/sec filter and then alter these values for 100 ohms and 100 kHz.

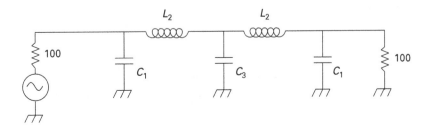

Problem 4.2. Use the results of Problem 4.1 to design a five-element bandpass filter with a Chebyshev 0.5-dB ripple shape. Let the input and output impedances remain at 100 ohms. The center frequency is to be 5 MHz and the total bandwidth (between outside 0.5-dB points) is to be 200 kHz.

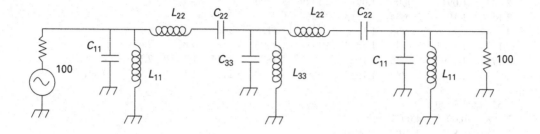

Problem 4.3. Convert the filter of Problem 4.2 to operate at 50 ohms by adding an L-section matching network at each end. Test the filter design using your ladder network analysis program, sweeping from 4.5 to 5.5 MHz in steps of 20 KHz.

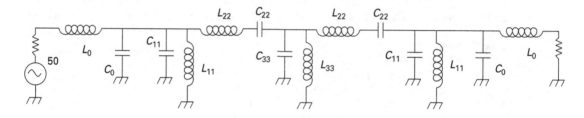

Problem 4.4. The one-section bandpass filter shown below uses a single parallel resonator. In its prototype lowpass filter, the resonator is a single shunt capacitor. Show that the frequency response of this filter is given by

$$\frac{P}{P_{max}} = \frac{1}{1 + Q^2 (f/f_0 - f_0/f)^2}$$

where f_0 is the resonant frequency of the LC combination and Q is defined as $R/(\omega_0 L)$, where R is the parallel combination of R_S and R_L.

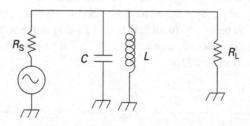

Problem 4.5. Highpass filters are derived from lowpass filters by changing inductors to capacitors and vice versa and replacing the component values in the prototype lowpass

tables by their reciprocals. (A 2-F capacitor, for example, would become a 0.5-H inductor.) The prototype highpass response at ω will be equal to the prototype lowpass response at 1/ω. Convert the lowpass filter of Figure 4.4(b) into a highpass filter. (Answer: 1 H, 0.5 F, 1 H.) Next, scale it to have a cutoff frequency of 5 kHz and to operate at 50 ohms. Finally, convert the scaled filter into a bandstop filter with a stopband 10 kHz wide, centered at 500 kHz.

Problem 4.6. Enhance your ladder network analysis program (Problem 1.3) to display not just the amplitude response of a network, but also the phase response (phase angle of the output voltage minus phase angle of the input voltage). Calculate the phase response of the Butterworth filter in Figure 4.4(a). Note: ladder networks belong to a class of networks ("minimum phase networks") for which the amplitude response uniquely determines the phase response and vice versa. In Chapter 12 we will encounter "allpass" filters which are not in this class; phase varies with frequency while amplitude remains constant.

Example answer: For the MATLAB program listing in Problem 1.3, simply insert the following two lines ahead of the last two lines in the original program.

```
figure(3);plot(-180/pi*angle(Vgen));
grid;xlabel('Frequency');ylabel('degrees');title('Phase response');
```

References

[1] Fink, D. G., *Electronic Engineers' Handbook*, New York: McGraw-Hill, 1975. See Section 12, Filters, Coupling Networks, and Attenuators by M. Dishal. Contains an extensive list references.

[2] Matthaei, G., Young, L. and Jones, E. M. T., *Microwave Filters, Impedance-Matching Networks, and Coupling Structures* New York: McGraw Hill, 1964, reprinted in 1980 by Artech House, Inc. Contains fully developed designs, comparing measured results with theory (spectacular fits, even at microwave frequencies) and has an excellent introduction and review of the theory.

5 Frequency converters

A common operation in RF electronics is frequency translation, whereby all the signals in a given frequency band are shifted to a higher frequency band or to a lower frequency band. Every spectral component is shifted by the same amount. Cable television boxes, for example, shift the selected cable channel to a low VHF channel (normally channel 3 or 4). Nearly every receiver (radio, television, radar, cell phone, …) uses the *superheterodyne* principle, in which the desired channel is first shifted to an intermediate frequency or "IF" band. Most of the amplification and bandpass filtering is then done in the fixed IF band, with the advantage that nothing in this major portion of the receiver needs to be retuned when a different station or channel is selected. The same principle can be used in frequency-agile transmitters; it is often easier to shift an already modulated signal than to generate it from scratch at an arbitrary frequency. Frequency translation is also called *conversion* and is even more commonly called *mixing*.[1]

5.1 Voltage multiplier as a mixer

A mixer takes the input signal or band of signals (segment of spectrum), which is to be shifted, and combines it with a reference signal whose frequency is equal to the desired shift in frequency. In a radio receiver, the reference or "L.O." signal is a sine-wave, generated within the receiver by a *local oscillator*.[2] Mixers, in order to produce new frequencies, must necessarily be nonlinear since linear circuits can change only the amplitudes and phases of a set of superposed sine waves. Multiplication is the nonlinear operation used in mixers

---

[1] In audio work "mixing" means addition, a linear superposition that produces no new frequencies. In RF work, however, mixing means multiplication; an RF mixer either directly or indirectly forms the product of the input signal voltage and a sinusoidal "local oscillator" (L.O.) voltage. Multiplication produces new frequencies.

[2] The earliest radio receivers employed no frequency conversion, so they had no "local" oscillator; the only oscillator was remote – at the transmitter location.

Figure 5.1. A voltage
multiplier used as a frequency
converter (mixer).

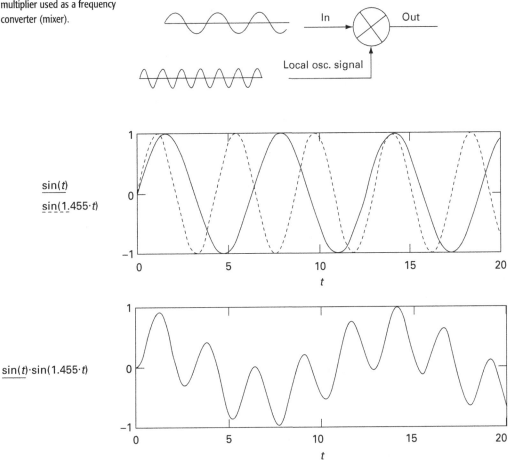

Figure 5.1. A voltage multiplier used as a frequency converter (mixer).

Figure 5.2. Multiplier input and output waveforms.

to produce new signals at the shifted frequencies. Figure 5.1 shows a voltage multiplier with its signal and L.O. inputs. A circumscribed X is the standard symbol for a mixer. The input signal port of the mixer is usually labeled "RF," while the other two ports are labeled "LO" and "OUT" (or "IF").

The output voltage from the multiplier is the product (or proportional to the product) of the two input voltages. In Figure 5.2 a sine-wave input signal is multiplied by an L.O. that is 1.455 times higher in frequency. These multiplicands are shown in the top graph. The bottom graph shows their product which can be seen to contain frequencies both higher and lower than the original frequencies.

The familiar "sin(a)sin(b)" trigonometric identity shows that, in this simple case, the multiplier output consists of just two frequencies: an up-shifted signal at $\omega_L + \omega_R$ and a down-shifted signal at $\omega_L - \omega_R$:

$$\sin(\omega_R t)\sin(\omega_L t) = \tfrac{1}{2}[\cos(\omega_R - \omega_L)t - \cos(\omega_R + \omega_L)t]. \qquad (5.1)$$

If we replace the single RF signal by $V_1\sin(\omega_{R1}t) + V_2\sin(\omega_{R2}t)$, a signal with two spectral components, you can confirm that the output will be

$$\tfrac{1}{2}V_1 \cos([\omega_{R1} - \omega_L]t) + \tfrac{1}{2}V_2 \cos([\omega_{R2} - \omega_L]t)$$
$$-\tfrac{1}{2}V_1 \cos([\omega_{R1} + \omega_L]t) - \tfrac{1}{2}V_2 \cos([\omega_{R2} + \omega_L]t). \qquad (5.2)$$

Just as this linear combination of two signals is faithfully copied into both an up-shifted band and a down-shifted band, any linear combination, i.e., any spectral distribution of signals, will be faithfully copied into these shifted output bands. With respect to signals applied to the RF port, you can see that the mixer is a linear device; all the components are translated (both up and down) in frequency, but their relative amplitudes are left unchanged and there is no interaction between them. Usually one wants only the up-shifted band or only the down-shifted band; the other is eliminated with an appropriate bandpass filter. Ideal analog (or digital) multipliers are being used more commonly as mixers in RF electronics as their speeds increase with improving technology.

5.2 Switching mixers

If the L.O. waveform is square, rather than a sinusoidal, the mixer output will contain not only the fundamental up-shifted and down-shifted outputs but also components at offsets corresponding to the third, fifth, and all other odd harmonics of the L.O. frequency, i.e., at offsets corresponding to all the frequencies in the Fourier decomposition of the square wave. You can confirm this by simply multiplying the Fourier series for the square-wave L.O. by the superposition of signals in the input at the RF port. These new components are usually very easy to filter out so there is no disadvantage in using a square-wave L.O. In fact, there is an advantage. Consider an L.O. signal that is a square wave with values ±1. In this case, since the multiplier multiplies the input signal only by either +1 or −1, it can be replaced by an electronic SPDT switch that connects the output alternately to the input signal and the negative of the input signal. This equivalence is shown in Figure 5.3.

The phase inversion needed for the bottom side of the switch can easily be done with a center-tapped transformer and the switching can be done with two

Figure 5.3. Switching mixer operation.

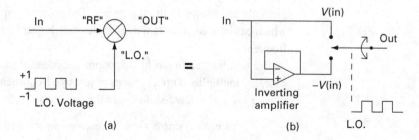

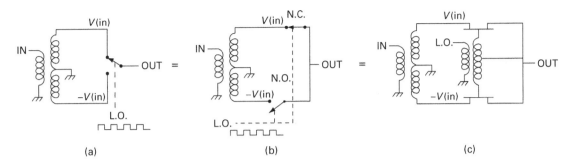

Figure 5.4. Active switching mixer using transistors.

Figure 5.5. Alternate active switching mixer.

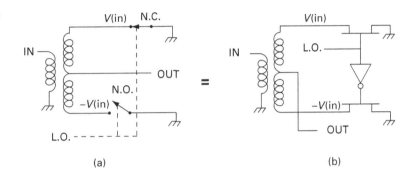

transistors, one for the high side and one for the low side. In the circuit of Figure 5.4 the switches are FETs. (Mixers based on transistors are called *active mixers*.) A second center-tapped transformer provides the L.O. phase inversion so that one FET is turned on while the other is turned off.

We could just as well have taken the signal from the center tap and used the FETs to ground one end of the secondary and then the other. With this arrangement, shown in Figure 5.5, it is easier to provide the drive signals to the transistors, since they are not floating.

Diodes are commonly used as the switching elements for the arrangement shown in Figure 5.5. This results in the *passive* switching mixer circuit shown in Figure 5.6.

Voltage from the L.O. transformer alternately drives the top diode pair and the bottom pair into conduction. The L.O. signal is made large enough that the conducting diodes have very low impedance (small depletion region) and the nonconducting diodes have a very large impedance (wide depletion region). The end of the input transformer connected between the turned-on diode pair is effectively connected to ground through the secondary of the L.O. transformer. Note that current uses both sides of the L.O. transformer on the way to ground, so no net flux is created in that transformer and it has zero impedance for this current. This circuit is usually drawn in the form shown in Figure 5.6(b) and is referred to as a *diode ring mixer*.

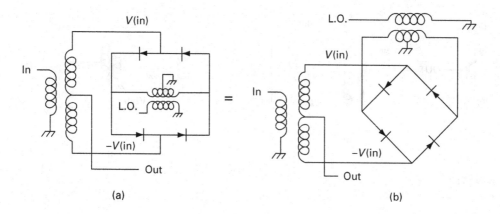

(a)   (b)

Figure 5.6. Diode ring mixer.

Figure 5.7. Unbalanced
switching mixers.

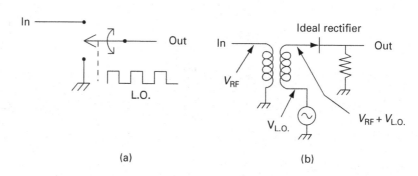

(a)   (b)

All the switching mixers shown above are "double-balanced" which means that no L.O. frequency energy appears at the RF or IF ports and no RF, except for the mixing products, appears at the OUT port. A balanced mixer is desirable, for example, when it is the first element in a receiver. An unbalanced mixer would allow L.O. energy to feed back into the antenna and the radiation could cause interference to other receivers (and could also reveal the position of the receiver). An unbalanced switching mixer is shown in Figure 5.7(a). It multiplies the signal by a square wave that goes from $+1$ to 0 (rather than $+1$ to -1) which is just a $+\frac{1}{2}$ to $-\frac{1}{2}$ square wave together with a bias of $\frac{1}{2}$. The square-wave term produces the up-shifted and down-shifted bands as before, but the bias term allows one-quarter of the RF input power to get through, unshifted in frequency, to the output.

A simple version of this mixer, using an ideal rectifier, is shown in Figure 5.7 (b). The L.O. and RF voltages are added here, by means of a transformer, and their sum is rectified. The voltage at the output is equal to the sum voltage when the sum is positive and is equal to zero when the sum is negative. If the L.O. voltage is large compared to the RF voltage, the rectifier effectively conducts when the L.O. voltage is positive and disconnects when the L.O. voltage is negative, allowing the resistor to pull down the output voltage to zero. Thus, the RF signal is switched to the output at the L.O. rate. Note, however, that this mixer is totally unbalanced; both the L.O. and RF signals appear at the mixer, together with the sum and difference frequencies.

Switching mixer loss

Let us consider the loss in the switching mixers in Figures 5.4, 5.5, and 5.6. Refer to Figure 5.4(a) and assume that the OUT port is terminated by a load resistor R_L whose value equals R_S, the source impedance of a sine-wave signal source connected to the IN port. Note that the source has no way of knowing that the load resistor is being reversed on half of every cycle of the L.O. The source just sees a matched load and therefore delivers its maximum power. Some of the power on the load is at the desired sum or difference frequency. The ratio of this desired power to the power available from the source is known as the conversion gain. For the diode ring mixer, the ratio is less than unity, i.e., a conversion loss. We can easily calculate this loss. The reversing switch presents the load with a voltage that is half the source voltage (since R_S and R_L form a voltage divider), multiplied by a dimensionless ± 1 square wave. Fourier analysis shows that the square wave is made of a sine wave at the square-wave frequency plus a sine wave at every odd multiple of this frequency. The amplitude of the fundamental sine wave is $4/\pi$. The amplitudes of the higher harmonics fall off as $1/n$. Evaluating the product of this square wave and the source voltage we have

$$
\begin{aligned}
V_{OUT} &= \tfrac{1}{2}V_S \cos(\omega_R t) \cdot [4/\pi](\cos(\omega_L t) + 3^{-1}\cos(3\omega_L t) + 5^{-1}\cos(5\omega_L t) + \ldots] \\
&= \tfrac{1}{2}V_S(4/\pi)\cos(\omega_R t)\cos(\omega_L t) + \ldots \\
&= \tfrac{1}{2}V_S(4/\pi)[(1/2)\cos(\omega_R - \omega_L)t + (1/2)\cos(\omega_R + \omega_L)t] + \ldots
\end{aligned}
$$

$$(5.3)$$

We see that the amplitude of the desired sum or difference frequency component is $\tfrac{1}{2}V_S(4/\pi)\cdot(1/2)$. The amplitude available from the source is $\tfrac{1}{2}V_S$, so the conversion gain is the ratio of the squares of these amplitudes:

$$
\text{Conversion gain} = \left(\frac{1}{2}\frac{4}{\pi}\frac{1}{2}\right)^2 \Bigg/ \left(\frac{1}{2}\frac{4}{\pi}\frac{1}{2}\right)^2 \left(\frac{1}{2}\right)\left(\frac{1}{2}\right)^2 = \left(\frac{4}{\pi}\frac{1}{2}\right)^2 = 0.4052
$$

$$(5.4)$$

or, in dB, $10 \log (0.4052) = -3.92$ dB. In practice, the loss is typically greater than this by a dB or so, due to loss in the diodes and in the transformers.

5.3 A simple nonlinear device as a mixer

Finally, let us consider a mixer that uses a single nonlinear device, but not as a switch. Figure 5.8 shows a single-diode mixer. The first op-amp is used to sum the RF and L.O. voltages. The sum is applied to the diode. The input of the second op-amp is a virtual ground so the full sum voltage is applied across the

Figure 5.8. Hypothetical single-diode mixer circuit.

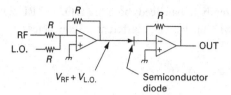

$V_{RF} + V_{L.O.}$ Semiconductor diode

diode. With its feedback resistor, the second op-amp acts as a current-to-voltage converter; it produces a voltage proportional to the current in the diode. The current, a nonlinear (exponential) function of the applied voltage, will contain mixing products at frequencies $N\omega_{RF} \pm M\omega_{L.O.}$ where N and M are simple integers. Note that this circuit is essentially the same as that of Figure 5.7(b), except that here we are considering low-level signals, where the diode cannot be treated as an ideal rectifier.

This op-amp circuit is intended to emphasize that the diode's nonlinearity operates on the *sum* of the RF and L.O. Commonly used circuits use passive components and the summing is not always obvious. Diodes are exponential devices; the current vs. applied voltage is given by

$$I = I_s(\exp(V/V_{th})-1), \tag{5.5}$$

where $V_{th} = V_{thermal} = kT/e$ (Boltzmann's constant × absolute temperature / electron charge) $= 26\,\mathrm{mV}$. The term I_s is a temperature-dependent "saturation current." In a small-signal situation, i.e., when $V \ll 26\,\mathrm{mV}$, we can expand the exponential to find the output of the above mixer:

$$V_{out} = I_s R[V/V_{th} + (V/V_{th})^2/2! + (V/V_{th})^3/3! + \ldots \tag{5.6}$$

Since $V = V_{RF} + V_{L.O.}$, the first term will give feedthrough (no balance) at both the RF and L.O. frequencies. The second term (the square law term) will produce the desired up-shifted and down-shifted sidebands since the square of $V_{L.O.} + V_{RF}$ contains the cross-product, $2V_{RF} V_{L.O.}$. This term also produces bias terms and double frequency components. The third-order term will give outputs at the third harmonics of the RF and L.O. frequencies and at $2\omega_{RF}+\omega_{L.O.}$, $2\omega_{RF}-\omega_{L.O.}$, $2\omega_{L.O.}+\omega_{RF}$, and $2\omega_{L.O.}-\omega_{RF}$. Normally these products are far removed from the desired output band and can be filtered out. If the input voltage is small enough, we do not have to continue the expansion. For larger signals, however, the next term (fourth-order) gives undesirable products within the desired output band. To see how this happens, consider an input signal with two components, $A_1\cos(\omega_1 t)$ and $A_2\cos(\omega_2 t)$. One of the fourth-order output terms will be, except for a constant,

$$\cos(\omega_L t)[A_1 \cos(\omega_1 t) + A_2 \cos(\omega_2 t)]^3. \tag{5.7}$$

You can expand this expression to show that it contains components with frequencies $\omega_L+2\omega_1+\omega_2$ and $\omega_L-2\omega_1-\omega_2$. When ω_1 and ω_2 are close to each other, $2\omega_1 + \omega_2$ and $2\omega_1 - \omega_2$ are nearby and can lie within the desired output band. In a radio receiver, this means that two strong signals will create an objectionable mixing product at a nearby, i.e., inband, frequency which will impede the reception of a weak signal at that frequency.

Simple mixers can also be made with a transistor. A bipolar transistor, if driven by the sum of the RF and L.O. voltages, will have a collector current containing the same set of frequency components as the diode mixer discussed above. Sometimes a dual-gate FET is used as a mixer; the L.O. voltage is applied to one gate and the RF voltage is applied to the other. This provides some isolation between L.O. and RF (which is provided automatically in a balanced mixer such as the diode ring mixer).

We will see later that multiplication, the basis of mixing, is also the operation needed to modulate the amplitude of a *carrier* sine wave, i.e., to produce amplitude modulation (AM). Multiplication, mixing, and AM modulation are all the same basic operation.

Problems

Problem 5.1. Sometimes two multipliers, two phase shifters, and an adder are used to build a mixer that has only one output band (a so-called single-sideband mixer). The design for an *upper* sideband mixer, for example, follows directly from the identity:

$$\cos([\omega_{RF} + \omega_{L.O.}]t) = \cos(\omega_{RF}t)\cos(\omega_{L.O.}t) - \sin(\omega_{RF}t)\sin(\omega_{L.O.}t).$$

Draw a block diagram for a circuit that carries out this operation.

Problem 5.2. The diode ring switching mixer also works when the L.O. and RF ports are interchanged. Explain the operation in this case.

Problem 5.3. Show that the diode ring switching mixer will work if the L.O. frequency is one-third of the nominal L.O. frequency. This is sometimes done for convenience if this $\frac{1}{3}\omega_{L.O.}$ frequency is readily available. Find the conversion gain (loss) for this situation. Why would this scheme not work if the L.O. frequency is half the nominal L.O. frequency?

Problem 5.4. Consider a situation where two signals of the same frequency but with a phase difference, θ, are separately mixed to a new frequency. Suppose identical mixers are used and that they are driven with the same L.O. signal. Show that the phase difference of the shifted signals is still θ.

Problem 5.5. In RF engineering, considerable use is made of the trigonometry identities $\cos(a+b) = \cos(a)\cos(b) - \sin(a)\sin(b)$ and $\sin(a+b) = \sin(a)\cos(b) + \cos(a)\sin(b)$. Prove these identities, either using geometric constructions or using the identity $e^{jx} = \cos(x) + j\sin(x)$.

6 Amplitude and frequency modulation

Modulation means *variation* of the amplitude or the phase (or both) of an otherwise constant sinusoidal RF *carrier wave* in order that the signal carry information: digital data or analog waveforms such as audio or video. In this chapter we look at pure amplitude modulation (AM) and pure frequency modulation (FM). Historically, these were the first methods to be used for communications and broadcasting. While still used extensively, they are giving way to modulation schemes, mostly digital, some of which amount to simultaneous AM and FM.

The simplest form of AM is *on/off keying*. This binary digital AM (full on is a data "1" and full off is a data "0") can be produced with a simple switch, originally a telegraph key in series with the power source or the antenna. The first voice transmissions used a carbon microphone as a variable resistor in series with the antenna to provide a continuous range of amplitudes. With AM, the frequency of the carrier wave is constant, so the zero crossings of the RF signal are equally spaced, just as they are for an unmodulated carrier. The simplest FM uses just two frequencies; the carrier has frequency f_0 for data "zero" and $f_0 + \Delta f$ for data "one." FM is usually generated by a VCO (voltage-controlled oscillator). For binary FSK (frequency shift keying), the control voltage has only two values: one produces f_0 and the other produces $f_0 + \Delta f$. FM broadcasting uses a continuous range of frequencies; the instantaneous frequency is determined by the amplitude of the audio signal. With FM, the amplitude of the carrier wave signal is constant. Figure 6.1 shows an unmodulated carrier wave, an AM-modulated wave, an FM-modulated wave, and a wave with simultaneous AM and FM modulation.

Phase modulation is a type of frequency modulation, since frequency is the time derivative of phase.

Amplitude and frequency (or phase) exhaust the list of carrier wave properties that can be modulated. The fractional bandwidth of RF signals is low enough that if one zooms in on a stretch of several cycles, the waveform is essentially sinusoidal and can be described by just its amplitude and frequency. Schemes such as single-sideband suppressed carrier (SSBSC), double-sideband

Figure 6.1. (a) Unmodulated wave; (b) with AM modulation; (c) with FM modulation; (d) with simultaneous AM and FM.

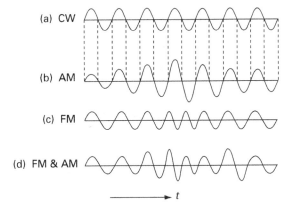

suppressed carrier AM (DSBSC), quadrature amplitude modulation (QAM), and others, produce simultaneous amplitude and frequency modulation. A sine-wave generator with provisions for both amplitude and phase modulation can, in principle, produce a radio signal with *any* specified type of modulation, for example, the comb of 52 individually modulated subcarriers used in Wi-Fi data links. Note: the term *digital modulation* means that the modulation uses a finite number of modulation states (often just two) rather than a continuum of states and that the time spent in any state is an integral multiple of a fundamental symbol time (baud).

6.1 Amplitude modulation

Amplitude modulation, the first scheme used for radio broadcasting, is still used in the long-wave, middle-wave, and short-wave broadcast bands.[1] Let us examine AM, first in the time domain and then in the frequency domain.

6.1.1 AM in the time domain

Without modulation (when the speech or music is silent) the voltage applied to the antenna is a pure sine wave at the carrier frequency. The power of a radio

---

[1] The long-wave (LW) band, used only outside the Western Hemisphere, extends from 153 to 179 kHz. The middle-wave (MW) band extends from 520 to 1700 kHz. Short-wave bands are usually identified by wavelength: 75 m, 60 m, 49 m, 41 m, 31 m, 25 m, 19 m, 16 m, 13 m and 11 m. The spacing between LW frequency assignments is 9 kHz. MW spacings are 10 kHz in the Western Hemisphere and 9 kHz elsewhere. Short-wave frequency assignments are less coordinated but almost all short-wave stations operate on frequencies which are an integral number of kHz.

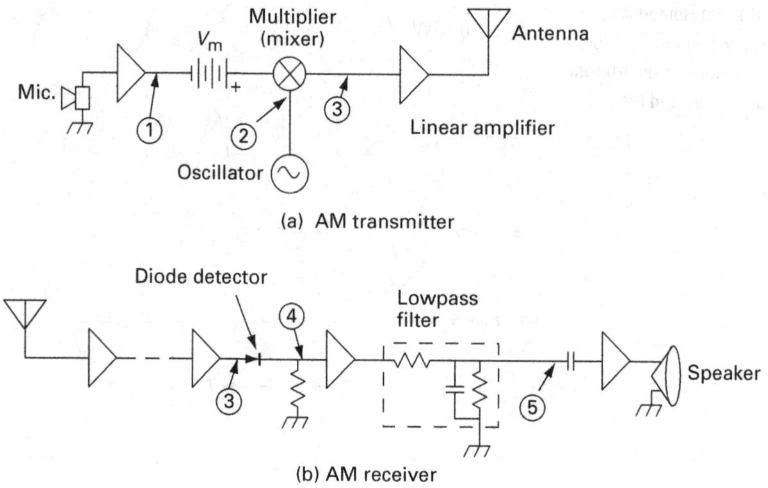

Figure 6.2. Hypothetical AM transmitter and receiver.

(a) AM transmitter

(b) AM receiver

station is defined as the transmitter output power when the modulation is zero. The presence of an audio signal changes the amplitude of the carrier. Figure 6.2 shows a hypothetical AM transmitter and receiver.

The audio signal (amplified microphone voltage) has positive and negative excursions but its average value is zero. Suppose the audio voltage is bounded by $+V_m$ and $-V_m$. A dc bias voltage of V_m volts is added to the audio voltage. The sum, $V_m + V_{audio}$, is always positive and is used to multiply the carrier wave, $\sin(\omega_c t)$. The resulting product is the AM signal; the amplitude of the RF sine wave is proportional to the biased audio signal. The simulation in Figure 6.3 shows the various waveforms in the transmitter and receiver of Figure 6.2 corresponding to a random segment of an audio waveform. The biased audio waveform is called the *modulation envelope*. Note that at full modulation where $V_{audio} = +V_m$, the carrier is multiplied by $2V_m$ whereas, at zero modulation, the carrier is multiplied by V_m (bias only). This factor of 2 in amplitude means the fully (100%) modulated signal has four times the power of the unmodulated signal (the carrier wave alone). It follows that the antenna system for a 50 000 W AM transmitter must be capable of handling 200 000 W peaks without breakdown. The average power of the modulated signal is determined by the average square of the modulation envelope. For example, in the case of 100% modulation by a single audio tone, the average power of the modulated signal is greater than the carrier by a factor of $\langle (1+\cos\theta)^2 \rangle = \langle 1+2\cos(\theta) + \cos(\theta)^2 \rangle = 1 + \langle \cos(\theta)^2 \rangle = 3/2$.

Figure 6.2 also shows how the receiver demodulates the signal, i.e., how it recovers the modulation envelope from the carrier wave. The detector is just an ideal rectifier, which eliminates the negative cycles of the modulated RF signal. A simple lowpass filter then produces the average voltage of the positive loops. (Usually this is a simple *RC* filter, but the lowpass filter used in Figure 6.3 is a

A piece of an arbitrary audio waveform
$$V1(t) := 3.6 \cdot \sin(2 \cdot \pi \cdot 21 \cdot t) - 5.3 \cdot \cos(2\pi \cdot 62 \cdot t)$$

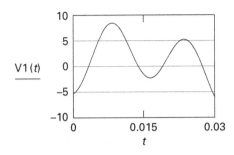

Sine–wave carrier
$$V2(t) := \sin(2 \cdot \pi \cdot 1000 \cdot t)$$

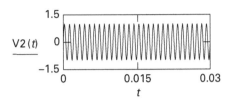

Modulated carrier
$$V3(t) := V2(t) \cdot (10 + V1(t))$$

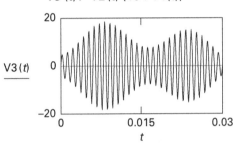

Received signal after rectification by detector diode
$$V4(t) := V3(t) \cdot (V3(t) > 0)$$

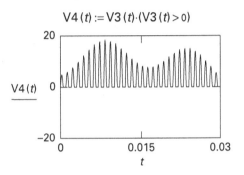

A lowpass filter (here a "box car" running average) recovers the modulation envelope

$$V5(t) := \frac{1}{40} \cdot \sum_{i=-20}^{20} V4(t + .0001 \cdot i)$$

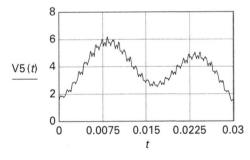

Figure 6.3. Waveforms in the AM system of Figure 6.2.

"boxcar" integrator that forms a running average.) Finally, ac coupling removes the bias, leaving an audio signal identical to the signal from the microphone.

6.1.2 AM in the frequency domain

Let us look at the spectrum of the AM signal to see how the power is distributed in frequency. This is easy when the audio signal is a simple sine wave, say $V_a \sin$

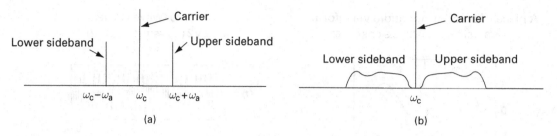

Figure 6.4. Spectrum of an AM power spectrum: (a) modulation by a single-tone modulation; (b) modulation by an audio spectrum.

$(\omega_a t)$. The voltage at point 3 in Figure 6.2 is then $\sin(\omega_c t)(V_m + V_a\sin(\omega_a t)) = V_m\sin(\omega_c t) + 1/2\ V_a\cos([\omega_c - \omega_a]t) - 1/2\ V_a\cos([\omega_c + \omega_a]t)$. These three terms correspond to the *carrier* at ω_c with power $V_m^2/2$, a *lower sideband* at $\omega_c - \omega_a$ with power $V_a^2/8$, and an *upper sideband* at $\omega_c + \omega_a$, also with power $V_a^2/8$. This spectrum is shown in Figure 6.4(a).

Note that the carrier, the component at ω_c, is always present and that its amplitude is independent of the modulation level. At 100% sine-wave modulation $V_a = V_m$ and the average power in each sideband is $\frac{1}{4}$ of the carrier power. The maximum average power is therefore $\frac{3}{2}$ times the carrier power, as we saw earlier from the time domain description. The sidebands, which carry all the information, account here for only $\frac{1}{3}$ of the total power transmitted. With speech waveforms the sidebands may contain even less power.

If, instead of a single sine wave, the audio signal is a superposition of sine waves with different frequencies (and amplitudes), the above analysis is readily extended to show that the upper and lower sidebands become continuous bands of frequency components straddling the carrier symmetrically, as shown in Figure 6.4(b). As the audio signal changes, the shape and size of the sidebands change. A typical audio signal has components at frequencies as high as 10 KHz. AM broadcast stations restrict their audio to not exceed 10 KHz but this still produces a 20 KHz-wide spectrum. The spacing between frequency assignments in the AM broadcast band is only 10 KHz. To prevent overlap, no two stations in a given listening area are assigned the same frequency or adjacent frequencies.

6.2 Frequency and phase modulation

In frequency modulation (FM) and phase modulation (PM) the audio voltage controls the phase angle of the sinusoidal carrier wave while the amplitude remains constant. Since frequency is the time derivative of phase, the two quantities cannot be varied independently; FM and PM are only slightly different methods of *angle modulation*. The advantage of FM (or PM) broadcasting over AM broadcasting is that, under strong signal conditions, the audio signal-to-noise ratio at the output of the receiver can be much higher for FM than for AM. The only price paid for this improvement is increased bandwidth (this topic is discussed in detail in Chapter 23).

6.2.1 FM in the time domain

A simple unmodulated carrier wave is given by $V(t) = \cos(\omega_c t + \phi_0)$ where ω_c and ϕ_0 are constants. The phase, $\phi(t) = (\omega_c t + \phi_0)$, of this cw signal increases linearly with time at a rate ω_c radians per second. In FM, the instantaneous frequency is made to shift away from ω_c by an amount proportional to the modulating audio voltage, i.e., $\omega(t) = \omega_0 + k_{osc} V_a(t)$. A linear voltage-controlled oscillator (VCO) driven by the audio signal, as shown in Figure 6.5a, makes an FM modulator.

The excursion from the center frequency, $k_{osc} V_a$, is known as the (radian/s) *deviation*. In FM broadcasting the maximum deviation, defined as 100% modulation, is 75 kHz, i.e., $k_{osc} V_{a_{max}} / 2\pi = 75 \cdot 10^3$. Consider the case of modulation by a single audio tone, $V_a(t) = A \cos(\omega_a t)$. The instantaneous frequency of the VCO is then $\omega(t) = \omega_c + k_{osc} A \cos(\omega_a t)$. And, since the instantaneous frequency is the time derivative of phase, the phase is given by the integral

$$\phi(t) = \int \omega(t)\mathrm{d}t = \int (\omega_c + k_{osc} A \cos(\omega_a t))\mathrm{d}t = \omega_c t + \frac{k_{osc} A}{\omega_a}\sin(\omega_a t) + \phi_0.$$

$$(6.1)$$

From Equation (6.1), we see that, in addition to the linearly increasing phase, $\omega_c t$, of the unmodulated carrier, there is a phase term that varies sinusoidally at the audio frequency. The amplitude of this sinusoidal phase term is $k_{osc} A/\omega_a$. This maximum phase excursion is known as the *modulation index*. The inverse dependence on the modulation frequency, ω_a, results from phase being the integral of frequency. Phase modulation (PM) differs from FM only in that it does not have this ω_a denominator; it could be produced by a fixed-frequency oscillator followed by a voltage-controlled phase shifter. An indirect way to produce a PM signal is to use a standard FM modulator (i.e., a VCO), after first passing the audio signal through a differentiator. The differentiator produces a factor, ω_a, which cancels the ω_a denominator produced by the VCO. This scheme is shown in Figure 6.6.

Similarly, a PM transmitter can be adapted to produce FM. A possible receiver for PM would be the receiver of Figure 6.5(b) with an integrator after

Figure 6.5. Basic FM system: (a) transmitter; (b) receiver.

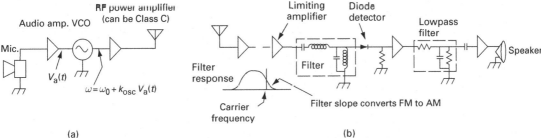

(a)   (b)

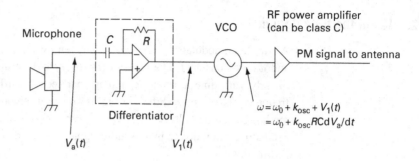

Figure 6.6. Phase modulation via frequency modulation.

the diode detector to undo the differentiation done in the transmitter of Figure 6.6. (See Chapter 18 for more FM and PM detectors.) In the simple FM receiver of Figure 6.5(b), a bandpass filter converts FM into AM because the carrier frequency is placed on the slope of the filter. When the carrier frequency increases, the filter response decreases the amplitude. After this filter, the rest of the receiver is an AM receiver. A limiting amplifier ahead of the filter is a refinement to eliminate amplitude noise, as well as to make the audio volume independent of signal strength. More refined AM and FM detectors are presented in Chapter 18.

6.2.2 Frequency spectrum of FM

Using the expression for phase from Equation (6.1) (neglecting ϕ_0) and using the usual expansion for $\cos(a+b)$, the complete FM signal (VCO output) for single-tone modulation becomes

$$V(t) = \cos(\phi(t)) = \cos(\phi_m \sin(\omega_a t)) \cos(\omega_c t) - \sin(\phi_m \sin(\omega_a t)) \sin(\omega_c t)$$

(6.2)

where $\phi_m = k_{osc}A/\omega_a$ is the modulation index. Let us first consider the case where the modulation index is small.

6.2.3 Narrowband FM or PM

At very low modulation levels, both FM and PM produce a power spectrum similar to AM, i.e., a carrier with an upper sideband ω_a above the carrier and a lower sideband ω_a below the carrier. To see this, we expand Equation (6.2) for very small ϕ_m, obtaining

$$V(t) \approx \left(1 - \frac{\phi_m^2 \sin^2(\omega_a t)}{2}\right) \cos(\omega_c t) - \phi_m \sin(\omega_a t) \sin(\omega_c t).$$

(6.3)

The first term is an almost constant amplitude spike at the carrier frequency, much like the carrier of an AM signal. You can verify directly that the amplitude of $V(t)$ remains constant by summing the squares of the coefficients of $\cos(\omega_c t)$ and $\sin(\omega_c t)$. Expanding the right-hand term gives the two sidebands,

$$V(t) \approx \cos(\omega_c t)\left(1 - \frac{\phi_m^2 \sin^2(\omega_a t)}{2}\right) + (\phi_m/2)\cos[(\omega_c + \omega_a)t]$$
$$- (\phi_m/2)\cos[(\omega_c - \omega_a)t]. \tag{6.4}$$

This differs from an AM signal in that the sidebands are equivalent to having multiplied the audio signal not by $\cos(\omega_c t)$, the carrier, but by $\sin(\omega_c t)$. Note also that, in this limit of small modulation index, the amplitude of the sidebands is small and the width of the spectrum stays fixed at $2\omega_a$, just as with AM.

6.2.4 Wideband FM spectrum

Looking at Equation (6.2), we see that the signal consists of the products of $\cos(\omega_c t)$ and $\sin(\omega_c t)$, multiplied respectively by the baseband signals $\cos[\phi_m \sin(\omega_a t)]$ and $\sin[\phi_m \sin(\omega_a t)]$. When ϕ_m is not small, these products are similar and each consists of upper and lower sidebands around ω_c, with frequency components spaced from the carrier frequency by integral multiples of ω_a. The $\cos(\omega_c t)$ and $\sin(\omega_c t)$ carriers are suppressed and the spectrum has no distinct central carrier spike. An exact analysis uses Fourier analysis of the cos (sin) and sin(sin) terms[2] to find the comb of sidebands, but a simpler argument can give an estimate of the bandwidth of the FM signal. Referring again to Equation (6.2), the baseband modulating signals produce sidebands. In a time equal to one-quarter of an audio cycle, the expression $\phi_m \sin(\omega_a t)$ changes by ϕ_m radians, so the phase of the baseband modulating signals changes by ϕ_m radians. Dividing this phase change by the corresponding time interval, $1/(4f_a) = \pi/(2\omega_a)$, the average sideband frequency is given by $2\phi_m \omega_a/\pi$. Substituting $\phi_m = k_{osc} A/\omega_a$, the average frequency is $2k_{osc} A/\pi = (2/\pi) \times$ deviation. Therefore, the one-sided bandwidth of the FM signal, when the modulation index is large, is roughly equal to the maximum deviation and the full bandwidth is about twice the maximum deviation. If the deviation is reduced, the bandwidth goes down proportionally at first but then, in the narrowband regime, stays constant at twice the audio bandwidth, as shown above.

---

[2] The terms in Equation (6.2) can be expanded into harmonics of the audio frequency by using the Bessel function identities:

$$\cos(\phi \sin(\omega t)) = J_0(\phi) + 2\sum_{n=1}^{n=\infty} J_{2n}(\phi)\cos(2n\omega t)$$

and

$$\sin(\phi \sin(\omega t)) = 2\sum_{n=0}^{n=\infty} J_{2n+1}(\phi)\sin((2n+1)\omega t).$$

6.2.5 Frequency multiplication of an FM signal

When a signal is passed through a times-N frequency multiplier its phase is multiplied by N. A square-law device, for example, can serve as a frequency doubler since $\cos^2(\phi) = 1/2 + 1/2\cos(2\phi)$. So if the phase of an FM signal, ϕ, (the right-hand expression in Equation 6.1), is multiplied by 2 by using a frequency doubler, both ω_c and kA/ω_a are multiplied by 2, i.e., the frequency and the modulation index are doubled. When a given VCO cannot be linearized well enough within the intended operating range, it can be operated with a low deviation and its signal can be frequency-multiplied to increase the deviation. If the resulting center frequency is too high, an ordinary mixer can shift it back down, preserving the increased deviation.

6.3 AM transmitters

The simple AM transmitter shown in Figure 6.2 is entirely practical. (Of course we would fix it up so that a battery would not be needed to supply the bias voltage.) The linear RF power amplifier would be the major part of this transmitter. We saw in Chapter 3 that a class-B linear amplifier has relatively high efficiency, $\pi/4$ or 78.5%, but that is for a maximum-amplitude sine wave. Let us find the efficiency for the transmitter of Figure 6.2, assuming the average modulation power is 10% of the peak modulation power. (The "crest" factor for speech is often taken to be 16 (12 dB), but broadcasters normally use dynamic range compression to increase "loudness.") Let the output signal be given by

$$V_{out} = V_{dc} \tfrac{1}{2}[1 + V_m(t)] \cos(\omega t) \tag{6.5}$$

where $|V_m(t)| \leq 1$. Note that the maximum output voltage is V_{dc}, the power supply voltage. (Refer to the amplifier circuit of Figures 3.7 or 3.8.) Remembering that $\langle \cos^2 \rangle = 1/2$ and assuming that the audio signal is an ac signal, i.e., $\langle V_m(t) \rangle = 0$, we can express the average output power:

$$\begin{aligned}
P_{out} &= \langle V_{out}^2 \rangle / R = \tfrac{1}{4} V_{dc}^2 \langle [1 + V_m(t)]^2 \rangle / (2R) \\
&= \tfrac{1}{4} V_{dc}^2 (1 + 0.1)/(2R) = 1.1 V_{dc}^2/(8R),
\end{aligned} \tag{6.6}$$

where R is the load (antenna) resistance and where we have used the fact that $\langle \cos(\omega t) \rangle = 0$. The average power delivered by the dc supply is given by

$$\begin{aligned}
\langle V_{dc} I_{supply} \rangle &= V_{dc} \langle |V_{out}(t)/R| \rangle \\
&= V_{dc} \cdot \tfrac{1}{2} V_{dc} \langle [1 + V_m(t)/R \rangle 2/\pi = V_{dc}^2/(\pi R),
\end{aligned} \tag{6.7}$$

where we have used the fact that the average of a positive loop $\sin(x)$ is equal to $2/\pi$. Calculating the efficiency, we find

$$\eta = \frac{\text{power out}}{\text{power in}} = (1.1)/V_{\text{dc}}{}^2/(8R) \div V_{\text{dc}}^2/\pi = 1.1\pi/8 = 43.2\%. \qquad (6.8)$$

Therefore, when running this AM transmitter, only 43.2% of the prime power gets to the antenna; the rest is dissipated as heat in the class-B amplifier.

Almost all AM transmitters obtain higher efficiency by using class-C, D, or E RF amplifiers. These amplifiers are not linear in the normal sense, that is, the output signal amplitude is not a constant multiple of the input signal amplitude. But, for a fixed input RF amplitude, the output amplitude *is* proportional to the supply voltage. These amplifiers can therefore be used as high-power *multipliers* that form the product of the power supply voltage times a unit sine wave at the RF frequency. Furthermore, the efficiency of these amplifiers, which is high, is essentially independent of the supply voltage. (These amplifiers are discussed in detail in Chapter 9.) Let us look at the overall efficiency of transmitters using these amplifiers.

6.3.1 AM transmitter using a class-C RF amplifier and a class-B modulator

In the traditional AM transmitter, audio voltage developed by a high-power class-B audio amplifier (the modulator) is added to a dc bias and the sum of these voltages powers a class-C RF amplifier. As explained above, a class-C RF amplifier acts as a high-power multiplier. In the traditional tube-type circuit of Figure 6.7, the dc bias is fed through the secondary of the modulation transformer. Audio voltage produced by the modulator appears across the secondary winding and adds to the bias voltage. With no audio present, the class-B audio amplifier consumes negligible power and the bias voltage supply provides power for the carrier. At 100% sine-wave modulation, the modulator must supply audio power equal to half the bias supply power. A 50 000 watt transmitter thus requires a modulator that can supply 25 000 W of audio power. (Again, this result is for a single tone, but is essentially the same for speech or music.)

Figure 6.7. Class-B plate-modulated AM transmitter.

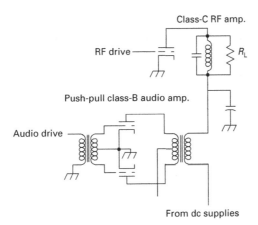

Let us find the efficiency of this transmitter. We will assume the class-C amplifier has an efficiency of 80% and that, as before, the average modulation power is 10% of the maximum possible modulation power. If the normalized output carrier power is 1 W, the dc bias supply must provide 1/0.8 W and the modulator must supply 0.1/0.08 W. To handle peaks, the maximum power from the modulator must be 1/0.08 W. To find the efficiency of the class-B modulator, we must know not only its peak output voltage, but also the mean of the absolute value of its output voltage. (Note that this last piece of information was not needed in the analysis of the class-B RF amplifier transmitter discussed above.) Let us just assume the modulating signal is a single audio sine wave whose power is 0.1/0.8 W. With that assumption, the modulator will have an efficiency of 0.35. The total input power, carrier plus modulation, will therefore be 1/0.8 +(0.1/0.35)/0.8 = 1.61 W. The output power will be 1 + 0.1, so the overall efficiency of this transmitter is 68%, a significant improvement over the transmitter using a linear class-B RF amplifier.

6.3.2 Class-C RF amplifier with a switching modulator

There are several newer methods to produce AM with even higher efficiency. All use switching techniques. The modulator shown in Figure 6.8(a) is just a high-power digital-to-analog converter. It uses solid-state switches to add the voltage of many separate low-voltage power supplies, rather than tubes or transistors to resistively drop the voltage of a single high-voltage supply. Modulators of this type can, in principle, be 100% efficient, since the internal switch transistors are either fully on or fully off. The modulator of Figure 6.8(b) is a pulse rate modulator whose output voltage is equal to the supply voltage multiplied by the duty factor of the switch tube.

This type of circuit, whose efficiency can also approach 100%, is discussed in detail with switching power supplies in Chapter 29. Again specifying an average modulation power of 10%, the combination of an 80% efficient class-C RF amplifier together with a switching modulator, makes an AM transmitter with an overall efficiency of 80%.

Figure 6.8. Class-C RF amplifier with (a) high-power D-to-A modulator and (b) duty cycle switching modulator.

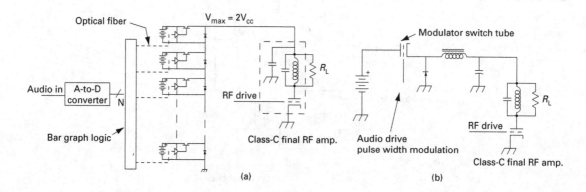

6.3.3 Class-D or E amplifiers with switching modulators

Class D and class-E RF amplifiers can approach 100% efficiency. Like the class-C amplifier, the amplitude of their output sine wave is proportional to the dc supply voltage. When one of these amplifiers is coupled with a switching modulator, the resulting AM transmitters can approach an overall efficiency of 100% for any modulation level.

6.4 FM transmitters

Since FM modulation is generated by a VCO at a low level, and the signal has a constant amplitude, there is no other modulator nor is there any requirement that the final power amplifier have a linear amplitude response. Class-C amplifiers have been used most often, with efficiencies around 80%.

6.5 Current broadcasting practice

Many $\frac{1}{2}$ MW to 2 MW AM stations operate in the long-wave, medium-wave, and short-wave bands. These superpower transmitters use vacuum tubes. Standard AM broadcast band transmitters in the U.S. are limited to 50 kW. For this power and lower powers, new AM transmitters manufactured in the U.S. use transistors. These transmitters combine power from a number of modular amplifiers in the 1 kW range. The advantages of solid state include lower (safer) voltages, indefinite transistor life rather than expensive tube changes every year or two, and better "availability;" a defective module only lowers the power slightly and can be replaced while the transmitter remains on the air. Most FM transmitters over about 10 kW still use vacuum tubes, but solid-state FM transmitters are available up to about 40 kW.

Stereophonic sound was added to FM broadcasting around 1960. Existing monophonic receivers operated as they had, receiving the left + right (L + R) audio signal. Thus the system is (backwardly) *compatible*, just as color television was compatible with existing black and white television receivers. The L − R signal information is in the demodulated signal, but clustered around 38 kHz, above the audible range. The L − R signal uses double-sideband suppressed carrier AM modulation, which is explained in Chapter 8. The demodulator to recover the L-R signal is discussed in Chapter 18. Stereo was added to AM radio broadcasting around 1980. Several systems competed to become the standard, but the public was largely indifferent, having already opted for FM stereo. Compatible digital broadcasting has recently been introduced, with versions for both the AM and FM bands. In this IBOC (In-Band On-Channel) system, a station's digital simulcast signal (which can contain different program material) shares the same channel with the standard AM or FM signal, mostly using the relatively empty spectral regions near (and somewhat past) the nominal edges of the channel. The digital signal, which can be produced by a separate transmitter and may even use a separate antenna, uses COFDM (Coded Orthogonal Frequency-Division Multiplexing), a modulation system, described in Chapter 22. This system is intended to be a stepping-stone to all-digital radio broadcasting in the traditional AM and FM bands. IBOC broadcasting equipment is equipped to transmit an all-digital mode, to be used if and when the traditional analog broadcasting is discontinued.

Problems

Problem 6.1.   An AM transmitter, 100% modulated with an audio sine wave, has sidebands whose total power is equal to half the carrier power. Consider 100% modulation by an audio square wave. What is the ratio of sideband power to carrier power?

Problem 6.2.   Suppose you are trying to listen to a distant AM station, but another station on the same frequency is coming in at about the same strength. Will you hear both programs clearly? If not, how will they interfere with each other?

Problem 6.3.   During periods where the audio signal level is low, the amplitude of an AM signal varies only slightly from the carrier level. The modulation envelope, which carries all the information, rides on top of the high-power carrier. If the average amplitude could be decreased without decreasing the amplitude of the modulation, power could be saved. Discuss how this might be accomplished at the transmitter and what consequences, if any, it might have at the receiver.

Problem 6.4.   Show how a PM transmitter can be used to generate FM.

Problem 6.5.   Consider an FM transmitter modulated by a single audio tone. As the modulation level is increased, the spectral line at the carrier frequency decreases. Find the value of the modulation index that makes the carrier disappear completely.

Problem 6.6.   One of the methods used for compatible AM stereo was a combined AM/PM system. The left + right (L + R) signal AM-modulated the carrier in the usual way, while the L − R signal was used to phase-modulate the carrier. Draw block diagrams of a transmitter and receiver for this system. Why would PM be preferable to FM for the L − R signal?

7 Radio receivers

In this chapter we will be mostly concerned with the sections of the receiver that come before the detector, sections that are common to nearly all receivers: AM, FM, television, cell phones, etc. Basic specifications for any kind of radio receiver are *gain, dynamic range, sensitivity* and *selectivity*, i.e., does a weak signal at the selected frequency produce a sufficiently strong and uncorrupted output (audio, video, or data) and does this output remain satisfactory in the presence of strong signals at nearby frequencies?

Sensitivity is determined by the noise power contributed by the receiver itself. Usually this is specified as an equivalent noise power at the antenna terminals. Selectivity is determined by a bandpass-limiting filter and might be specified as "3 dB down at 2 kHz from center frequency and 20 dB down at 10 kHz from center frequency." (Receiver manufacturers usually do not specify the exact bandpass shape.)

7.1 Amplification

Let us consider how much amplification is needed in ordinary AM receivers. One milliwatt of audio power into a typical earphone produces a sound level some 100 dB above the threshold of hearing. A barely discernable audio signal can therefore be produced by -100 dBm (100 dB below 1 mW or 10^{-13} watts). Let us specify that a receiver, for comfortable earphone listening, must provide 50 dB more than this threshold of hearing, or 10^{-8} watts. You can see that, with efficient circuitry, the batteries in a portable receiver could last a very long time! (Sound power levels are surprisingly small; you radiate only about 1 mW of acoustic power when shouting and about 1 nW when whispering.) How much signal power arrives at a receiver? A simple wire antenna could intercept 10^{-8} watts of RF power at a distance of about 20 000 km from a 10 kW radio station at 1 MHz having an omnidirectional transmitting antenna, so let us first consider "self-powered" receivers.

7.2 Crystal sets

The earliest radios, crystal sets, were self-powered. A crystal diode rectifier recovered the modulation envelope, converting enough of the incoming RF power into audio power to drive the earphone. A simple *LC* tuned circuit served as a bandpass filter to select the desired station and could also serve as an antenna matching network. The basic crystal set receiver is shown in Figure 7.1.

Figure 7.1. Self-powered crystal set receiver.

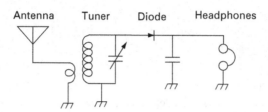

The considerations given above show that a self-powered receiver can have considerable range. But when the long-wire antenna is replaced by a compact but very inefficient loop antenna and the earphone is replaced by a loud-speaker, amplification is needed. In addition, we will see later that the diode detector, when operated at low signal levels, has a square-law characteristic, which causes the audio to be distorted. For proper envelope detection of an AM signal, the signal applied to a diode detector must have a high level, several milliwatts. The invention of the vacuum tube, followed by the transistor, provided the needed amplification. Receivers normally contain both RF and audio amplifiers. RF amplification provides enough power for proper detector operation, while subsequent audio amplification provides the power to operate loudspeakers.

7.3 TRF receivers

The first vacuum tube radios used a vacuum tube detector instead of a crystal, and added RF preamplification and audio post-amplification, as described above. These *TRF* (Tuned Radio Frequency) sets[1] had individual tuning adjustments for each of several cascaded RF amplifier stages. Changing stations

---

[1] Early radios were called "radio sets" because they were literally a set of parts including one or more tubes and batteries (or maybe just a crystal detector), inductors, "condensers," resistors, and headphones. Many of these parts were individually mounted on wooden bases, and, together with "hook-up" wires, would spread out over a table top.

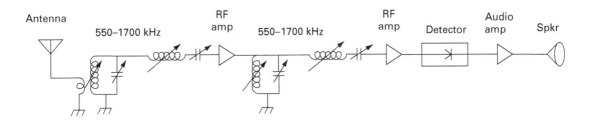

Figure 7.2. TRF receiver.

required the user to adjust several dials (often with the aid of a tuning chart or graph).

Figure 7.2 shows a hypothetical TRF receiver with cascaded amplifiers and bandpass filters.

Note that all the inductors and capacitors should be variable in order to tune the center frequency of the bandpass filters and also maintain the proper bandwidth, which is about 10 kHz for AM. (In a practical circuit, the bandpass filters would use a coupled-resonator design rather than the straightforward lowpass-to-bandpass conversion design shown here.)

7.4 The superheterodyne receiver

The disadvantages of TRF sets were the cost and inconvenience of having many tuning adjustments. Most of these adjustments were eliminated with the invention of the superheterodyne circuit by Edwin H. Armstrong in 1917. Armstrong's circuit consists of a fixed-tuned, i.e., single-frequency, TRF back-end receiver, preceded by a frequency converter front-end (mixer and local oscillator) so that the signal from any desired station can be shifted to the frequency of the TRF back-end. This frequency is known as the intermediate frequency or IF. The superheterodyne is still the circuit used in nearly every radio, television, and radar receiver. Among the few exceptions are some toy walkie-talkies, garage-door openers, microwave receivers used in radar-controlled business place door openers, and highway speed trap radar detectors. Figure 7.3 shows the classic broadcast band "superhet."

Selectivity is provided by fix-tuned bandpass filters in the IF amplifier section. The AM detector here is still a diode, i.e., the basic envelope detector. (In Chapter 18 we will analyze this detector, among others.) All the RF gain can be contained in the fixed-tuned IF amplifier, although we will see later that there are sometimes reasons for having some amplification ahead of the mixer as well. Figure 7.3 also serves as the block diagram for FM broadcast receivers, where the IF frequency is usually 10.7 MHZ, and for television receivers, where the IF center frequency is commonly around 45 MHZ.

Note: There was indeed a *heterodyne* receiver that preceded the superheterodyne. Invented by radio pioneer Reginald Fessenden, the heterodnye receiver

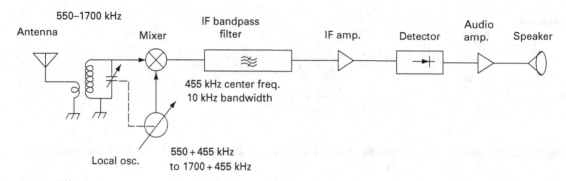

Figure 7.3. Standard superheterodyne receiver for the AM broadcast band.

converted the incoming RF signal directly to audio. This design, known as a "direct-conversion receiver," is disucussed below.

7.4.1 Image rejection

The superhet has some disadvantages of its own. With respect to signals at the input to the mixer, the receiver will simultaneously detect signals at the desired frequency and also any signals present at an undesired frequency known as the *image frequency*. Suppose we have a conventional AM receiver with an IF frequency of 455 kHz and suppose the local oscillator is set at 1015 kHz in order to receive a station broadcasting at $1015 - 455 = 560$ kHz (560 kHz is near the lower edge of the AM broadcast band). All the mixers we have considered will also produce a 455-kHz IF signal from any input signal present at 1470 kHz, i.e., 455 kHz above the local oscillator. If the receiver has no RF filtering before the mixer and if there happens to be a signal at 1470 kHz, it will be detected along with the desired 560 kHz signal. A bandpass filter ahead of the mixer is needed to pass the desired frequency and greatly suppress signals at the image frequency. Note that in this example (the most common AM receiver design), this anti-image bandpass filter must be tunable and, for the receiver to have single-dial tuning, the tuned filter must always "track" 455 kHz below the L.O. frequency. In this example, the tracking requirement is not difficult to satisfy; since the image frequency is more than an octave above the desired frequency, the simple one-section filter shown in Figure 7.2 can be fairly broad and still provide adequate image rejection, maybe 20 dB. (Note, though, that 20 dB is not adequate if a signal at the image frequency is 20 dB stronger than the signal at the desired frequency.)

What if a receiver with the same 455 kHz IF frequency is also to cover the short-wave bands? The worst image situation occurs at the highest frequency, 30 MHz, where the image is only about 3% higher in frequency than the desired frequency. A filter 20 dB down at only 3% from its center frequency will need to have many sections, all of which must be tuned simultaneously with a mechanical multisection variable capacitor or voltage-controlled varicaps. As explained above, the center frequency of the filter must be track with a 455 kHz offset from the L.O. frequency in order that the desired signal fall within the narrow IF passband. Image rejection is not simple when the IF frequency is much lower than the input frequency.

Solving the image problem

A much higher IF frequency can solve the image problem. If the AM broadcast band receiver discussed above were to have an IF of 10 MHz rather than 455 kHz, the L.O. could be tuned to 10.560 MHz to tune in a station at 560 kHz. The image frequency would be 20.560 MHz. As the radio is tuned up to the top end of the AM broadcast band, 1700 kHz, the image frequency increases to 21.700 MHz. In this case, a fixed-tuned bandpass filter, wide enough to cover the entire broadcast band, can be placed ahead of the receiver to render the receiver insensitive to images. This system is shown in Figure 7.4. Only the local oscillator needs to be changed to tune this receiver. Of course, the 10 MHz IF filter must still have a narrow 10 kHz passband to establish the receiver's basic selectivity.

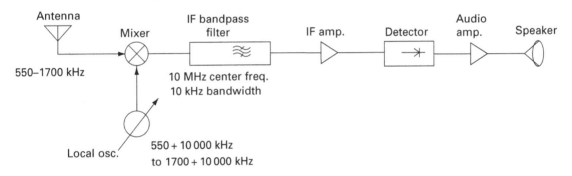

Figure 7.4. Image-free broadcast receiver using a 10 MHz IF.

The circuit of Figure 7.4 is entirely practical although it is more expensive to make the necessary narrowband filters at higher frequencies; quartz crystal resonator elements must be used in place of lumped *LC* elements. If the input band is wider, e.g., 3–30 MHz for a short-wave receiver, the IF frequency would have to be much higher, and narrowband filters become impractical. (Even at 10 MHz, a bandwidth of 10 kHz implies a very narrow fractional bandwidth, 0.1%.) A solution to both the image problem and the narrow fractional bandwidth problem is provided by the *double-conversion superhet*.

Double conversion superhet

Figure 7.5 shows how a second frequency converter takes the first IF signal at 10 MHz and converts it down to 455 kHz, where it can be processed by the standard IF section of the receiver of Figure 7.3. The 10-MHz first IF filter can be wider than the ultimate passband.

Suppose, for example, that the first IF section has a bandwidth of 500 kHz. The second L.O. frequency is at 10.455 MHz, so the second mixer would

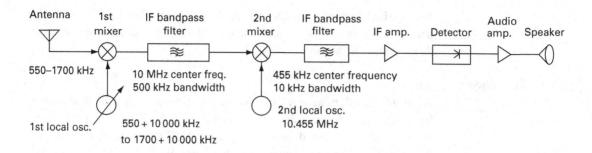

Figure 7.5. Double conversion superhet.

produce an image from a signal at 10.910 MHz. But note that our first IF filter cuts off at 10 MHz + 0.500 MHz /2 = 10.25 MHz, so there will be no signals at 10.910. This system has its own special disadvantages: the receiver usually cannot be used to receive signals in the vicinity of its first IF, since it is difficult to avoid direct feedthrough into the IF amplifier. Multiple conversions require multiple local oscillators and various sum and difference frequency combinations inevitably are produced by nonlinearities and show up as spurious signals known as "birdies."

Present practice for communications receivers is to use double or triple conversion with a first IF frequency at, say, 40 MHz. The front-end image filter is usually a 30 MHz lowpass filter. In as much as modern crystal filters can have a fairly small bandwidth even at 40 MHz, the output of the first IF section can be mixed down to a second IF with a much lower frequency. Sometimes triple conversion is necessary when the final IF frequency is very low, e.g., 50 kHz. The use of first IF frequencies in the VHF region requires very stable local oscillators but crystal oscillators and frequency synthesizers provide the necessary stability. (Oscillator phase noise was a problem in the first generation of receivers with synthesized local oscillators; the oscillator sideband noise was shifted into the passband by strong signals near the desired signal but outside the nominal passband.)

Image rejection mixer

Another method of solving the image problem is to use an image rejection mixer, such as the circuit shown in Figure 7.6.

This circuit uses two ordinary mixers (multipliers). The lower multiplier forms the product of $\cos(\omega_{L.O.}t)$ and an input signal, $\cos[(\omega_{L.O.}\pm\omega_{IF})\,t]$, depending on whether the input signal is above or below the L.O. frequency. The upper multiplier has a 90° delay in its connection to the L.O., so it forms the product of $\sin(\omega_{L.O.}t)$ and $\cos[(\omega_{L.O.}\pm\omega_{IF})t]$. Neglecting the sum frequency terms, the outputs of the upper and lower multipliers are, respectively, $\mp\sin(\omega_{IF}t)$ and $\cos(\omega_{IF}t)$. The output from the lower multiplier is delayed by 90°, so the upper and lower signals at the input to the adder are $\mp\sin(\omega_{IF}t)$ and $\sin(\omega_{IF}t)$. Thus, for an input frequency of $\omega_{L.O.} - \omega_{IF}$, the output of the adder is $2\sin(\omega_{IF}t)$, but when the input frequency is $\omega_{L.O.} + \omega_{IF}$, the output of the adder is

Figure 7.6. Image rejection mixer.

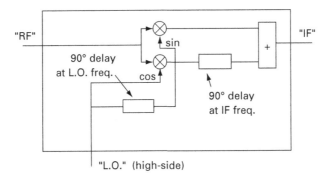

zero. Thus, this mixer rejects signals above the L.O. frequency. The same circuit crops up in Chapter 8, as a *single-sideband generator*. In practice this circuit might provide 20–40 dB of image rejection. It can be used together with the standard filtering techniques to get further rejection.

Zero IF frequency – direct conversion receivers

The evolution of the superhet, which was always toward higher IF frequencies and multiple conversions, has taken a new twist with the advent of the nearly limitless signal processing power available from DSP chips. Direct conversion receivers, in a reversion to the heterodyne architecture, shift the center frequency of the desired signal, ω_C, all the way to zero Hz ("baseband"). Generally this requires that the signal be mixed separately with local oscillator signals $\cos(\omega_C t)$ and $\sin(\omega_C t)$ to preserve all the signal information. To see this, note that the input signal, which is sinusoidal, will, in general, be out of phase at times with $\cos(\omega_C t)$ and at other times with $\sin(\omega_C t)$. Thus, the outputs of the "cosine mixer" or "sine mixer" can go to zero but, together, these "I" (in-phase) and "Q" (quadrature) signals contain all the signal information. To see this, note that the original signal could be easily reconstructed from the I and Q signals. Lowpass filtering of the I and Q signals determines the passband of the receiver, e.g., 5-kHz rectangular lowpass filters on the I and Q signals would give the receiver a flat passband extending 5 kHz above and 5 kHz below the L.O. frequency. The classic image problem, severe at low IF frequencies, disappears when the IF frequency is zero. The low-frequency I and Q signals can be digitized directly for subsequent digital processing and demodulation (see Problem 7.8). No bulky IF bandpass filters are required. Even tiny surface acoustic wave (SAW) bandpass filters are huge, compared to the real estate on DSP chips. With this architecture, nearly 100% of a receiver can be incorporated on a chip, including a tunable frequency synthesizer to produce the L.O. signals. Figure 7.7 shows a block diagram of a direct-conversion receiver. Everything on this diagram plus an L.O. synthesizer is available on a single chip intended for use in HD television receivers.

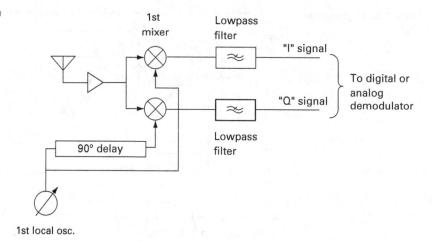

Figure 7.7. Direct conversion receiver.

7.4.2 Automatic gain control

Nearly every receiver has some kind of automatic gain control (AGC) to adjust the gain of the RF and/or IF amplifiers according to the strength of the input signal. Without this feature the receiver will overload; overdriven amplifiers go nonlinear ("clip") and the output will be distorted as well as too loud. The output sound level of an FM receiver, depending on the design of the demodulator, may not vary with signal level but overloading the IF amplifier stages will still produce distortion, so FM receivers also need AGC. Television receivers need accurate AGC to maintain the correct contrast level (analog TV) or threshold levels (digital TV). Any AGC circuit is a feedback control system. In simple AM receivers the diode detector provides a convenient dc output voltage that can control the bias current (and hence gain) of the RF amplifiers. The controlled bias current can also be used to drive a signal strength indicator.

7.5 Noise blankers

Many receivers, including most television receivers, have a noise blanker circuit to reduce the effects of impulse noise such as the spiky noise produced by automobile ignition systems. Here the interfering pulses are of such short duration that the IF stages can be gated off briefly while the interference is present. The duty cycle of the receiver remains high and the glitch is all but inaudible (or invisible). An important consideration is that the gating must be done before the bandwidth is made very narrow since narrow filters elongate pulses.

7.6 Digital signal processing in receivers

Advances in digital electronics, notably fast A-to-D conversion and processors, make it possible to do all-digital filtering and detection in a receiver. Any desired filter amplitude and phase response can be realized.

Besides direct-conversion receivers on a chip, there are many single-chip superhet chips, usually using image cancelling mixers followed by digital bandpass filters operating at low IF frequencies. Adaptive digital filters can correct for propagation problems such as multipath signals. Digital demodulators allow the use of elaborate signal encoding which provides high spectral efficiency (bits/sec/Hz) as well as error correction through the use of redundant bits. Digital modulation techniques are discussed in Chapter 22.

Problems

Problem 7.1. The FM broadcast band extends from 88 to 108 MHz. Standard FM receivers use an IF frequency of 10.7 MHz. What is the required tuning range of the local oscillator?

Problem 7.2. Why are airplane passengers asked not to use radio receivers while in flight?

Problem 7.3. Two sinusoidal signals that are different in frequency, if simply added together, will appear to be a signal at a single frequency but amplitude modulated. This "beat" phenomenon is used, for example, to tune two guitar strings to the same frequency. When they are still at slightly different frequencies, the sound seems to pulsate slowly at a rate equal to their frequency difference. Show that

$$\sin([\omega_0 - \delta\omega]t) + \sin([\omega_0 + \delta\omega]t) = A(t)\sin(\omega_0 t)$$

where

$$A(t) = 2\cos(\delta\omega t).$$

(Note that this addition is a linear process; no new frequencies are generated.)

Problem 7.4. Using an AM receiver in an environment crowded with many stations, you will sometimes hear an annoying high-pitched 10 kHz tone together with the desired audio. If you rock the tuning back and forth the pitch of this tone does not change. What causes this?

Problem 7.5. When tuning an AM receiver, especially at night, you may hear "heterodynes" or whistling audio tones that change frequency as you slowly tune the dial. What causes this? Can it be blamed on the receiver? (Answer: Yes.)

Problem 7.6. Using modern components and digital control, we could build good TRF radios. What advantages would such a radio have over a superheterodyne? What disadvantages?

Problem 7.7. You may have observed someone listening to distorted sound from an AM radio whose tuning is not centered on the station. Often this mistuning is done deliberately when the listener has impaired high-frequency hearing and/or the radio has insufficient bandwidth. What is going on here? Why would a radio not have sufficient bandwidth and why would insufficient bandwidth cause some listeners to tune slightly off station?

Problem 7.8. Design a direct-conversion AM broadcast receiver using digital processing of the I and Q signals. Assume that the L.O. frequency may not be set exactly equal to the frequency of the desired station. Hint: compute the input signal amplitude from the digitized I and Q signals. Then feed this stream of amplitudes into a D-to-A converter.

References

[1] The American Radio Relay League, *The ARRL Handbook for Radio Communications*, 2008 Edition. Almost five pounds of practical circuits, explanations, and construction information.

[2] Gosling, W., *Radio Receivers*, London: Peter Peregrinus, 1986. Good concise discussion of receivers.

[3] Rohde, U. and Bucher, T., *Communications Receivers, Principles & Design*, Second Edition, New York: McGraw-Hill, 1988. A whole course in itself.

[4] Rohde, U. and Whitaker, J., *Communications Receivers*, Third Edition, New York: McGraw-Hill, 2001. The third edition includes material on digital processing in receivers.

8 Suppressed-carrier AM and quadrature AM (QAM)

Viewed just in the time domain, ordinary AM, as used in broadcasting, seems so obvious that one would scarcely imagine how it might be done otherwise. But viewed in the frequency domain, as in Chapter 6, this system shows some obvious inefficiencies. First, most of the average transmitted power (about 95% when transmitting typical audio material) is in the carrier, which is a spike or "delta function" in the frequency domain. Since its amplitude and frequency are constant, it carries virtually no information. The information is in the sidebands. Could broadcasters just suppress the carrier to reduce their electric power costs by 95%? Second, the upper and lower sidebands are mirror images of each other, so they contain the same information. Could they not suppress (filter away) one sideband, making room for twice as many stations on the AM band? The answer to both questions is yes but, in both cases, the simple AM receiver, with its envelope detector, will no longer work properly. Economics favored the simplicity of the traditional AM receiver until it because possible to put all the receiver signal processing on an integrated-circuit chip, where the additional complexity can have negligible cost. In this chapter we examine alternate AM systems that remove the carrier and then at AM systems that reduce the signal bandwidth or double the information carried in the original bandwidth.

8.1 Double-sideband suppressed-carrier AM

Let us look at a system that removes the carrier at the transmitter and regenerates it at the receiver. It is easy enough to modify the transmitter to eliminate the carrier. Review the circuit diagram given previously for an AM transmitter (Figure 6.2a). If we replace the bias battery by a zero-volt battery (a wire), the carrier disappears. The resulting signal is known as *Double-Sideband Suppressed-Carrier* (DSBSC). To restore the missing carrier we might try the receiver circuit shown in Figure 8.1. The locally generated carrier, from a *beat frequency oscillator* (BFO), is simply added back into the IF signal, just ahead

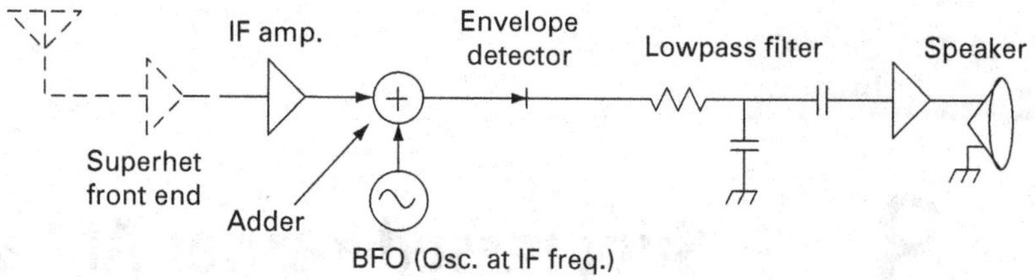

Figure 8.1. Hypothetical DSBSC receiver: additive carrier reinsertion and envelope detection.

of the envelope detector. The energy saved by suppressing the carrier can increase battery lifetime in walkie-talkies by a factor of maybe 20. The modulation schemes used in many cell phones (after the first generation) likewise do away with battery-draining constant carriers.

In a superheterodyne receiver, we only have to generate this carrier at one frequency, the intermediate frequency (IF). As you would expect, the added carrier must have the right frequency. But it must also have the right phase. Suppose, for example, that the modulation is a single audio tone at 400 Hz. If the replacement carrier phase is off by 45°, the output of the envelope detector will be severely distorted. And if the phase is off by 90° the output of the detector will be a tone at 800 Hz (100% distortion!). These example waveforms are shown in Figure 8.2.

Correct carrier reinsertion is fairly easy if the transmitter provides a low-amplitude "pilot" carrier to provide a phase reference. In the model transmitter of Figure 6.2(a) we would go back to using a bias battery, but it would have a low voltage. In principle, the receiver could have a tuned IF amplifier with a narrowband gain peak at the center frequency to bring the pilot carrier up to full level. The most common method, however, is to use a VCO for the BFO in Figure 8.1, together with a feedback circuit to phase lock it to the pilot carrier.

8.2 Single-sideband AM

The requirement for correct BFO phasing is not critical if only one of the two sidebands is transmitted. The sidebands are mirror images of each other so they carry the same information and one will do. We will discuss later three methods to eliminate one of the sidebands from a DSBSC signal, but the first and most obvious method is to use a bandpass filter to select the desired sideband. The resulting signal is known as *Single-Sideband Suppressed-Carrier* (SSBSC) or simply as *Single-Sideband* (SSB). Take the previous example of a single 400-Hz audio tone. The SSB transmitter will put out a single frequency, 400 Hz above the frequency of the (suppressed) carrier if we have selected the upper sideband. This signal will appear in the receiver 400 Hz away from the IF center frequency. Suppose we are still using the BFO and envelope detector. The

Demonstration that Detection of Double Sideband A.M. Requires the Correct Carrier Phase

$t := 0, 0.1 .. 300$

carrier(t) := sin(t)

audio(t) := 0.8·sin(0.05·t) Audio is sine wave at 1/20 carrier frequency

dssc(t) := audio(t)·carrier(t) Double sideband suppressed carrier (sidebands alone)

Note that dssc(t) = 1/2(cos(0.95t) − cos(1.05)t)

signal(t) := carrier(t) + dssc(t) Carrier plus sidebands = (1 + audio)(carrier) = normal AM

(a) Sidebands plus correct carrier. (note that the envelope is the sinewave audio)

$\underline{\text{signal}(t)}$

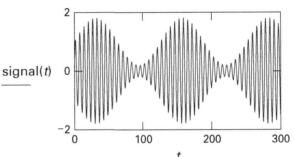

alt(t) := dssc(t) + cos(t + $\frac{\pi}{4}$)

Give the re-injected carrier a 45° phase error.

(b) Sidebands with carrier 45° out of phase

(Note envelope distortion)

$\underline{\text{alt}(t)}$

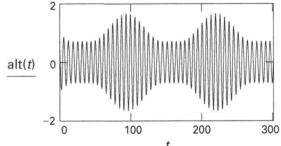

alt2(t) := dssc(t) + cos(t)

Here the re-injected carrier has a 90° phase error.

(c) Sidebands with carrier 90° out of phase

(Note that the envelope has twice the audio frequency − total distortion)

$\underline{\text{alt2}(t)}$

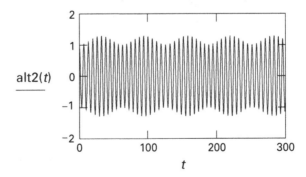

Figure 8.2. Detection of DSSC by adding a local carrier.

envelope of the signal-plus-BFO will indeed be an undistorted sine wave at 400 Hz. So far, so good. But suppose the signal had been two tones, one at 400 Hz and one at 600 Hz. When the BFO carrier is added, the resulting envelope will have tones at 400 and 600 Hz but it will also have a component at 200 Hz (600 − 400) which is distortion. Unwanted signals are produced by the IF components beating with each other; their strength is proportional to the product of the two IF signals. The strength of each wanted component, on the other hand, is proportional to the product of its amplitude times the amplitude of the BFO. If the BFO amplitude is much greater than the IF signal the distortion can be reduced.

8.3 Product detector

The distortion described above can be avoided by using a *product detector* instead of an envelope detector. In the receiver shown in Figure 8.3, the product detector is just the familiar multiplier (a.k.a. mixer).

The output of this detector is, by the distributive law of multiplication, the sum of the products of each IF signal component times the BFO signal. There are no cross-products of the various IF signal components. Product detectors are also called "baseband mixers" or "baseband detectors," because they translate the sideband components all the way down to their original audio frequencies.

While a product detector does not produce cross-products (intermodulation distortion) of the IF signal components, the injected carrier should ideally have the correct phase. The wrong phase is of no consequence in our single-tone example. But when the signal has many components, their relative phases are important. A waveform will become distorted if all its spectral components are given an identical phase shift (see Problem 8.3). It is only when every component is given a phase shift proportional to its frequency that a waveform is not distorted but only delayed in time. Therefore a single-sideband transmitter, like the double-sideband transmitter, should really transmit a pilot carrier and the receiver should lock its BFO phase to this pilot. Some SSB systems do just that. But for voice communications it is common to use no pilot. Speech remains intelligible and almost natural-sounding even when the BFO phase is

Figure 8.3. SSBSC receiver using a product detector.

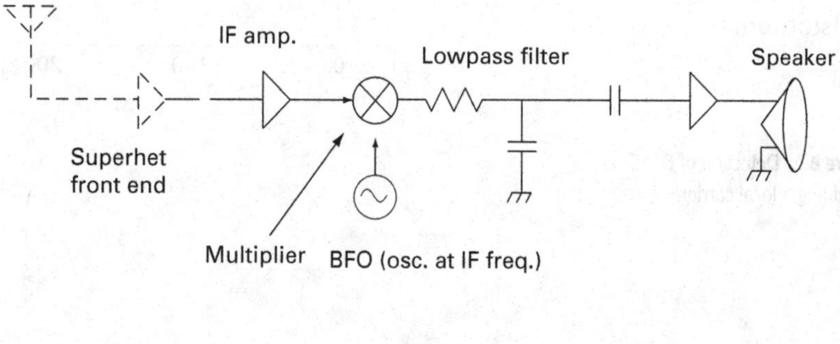

wrong.[1] The BFO can even be slightly off frequency. When the frequency is too high, a demodulated upper sideband (USB) voice signal has a lower than natural pitch. When it is too low, the pitch is higher than natural.

Finally, what happens if we try to use a product detector with a free-running (not phase locked) BFO to receive DSBSC? If the BFO phase is off by 90°, the detector produces no output. If the BFO phase happens to be correct, the output will have maximum amplitude. For an intermediate-phase error the amplitude will be reduced.

Advantages of SSB

Single-sideband, besides not wasting power on a carrier, uses only half the bandwidth, so a given band can hold twice as many channels. Halving the receiver bandwidth also halves the background noise so there is a 3-dB improvement over conventional AM in signal-to-noise ratio. When a spectrum is crowded with conventional AM signals, their carriers produce annoying beat notes in a receiver (a carrier anywhere within the receiver passband appears to be a sideband belonging to the spectrum of the desired signal). With single-sideband transmitters there are no carriers and no beat notes. Radio amateurs, the military, and aircraft flying over oceans use SSB for short-wave (1.8–30 MHz) voice communication.

8.4 Generation of SSB

There are at least three well-known methods to generate single-sideband. In the filter method of Figure 8.4, a sharp bandpass filter removes the unwanted sideband. This filter usually uses crystal or other high-Q mechanical resonators and has many sections. It is never tunable, so the SSB signal is generated at a single frequency (a transmitter IF frequency) and then mixed up or down to the desired frequency. The filter, in as much as it does not have infinitely steep skirts, will cut off some of the low-frequency end of the voice channel.

The phasing method uses two multipliers and two phase-shift networks to implement the trigonometric identity:

$$\cos(\omega_c t) \cos(\omega_a t) \pm \sin(\omega_c t) \sin(\omega_a t) = \cos([\omega_c \mp \omega_a]t). \qquad (8.1)$$

The subscripts "a" and "c" denote the audio and (suppressed) carrier frequencies. If the audio signal is $\cos(\omega_a t)$, we have to generate the $\sin(\omega_a t)$ needed for the second term. A 90° audio phase shift network can be built with passive components or digital signal processing techniques. The audio phase-shift network is usually implemented with two networks, one ahead of each mixer. Their phase *difference* is close to 90° throughout the audio band. The second term also

---

[1] Human perception of speech and even music seems to be remarkably independent of phase. When an adjustable allpass filter is used to produce arbitrary phase vs. frequency characteristics (while not affecting the amplitude), test subjects find it difficult or impossible to distinguish phase-modified audio from the original source.

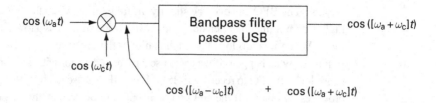

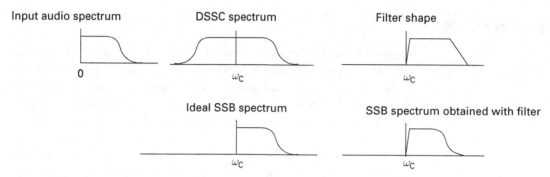

Figure 8.4. SSB generator – filter method.

Figure 8.5. SSB Generator – phasing method.

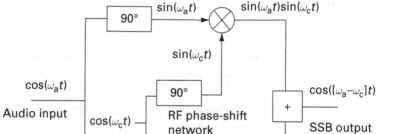

requires that we supply $\sin(\omega_c t)$ but, since ω_c is fixed, anything that provides a 90° shift at this one frequency will suffice. Figure 8.5 illustrates the phasing method used to generate an upper sideband (USB) signal.

If the adder is changed to a subtractor (by inverting the polarity of one input), the output will be the upper rather than the lower sideband.

A third method [2] needs neither a sharp bandpass filter nor phase-shift networks. Weaver's method, shown in Figure 8.6, uses four multipliers and two lowpass filters. The trick is to mix the audio signal with a first oscillator whose frequency, ω_0, is in the center of the audio band. The outputs of the first

Figure 8.6. SSB generator –
Weaver method.

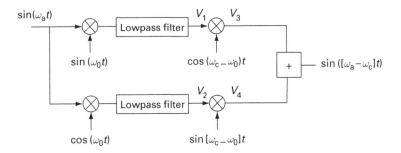

set of mixers have the desired 90° phase difference. The second set of mixers
then works the same way as the two mixers in the phasing method.

Referring to Figure 8.6,

$$V_1(t) = \text{lowpass} \; [\sin(\omega_a t)\sin(\omega_0 t)] = {}^1\!/_2 \cos([\omega_a - \omega_0]t) \qquad (8.2)$$

$$V_2(t) = \text{lowpass} \; [\sin(\omega_a t)\cos(\omega_0 t)] = {}^1\!/_2 \sin([\omega_a - \omega_0]t) \qquad (8.3)$$

$$\begin{aligned} V_3(t) &= {}^1\!/_2 \cos[(\omega_c - \omega_0)t]\cos[(\omega_a - \omega_0)t] \\ &= {}^1\!/_4 \cos([\omega_a - \omega_c]t) + {}^1\!/_4 \cos([\omega_a + \omega_c - 2\omega_0]t) \end{aligned} \qquad (8.4)$$

$$\begin{aligned} V_4(t) &= {}^1\!/_2 \sin[(\omega_c - \omega_0)t]\sin[(\omega_a - \omega_0)t] \\ &= {}^1\!/_4 \cos[(\omega_a - \omega_c)t] - {}^1\!/_4 \cos[(\omega_a + \omega_c - 2\omega_0)t] \end{aligned} \qquad (8.5)$$

$$V_{\text{out}}(t) = V_3(t) + V_4(t) = {}^1\!/_2 \cos[(\omega_c - \omega_a)t]. \qquad (8.6)$$

We see from Equation (8.6) that this particular arrangement generates the lower
sideband.

8.5 Single-sideband with class C, D, or E amplifiers

The three methods described above for generating a single-sideband signal are
done in low-level circuitry. Linear amplifiers then produce the required power
for the antenna. There is, however, a way to use a class-C or class-D amplifier
with simultaneous phase modulation and amplitude modulation to produce
SSB. This follows from the fact that any narrowband signal (CW, AM, FM,
PM, SSB, DSB, narrowband noise, etc.) is essentially a sinusoid at the center
frequency, ω_0, whose phase and amplitude vary on a time-scale longer than
$(\omega_0)^{-1}$. Suppose we have generated SSB at low level. We can envelope-detect
it, amplify it, and use the amplified envelope to AM-modulate the class-C or
class-D amplifier. At the same time we can phase-modulate the amplifier with
the phase determined from the low-level SSB signal. The low-level signal is
simply amplitude-limited and then used to drive the modulated amplifier. The
point is that the high-power SSB signal is produced with an amplifier that has
close to 100% efficiency. Moreover, an existing AM transmitter can be converted to single-sideband by making only minor modifications.

8.6 Quadrature AM (QAM)

We noted above that a product detector used for DSBSC will produce no output if the BFO phase is wrong by 90°. It follows that two independent DSBSC signals can be transmitted over the same channel. At the receiver, the phase of the BFO selects one or the other. This is known as quadrature AM (QAM). Note that the carrier can be suppressed, but retaining at least a low-power pilot carrier provides a phase reference for the receiver. Figure 8.7 shows the transmitter (a) and receiver (b) for a QAM system. Here, the receiver uses a narrowband filter to isolate the pilot carrier and then amplifies it to obtain a reconstituted BFO or LO, depending on whether the receiver is a superheterodyne or uses direct conversion to baseband. In practice, carrier regeneration is almost always done with a phase lock loop.

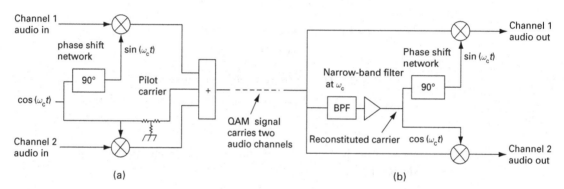

Figure 8.7. QAM:
(a) transmitter; (b) receiver.

The U.S. NTSC and European PAL standards for color television use QAM to transmit a pair of video color difference signals. In the NTSC system, these so-called I and Q signals (In-phase and Quadrature) are multiplied by sine and cosine versions of a 3.579-MHZ oscillator signal and the resulting DSBSC signals are added to the normal video ("luminance") signal. Other uses of QAM include delivery of digital TV signals to set-top converters in some cable television systems and data distribution in some LANs.

The QAM system of Figure 8.7 could also be used to broadcast AM stereo sound, with both the $L+R$ (left plus right) and $L-R$ signals fitting into the channel used traditionally for $L+R$ alone. However, such a QAM stereo signal would not be compatible with the envelope detector in standard AM receivers. Several compatible analog systems were developed and used for AM stereo, but at this writing, both AM and FM stations are adopting compatible *hybrid digital* (HD) systems in which the digital signals occupy spectral space on either side of the conventional analog signal. This system, which uses *COFDM* modulation, is discussed in Chapter 22.

Problems

Problem 8.1. Demonstrate for yourself the kind of phase distortion that will occur when the BFO in a product detector does not have the same phase as the suppressed carrier. Use the Fourier decomposition of a square wave: $V(t) = \sin(\omega t) + 1/3 \sin(3\omega t) + 1/5 \sin(5\omega t) + \dots$. Have your computer plot $V(\theta) = \sin(\theta) + 1/3 \sin(3\theta) + 1/5 \sin(5\theta) + \dots + 1/9 \sin(9\theta)$ for θ from 0 to 2π. Then plot $V'(\theta) = \sin(\theta+1) + 1/3 \sin(3\theta+1) + 1/5 \sin(5\theta+1) + \dots + 1/9 \sin(9\theta+1)$, i.e., the same function but with every Fourier component given an equal phase shift (here 1 radian). Would you expect these waveforms to sound the same? Consider the case in which the phase shift is 180°. This just inverts the waveform. Is this a special case or is an inverted audio waveform actually distorted?

Problem 8.2. The phase-shift networks used in the phasing method of single-sideband generation have a flat amplitude vs. frequency response – they are known as *allpass* filters. Allpass filters have equal numbers of poles (all in the left-hand plane) and zeros (all in the right-hand plane). Find the amplitude and phase response of the network shown below, which is a first-order allpass filter. (Note that the inverting op-amp is used only to provide an inverted version of the input.)

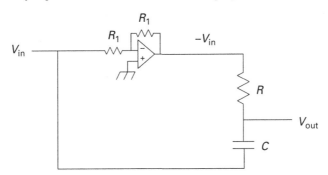

Problem 8.3. The most general allpass filter can be obtained by cascading first-order allpass sections (Problem 8.2) with second-order allpass sections. Find the amplitude and phase response of the network shown below, a second-order allpass filter.

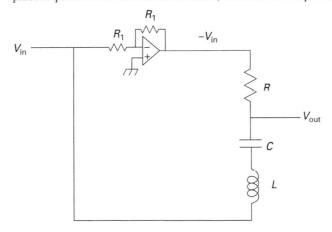

References

[1] Sabin, W. E., and Schoenike, E. O., Editors, *Single-Sideband Systems and Circuits*, New York: McGraw- Hill, 1987.

[2] Weaver, D. K. Jr., A third method of generation and detection of single-sideband signals, *Proceedings of the IRE*, pp. 1703–1706, December 1956.

9 Class-C, D, and E power RF amplifiers

Class-C, D, and E RF power amplifiers are all about high efficiency. They are used in large transmitters and industrial induction heaters, where high efficiency reduces the power bill and saves on cooling equipment, and also in the smallest transmitters, such as cell phones, where high efficiency increases battery life. These amplifiers are so nonlinear (the output signal amplitude is not proportional to the input signal amplitude), they might better be called synchronized sine wave generators. They consist of a power supply, at least one switching element (a transistor or vacuum tube), and an *LCR* circuit. The "*R*" is the load, R_L, often the radiation resistance of an antenna, equivalent to a resistor. The *LC* network is resonant at the operating frequency. The output sine-wave amplitude, while not a linear function of the input signal amplitude, *is* proportional to the power supply voltage. Thus, these amplifiers can be amplitude modulated by varying the supply voltage. Of course they can also be frequency modulated by varying the drive frequency (within a restricted bandwidth, determined by the *Q* of the *LC* circuit). Finally, they can be used as frequency multipliers by driving them at a subharmonic of the operating frequency.

9.1 The class-C amplifier

Figure 9.1 shows a class-C amplifier (a), together with an equivalent circuit (b). The circuit looks no different from the class-B amplifier of Figure 3.15 or a small-signal class-A amplifier. But here the active device (transistor or tube) is used not as a continuously variable resistor, but as a switch. To simplify the analysis, we consider the switch to have a constant on-resistance, *r*, and infinite off-resistance. This model is a fairly good representation of a power FET, when used as a switch.

The switch is closed for less than half the RF cycle, during which time the power supply essentially "tops off" the capacitor, restoring energy that the load has sapped from the resonant circuit during the cycle. The switch has internal resistance (i.e., loss), so in recharging the capacitor, some energy is lost in the

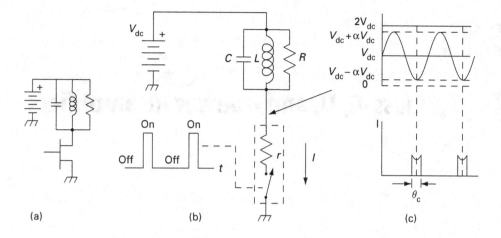

Figure 9.1. Class-C amplifier operation.

switch. Normally the *LC* circuit has a high *Q* (at least 5) so its flywheel action minimizes distortion of the sine wave caused by the abrupt pull-down of the switch and by the damping caused by the load. The drive is shown as a rectangular pulse but is often a sine wave, biased so that conduction takes place just around the positive tips. Class-C amplifiers are normally run in saturation, meaning that the switch, when on, always has its lowest possible resistance, ideally much less than the load resistance.

9.1.1 Simplified analysis of class-C operation

The simplified analysis, which we will also use later to analyze the class-E amplifier, is based on the assumption that the *LC* circuit provides enough flywheel effect to maintain a perfect sine wave throughout the cycle, including the interval when the resistive switch is closed. With this assumption, let us analyze the circuit of Figure 9.1 to find the output voltage and, thereby, the power and the efficiency. Referring to the figure, θ_c is the conduction angle, V_{dc} is the supply voltage, r is the on-resistance of the switch, and αV_{dc} is the peak voltage of the sine wave. We can find α as follows. The input power (the power supplied by the battery) must be equal to the sum of the output power (the power dissipated in R) plus the power dissipated in the switch resistance r. These terms are given respectively by the average of the battery voltage times the current, $(\alpha V_{dc})^2/(2R)$, and the average of I^2/r. The power equation becomes

$$\frac{1}{2\pi} \int_{\theta_c/2}^{\theta_c/2} V_{dc} \left(\frac{V_{dc} - \alpha V_{dc} \cos\theta}{r} \right) d\theta = \frac{(\alpha V_{dc})^2}{2R}$$

$$+ \frac{1}{2\pi} \int_{-\theta_c/2}^{\theta_c/2} \left(\frac{V_{dc} - \alpha V_{dc} \cos\theta}{r} \right)^2 r \, d\theta.$$

$$(9.1)$$

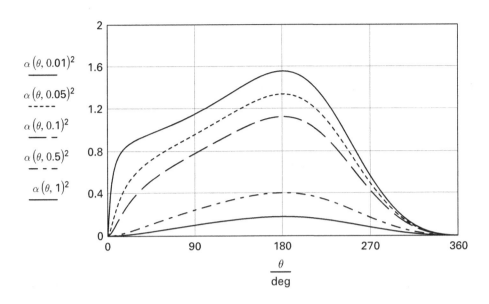

Figure 9. 2. Class-C output power ($\alpha 2$) vs. conduction angle for five values of r/R. The efficiency is given by the output power divided by the power supplied by the battery.

Rearranging this equation, we get

$$\frac{1}{\pi r/R} \int_0^{\theta_c/2} (1 - \alpha \cos \theta)(\alpha \cos \theta)\mathrm{d}\theta - \frac{\alpha^2}{2} = 0. \tag{9.2}$$

Carrying out the integral in Equation (9.2) and solving for α, we find

$$\alpha(\theta_c, r/R) = \frac{2 \sin(\theta_c/2)}{\theta_c/2 + \sin(\theta_c)/2 + \pi r/R}. \tag{9.3}$$

The quantity α^2, proportional to the output power, is plotted in Figure 9.2 for five values of r/R, the ratio of the switch resistance to the load resistance.

The middle value, $r/R = 0.1$, is typical in actual practice. Maximum power is produced for a conduction angle of 180°, i.e., if the switch is closed during the entire negative voltage loop. If the conduction angle exceeds 180°, the incursions into the positive loop extract energy from the tuned circuit and the power is reduced. Note that α^2, and therefore α, can be greater than unity, especially when the conduction angle is 180°. We will see below, however, that much higher efficiency is obtained for conduction angles substantially less than 180°. The efficiency is given by the output power divided by the power supplied by the battery:

$$\eta(\theta_c, r/R) = \frac{(\alpha V_{\mathrm{dc}})^2/2R}{\frac{1}{2\pi} \int_{-\theta_c/2}^{\theta_c/2} V_{\mathrm{dc}}(V_{\mathrm{dc}} - \alpha V_{\mathrm{dc}} \cos \theta'/r)\mathrm{d}\theta'}. \tag{9.4}$$

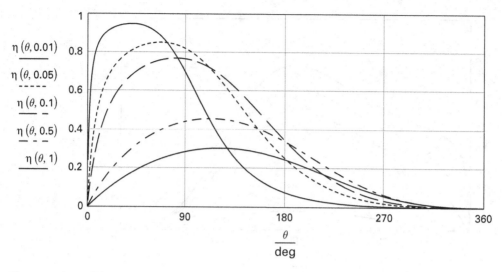

Figure 9.3. Class-C efficiency vs. conduction angle for five values of r/R.

Evaluating the integral in Equation (9.4) we find

$$\eta(\theta_c, r/R) = \frac{(\pi r/R)\alpha^2}{\theta_c - 2\alpha\sin(\theta_c/2)},$$ (9.5)

where α is given by Equation (9.3). This expression for efficiency is plotted in Figure 9.3 for the same five values of r/R. For $r/R = 0.1$, the efficiency is a maximum at about 90°.

Note that this amplifier model, assuming a constant resistance in the switch, can be solved exactly by finding the general solution for the transient waveform during the switch-off period, as well as the transient waveform during the switch-on period. Because the differential equations are of second order, these general solutions will each have two adjustable parameters. The parameters are found by imposing the boundary conditions that, across the switch openings and closings, the voltage on the capacitor is continuous and the current through the inductor is continuous.

9.1.2 General analysis of a class-C operation with a nonideal tube or transistor

The above analysis will not give accurate results for class-C amplifiers made with tubes or bipolar transistors, because these devices do not have the simple constant on-resistance characteristic of a FET. Nevertheless, their nonlinear characteristics are specified graphically on data sheets, and accurate class-C analyses can be done numerically. The method is basically the same as the simplified analysis; one assumes the resonant LC circuits at the input and output have enough Q to force the input and output waveforms to be sinusoidal. For this analysis, the device characteristics are plotted in a "constant current" format: in the case of a tube, curves of constant plate current are plotted on a graph whose axes are plate voltage and grid voltage. Sinusoidal plate and grid voltages are assumed and a numerical integration of plate current × plate voltage, averaged through one complete cycle, gives the power dissipated in the device. The power supply voltage times the *average* current gives the total input power. The difference is the power delivered to the load. The designer selects a device and a power supply voltage and assumes trial waveforms with different bias points and sine-wave amplitudes. Usually several trial designs are needed to maximize output power with the given device or to maximize efficiency for a specified output power. A class-C amplifier can approach 100% efficiency, but only in the limit that the output power goes to zero. We will see below that class-D and E amplifiers can approach 100% efficiency and still produce considerable power.

9.1.3 Drive considerations

Class-C amplifiers using vacuum tubes[1] nearly always drive the control grid positive when the tube is conducting. The grid draws current and dissipates power. Data sheets include grid current curves so that the designer can use the procedure outlined above to verify that the chosen operating cycle stays within both the maximum plate dissipation rating and the maximum grid dissipation rating. When tetrodes are used, a third analysis must be done to calculate the screen grid dissipation.

9.1.4 Shunt-fed class-C amplifier

In Figure 9.1, since the switch and the *LCR* tank circuit are in series, they can be interchanged, allowing the bottom of the tank to be at ground potential, which is often a convenience. But the usual way to put the tank at dc ground is to use the shunt-fed configuration shown in Figure 9.4(c).

In the shunt-fed circuit, an RF choke connects the dc supply to the transistor and a blocking capacitor keeps dc voltage off the tank circuit. The RF choke has

Figure 9.4. Equivalence of series-fed and shunt-fed circuits.

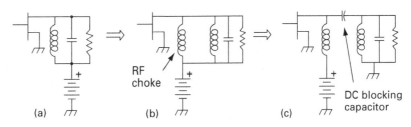

(a)   (b)   (c)

RF choke

DC blocking capacitor

---

[1] A triode vacuum tube is analogous to an NPN transistor. The tube's plate, control grid, and cathode correspond respectively to the transistor's collector, base, and emitter. Tetrode tubes have an additional grid, the screen grid, between the control grid and the plate. The screen grid is usually run at a fixed bias voltage and forms an electrostatic shield between the control grid and the plate [5].

a large inductance, so the current through it is essentially constant. The switch pulls pulses of charge from the blocking capacitor. This charge is replenished by the current through the choke. Figure 9.4 shows, from right to left, the equivalence of the shunt-fed and series-fed circuits. Note that the series-fed and shunt-fed equivalence applies as well to amplifiers of class-A, B, or C and for large-signal or small-signal operation.

9.1.5 The class-C amplifier as a voltage multiplier

An important property of the saturated class-C amplifier is that the peak voltage of the output RF sine wave is directly proportional to the supply voltage ($V_{pk} = \alpha V_{dc}$). In the standard saturated operation, the proportionality constant, α, is about 0.9. The class-C amplifier is therefore equivalent to a voltage multiplier which forms the product of a nearly unit-amplitude sine wave times the power supply voltage. Modulating (varying) the power supply voltage of a class-C amplifier is the classic method used in AM transmitters. (Note that this useful property of a class-C (or D or E) amplifier would be considered a defect for an op-amp circuit – poor power supply rejection.)

9.1.6 The class-C amplifier as a frequency multiplier

If the class-C amplifier drive circuit furnishes a turn-on pulse only every other cycle, you can see that the decaying oscillation of the tank circuit voltage will execute two cycles between refreshes. Of course there will be a greater voltage droop, producing less output power, but note that the circuit becomes a frequency doubler. If the circuit is pulsed only on every third cycle, it becomes a tripler, etc. It is common to use a cascade of frequency multipliers to produce a high RF frequency that is a multiple of the frequency of a stable low-frequency oscillator. Class-D and class-E amplifiers can also be used as frequency multipliers.

9.2 The class-D RF amplifier

Class-D amplifiers can, in principle, achieve 100% efficiency. At least two switches are required, but neither is forced to support simultaneous voltage and current. The *class-D series resonant amplifier* is shown in Figure 9.5(a).

A single-pole double-throw switch produces a square-wave voltage. The series LC filter lets the fundamental sine-wave component reach the load, R_L. The bottom of the switch could be connected to a negative supply but ground will work since the resonating capacitor provides ac coupling to the load resistor. A real circuit is shown in Figure 9.5(b); two transistors form the switch. This is a push–pull circuit, i.e., the transistors are driven out-of-phase so that when one is on the other is off. Let us find the voltage on the load. Since the capacitor also acts as a dc block, we can consider this square wave to be

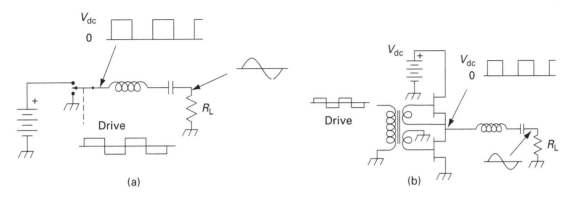

Figure 9.5. Class-D series
resonant amplifier.

symmetric about zero, swinging from $-V_{dc}/2$ to $+V_{dc}/2$. The square wave is equivalent to a Fourier series, i.e., a sum of sine waves whose frequencies are the fundamental, and odd multiples of the fundamental. The series resonant LC filter passes the fundamental sinusoidal component of the square wave. We can find the amplitude of the output sine wave by equating the dc input power to the sine-wave output power. They must be equal, since the circuit has no lossy element other than R_L, the load. Since both the voltage and current change sign during the negative half of the cycle, the power is the same in the negative half-cycle as in the positive half-cycle. The power delivered by the dc supply is therefore the product of the square-wave voltage, $V_{dc}/2$, times the average of the positive current in the series LCR. The current can be written as $I(t) = I_{pk}\sin(\omega t)$ and the average positive current is $\langle |I| \rangle = (2/\pi) I_{pk} = (2/\pi) V_{pk}/R_L$, where V_{pk} is the peak value of the sine-wave voltage on the load. Equating the power from the supply to the sine-wave power on the load, we have $(V_{dc}/2)\cdot(2/\pi)(V_{pk}/R_L) = 2V_{dc}^2/(R_L)$. Solving for V_{pk}, we find $V_{pk} = 2V_{dc}/\pi$.

To estimate the loss, we will assume the FETs have constant on-resistance, r. The current through the load passes through one or the other of the switches, so the ratio of output power to switch power is $(I^2 R_L)/(I^2 r)$ and the efficiency is $\eta = R_L/(R_L+r)$. You can see that, inas much as $r \ll R_L$, the efficiency can approach 100%. However, there is another, more important, source of loss in this class-D amplifier. Each switching transistor has parasitic capacitance which is abruptly charged and discharged through the transistor's resistance once per cycle. The energy lost is $\frac{1}{2}CV^2$ per transition so, for the circuit of Figure 9.5, the switching losses would be $4 \times \frac{1}{2}CV_{dc}^2 \times f$. Suppose the frequency, f, is 10 MHz and the FET switches each have, say, 200 pF of parallel capacitance. With a supply voltage of 200 volts, this would produce a loss of 160 watts! This loss can be avoided by using an alternate topology, the parallel resonant class-D amplifier.

9.2.1 Parallel-resonant class-D RF amplifiers

The class-D amplifier of Figure 9.6 consists of a square-wave *current source* driving a parallel *LCR* circuit. In this circuit, a large inductor (RF choke)

Figure 9.6. Class-D parallel-resonant RF amplifier operation.

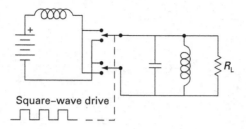

Figure 9.7. Practical class-D parallel-resonant RF amplifiers.

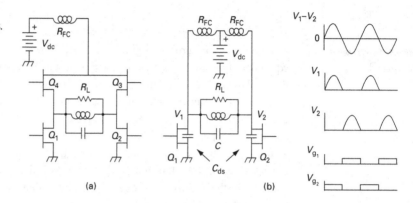

provides the constant current. A DPDT switch commutates the load, effectively forming a square-wave current source.

Two practical versions of this circuit are shown in Figure 9.7. In (a) the peak sine-wave voltage on the load, R_L, will be $\pi V_{dc}/2$ and the dc supply current will be $\pi^2 V_{dc}/(8R_L)$. For the appealing circuit in (b), these quantities are πV_{dc} and $\pi^2 V_{dc}/(2R_L)$. (See problem 9.4.) But the beauty of the parallel class-D amplifier is that the voltage on the parasitic capacitances of the transistors (drain-to-source capacitance, shown in dotted lines in Figure 9.7b) is zero at the instants the switches open or close. Thus, there is no lossy abrupt charging or discharging of these parasitic capacitors. (Refer to the waveforms in Figure 9.7b. Note that the value of C is effectively increased by C_{ds}, the parasitic capacitance of one transistor – first one, then the other.)

9.3 The class-E amplifier

We have seen that a class-C amplifier can be built with a single transistor and that its efficiency is high, relative to class-B and class-A amplifiers. The class-D switching amplifier, on the other hand can, in principle, achieve 100% efficiency, but requires two transistors. The *class-E amplifier* [4], has both virtues: single transistor operation and up to 100% efficiency. Figure 9.8 shows the circuit.

Figure 9.8. Class-E amplifiers.

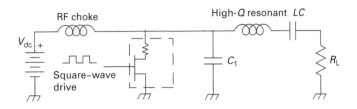

Figure 9.9. Class-E amplifier equivalent circuit.

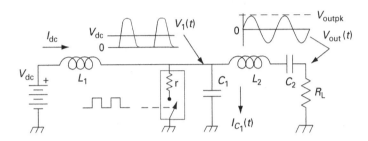

The transistor is used as a switch: fully on or fully off. An equivalent circuit is shown in Figure 9.9, where the transistor is represented as a switch. The switch operates with a 50% duty factor, so the voltage at the switch, $V_1(t)$, is forced to be zero for half the cycle (or almost zero if we consider the switch to have a small series resistance, r).

During the other half-cycle, $V_1(t)$ consists of a rounded positive pulse. The power supply is connected through L_1, an RF choke (a high-value inductor). The choke and the switch form a *flyback* circuit in which the power supply pumps energy into the inductor while the switch is closed and the inductor pumps energy into the rest of the circuit while the switch is open. Since there can be no dc drop across the choke, you can see that the pulses in $V_1(t)$ at the switch must have an average amplitude equal to twice the power supply voltage. The key to the efficiency of this circuit is that it can be designed so that the voltage $V_1(t)$ has fallen to zero at precisely the instant the switch closes, i.e., the capacitor C_1 is shorted out without a sudden lossy discharge. Note that C_1 includes the transistor's parasitic capacitance. The RF choke, L_1, is large enough to ensure that the dc supply current, I_{dc}, has essentially no ac component, i.e., the inductor current's increase and decrease during the flyback cycle are much smaller than the average current.

A simplified description of the circuit operation is as follows. The pulses at C_1 are not square, but must have an average voltage of $2V_{dc}$. The waveform is a rough approximation to a sine wave with a peak voltage of V_{dc} plus an equal dc offset. The bandpass filter formed by L_2 and C_2 passes a good sine wave to the load, R_L. Let us now look in detail at the circuit operation and design.

9.3.1 Class-E amplifier design procedure

We have seen that we can expect this amplifier to furnish the load with a sine wave whose amplitude is something like V_{dc}, so the RF power output will be approximately $V_{dc}^2/(2R_L)$. If a different power is desired, the resistance of the load, R_L, can be transformed to a different value using any of the techniques of Chapter 2. Next, we can simply pick a value for L_2 that will give a reasonably high Q, maybe 10, so that the waveform at the load will be a good sine wave, i.e., have minimal harmonic content. We are left with finding values for C_1 and C_2. So far, the only constraint on the circuit is the condition that $V_1(t)$ have fallen to zero at $\tau/2$, the instant of switch closure. It turns out that this can be satisfied over a range of combinations of C_1 and C_2 and that this provides a way to set the output power. Combinations with lower values of C_1 reduce the output power. However, it is possible and beneficial to impose a second constraint which will require a unique combination of C_1 and C_2. This constraint is that dV_1/dt, as well as $V_1(t)$, be zero at the instant the switch is closed. If this condition is met, the frequency of the amplifier can be shifted somewhat without seriously violating the condition that $V_1(\tau/2)$ be zero. Moreover, at high frequencies, where the transistor's switching time is relevant, $V_1(t)$ will at least remain close to zero during the switching process. Before outlining the analysis, we summarize the resulting design procedures as follows:

1. Pick a value for Q, say $Q = 10$. Then $\omega L_2 = QR_L$.

$$\text{Pick } \omega L_1 \text{ to be, say, } 100R_L. \tag{9.6}$$

2. Calculate C_1 as follows:

$$\frac{1}{\omega C_1} = \frac{\pi}{2}\left(1 + \frac{\pi^2}{4}\right)R_L = 5.45R_L. \tag{9.7}$$

3. Calculate C_2 as follows:

$$\frac{1}{\omega C_2} = \omega L_2 - \frac{\left(\dfrac{\pi^2}{8} - \dfrac{1}{2}\right)}{\left(1 + \dfrac{\pi^2}{4}\right)}\frac{1}{\omega C_1} = \omega L_2 - 0.212\frac{1}{\omega C_1}. \tag{9.8}$$

4. The amplitude of the output sine wave (the voltage on R_L) is

$$V_{out_{pk}} = V_{dc}\frac{2}{\sqrt{1 + \frac{\pi^2}{4}}} = 1.07V_{dc}. \tag{9.9}$$

5. The output power is given by

$$P_{out} = \frac{(V_{out_{pk}})^2}{2R_L} = 1.154\frac{V_{dc}^2}{2R_L}. \tag{9.10}$$

6. The average current drawn from the power supply is

$$I_{dc} = \frac{V_{out_{pk}}}{2R_L} = 0.577\frac{V_{dc}}{R_L}. \tag{9.11}$$

9.3.2 Class-E circuit analysis

This section is included for the reader who wants to understand the derivation of the above design formulas. We will analyze the circuit assuming $r = 0$ and afterward estimate the loss when r is small compared with R_L. The switch will be open whenever $\sin(\omega t)$ is positive, i.e., it opens at $t = 0$, closes at $\omega t = \pi$, etc. Because of the high Q of the series combination, L_2C_2, we will again use the simplifying assumption that the load voltage is a perfect sine wave, $V_{RL} = V_{out_{pk}}(t)\sin(\omega t + \phi)$ where ϕ is a phase shift to be determined [3]. As with the class-C amplifier, an exact analysis requires finding the properly connected switch-open and switch-closed transient solutions. Assuming the sine wave, the current into the load is given by $I_{RL}(t) = (V_{out_{pk}}/R_L)\sin(\omega t + \phi)$.

By inspection of Figure 9.9, we can immediately write an equation for the current through the capacitor C_1 while the switch is open:

$$I_{C_1}(t) = C_1\frac{dV_1}{dt} = I_{dc} - \frac{V_{out_{pk}}}{R_L}\sin(\omega t + \phi). \tag{9.12}$$

Imposing the condition that I_{C_1} be zero at $\omega t = \pi$, we find that

$$\sin(\phi) = \frac{-I_{dc}R_L}{V_{out_{pk}}}. \tag{9.13}$$

We can integrate Equation (9.12) to get $V_1(t)$, the voltage on the capacitor while the switch is open:

$$V_1(t) = \frac{1}{C_1}\left(I_{dc}t + \frac{V_{out_{pk}}}{\omega R_L}\cos(\omega t + \phi) - \frac{V_{out_{pk}}}{\omega R_L}\cos(\phi)\right), \tag{9.14}$$

where the last term is a constant of integration, added to satisfy the condition $V_1(0) = 0$, imposed by the switch having been in its closed state before $t = 0$. Next, imposing the condition that V_1 be zero also at $\omega t = \pi$, Equation (9.14) gives us

$$\cos(\phi) = \frac{\pi}{2}\frac{I_{dc}R_L}{V_{out_{pk}}}. \tag{9.15}$$

We can combine Equations (9.13) and (9.15), using $\cos^2\theta + \sin^2\theta = 1$, to find

$$\frac{V_{out_{pk}}}{I_{dc}R_L} = a \quad \text{and} \quad \sin(\phi) - \frac{-1}{a}, \quad \text{where} \quad a = \sqrt{\frac{\pi^2}{4} + 1}. \tag{9.16}$$

From this point, it remains to express $V_1(t)$ and the load current as complex constants times $e^{j\omega t}$ (the actual voltage and current are, of course, the real parts) and calculate the ratio, which must be equal to the complex impedance looking into the $L_2 C_2 R_L$ series circuit, $Z = R_L + j(\omega L_2 - 1/(\omega C_2))$.

The load current, $(V_{pk}/R_L)\sin(\omega t + \phi) = a\, I_{dc} \sin(\omega t + \phi)$, corresponds to the complex current $I(t) = -j\, a\, I_{dc}\, e^{j(\omega t + \phi)}$, since $\mathrm{Re}(-j\, a\, I_{dc}\, e^{j(\omega t + \phi)}) = a\, I_{dc} \sin(\omega t + \phi)$. Since the voltage $V_1(t)$ is not a sine wave, we have to find the complex representation of its fundamental Fourier component by evaluating the integral

$$\text{complex } V(t) = e^{j\omega t} \frac{2}{\tau} \int_0^{\pi/\omega} V(t) e^{-j\omega t} \mathrm{d}t. \tag{9.17}$$

Note that the upper limit of the integral would normally be $2\pi/4$ but by integrating over only the first half-cycle, we account for the fact that the closed switch forces $V_1(t)$ to be zero during the second half-cycle. We can rewrite Equation (9.14) in terms of a and I_{dc}:

$$V_1(t) = \frac{I_{dc}}{C_1}\left(t + \frac{a}{\omega}\cos(\omega t + \phi) - \frac{a}{\omega}\cos(\phi)\right). \tag{9.18}$$

Evaluating the integral in Equation (9.17), and then setting the ratio *complex $V_1(t)$ / complex I(t)* equal to the impedance, $R_L + j(\omega L - 1/(\omega C_2))$, results in the formulas for C_1 and C_2 (Equations 9.7 and 9.8).

9.3.3 Efficiency of the class-E amplifier

We will now assume that r is not zero, but still small enough that we can use the currents derived above, where r was taken as zero. Power is dissipated in r when the switch is closed. The current through r is just

$$I_r(t) = I_{dc} - \frac{V_{outpk}}{R_L}\sin(\omega t + \phi), \tag{9.19}$$

which is the same as Equation (9.12), except that now we assume any current flowing in the capacitor is negligible compared with the current flowing through the closed switch.

The instantaneous power dissipated in r is

$$(I_r(t))^2 r = \left(I_{dc} - \frac{V_{outpk}}{R_L}\sin(\omega t + \phi)\right)^2 r \tag{9.20}$$

and the average power dissipation in r is, therefore, given by

$$\langle (I_r(t))^2 r \rangle = \left\langle \left(I_{dc} - \frac{V_{outpk}}{R_L}\sin(\omega t + \phi)\right)^2 \right\rangle r$$

$$= \frac{1}{\tau} \int_{\tau/2}^{\tau} \left(I_{dc} - \frac{V_{outpk}}{R_L}\sin(\omega t + \phi)\right)^2 r\, \mathrm{d}t, \tag{9.21}$$

where the period, τ, is just $2\pi/\omega$. Evaluating this integral and using Equations (9.15) and (9.16), we find

$$\langle (I_r(t))^2 r \rangle = \frac{r}{R_L^2} \left(\frac{3 V_{\text{out}_{pk}}^4}{8 V_{dc}^2} + \frac{V_{\text{out}_{pk}}^2}{4} \right). \tag{9.22}$$

To get the fractional power loss, just divide this by the power out, $V_{\text{out}_{pk}}/(2R_L)$.

9.3.4 Class-E design example

The formulas above were used to design a class-E amplifier designed to produce a 1-MHz, 12-V peak sine wave on a 50-ohm load, using a 12-V power supply. The component values were as follows: $L_1 = 1$ mH, $C_1 = 584$ pF, $L_2 = 79.6$ μH, $C_2 = 360$ pF, and $R_L = 50$ ohms. If the transistor has an on-resistance of 0.5 ohms, the fractional loss will be 1.9% (98% efficiency). A SPICE analysis of this circuit shows the second harmonic power at the load to be 26 dB below the fundamental power. The 3-dB bandwidth is about 15%.

9.4 Which circuit to use: class-C, class-D, or class-E?

Remember that the high-efficiency amplifiers discussed in this chapter are narrowband and nonlinear. They are ideal for producing conventional signals with AM, pulse, or phase modulation, such as radio broadcast signals and signals from cell phones. Lower efficiency linear class-A, AB, or B amplifiers are used with more complicated signals such as the superposition of many individual signals from a cellular base station or from a direct broadcast television satellite.[2] AM and FM broadcast transmitters have historically used class-C vacuum tube amplifiers. This type of transmitter is still widely used. Tubes are available with maximum plate dissipations of up to more than 1 MW. Typical class-C amplifiers have efficiencies of 75–85%, so a single large tube can produce in excess of 6 MW of RF power. The highest power used in radio broadcasting is about 2 MW, but amplifiers (or class-C oscillators) for much higher power are found in industrial heating applications such as curing plywood and welding. Many new designs use class-D and class-E solid-state amplifier modules, with the power from multiple modules being combined to achieve the desired total power. The complexity of using many modules is sometimes offset by the ability to "hot swap" defective modules without interrupting operations. The switching amplifiers, class-D and class-E, benefit from advances in transistor technology and have reached GHz frequencies.

---

[2] In principle, class-C, D, and E amplifiers can reproduce any type of band-limited signal by using simultaneous amplitude and phase modulation, as in the efficient amplification of single-sideband signals discussed in Chapter 8.

Problems

Problem 9.1. For the shunt-fed class-C amplifier of Figure 9.4(c), sketch the waveforms of the voltage on each side of the blocking capacitor and of the current through the blocking capacitor.

Problem 9.2. Suppose the simple series class-D amplifier of Figure 9.5(b) is driving a 50-ohm resistive load at a frequency of 1 MHz. The values of the inductor and capacitor are 9.49 μH and 2.67 nF (equal and opposite reactances at 1 MHz). This resonant circuit passes the 1-MHz component of the square wave and greatly reduces the harmonics. By what factor is the 3-MHz (i.e., third harmonic) power delivered to the load lower than the fundamental (1-MHz) power? Hints: in a square-wave, the amplitude of the third harmonic is one-third of the amplitude of the fundamental. Remember that at 1 MHz, $X_L = X_C$ while at 3 MHz, X_L is increased by a factor of 3 and X_C is decreased by a factor of 3.

Problem 9.3. A single-tube class-C amplifier with 75% efficiency is providing 500 kW of continuous cw output power. The supply voltage is 60 kV and the conduction angle is 90°. What is the average current drawn from the power supply? Answer: 11.1 amperes. What is the average current when the tube is on? Answer: 44.4 amperes.

Problem 9.4. (a) Consider the class-D amplifier of Figure 9.7(a). Show that the peak sine-wave voltage on the load, R_L, will be $\pi V_{dc}/2$ and the dc supply current will be $\pi^2 V_{dc}/(8R_L)$. Hint: since the efficiency is 100% (there are no lossy components except the load), the power supplied by the dc source will be $I \cdot V_{dc}$ where I is the constant current flowing through the RF choke. But the power from the supply can also be written as V_{dc} times the average of the absolute value of the sine wave on the load. (The average of $|A\sin(\theta)|$ is $2A/\pi$.)

(b) For the amplifier of Figure 9.7(b), show that the peak sine-wave voltage on the load will be πV_{dc} and the dc supply current will be $\pi^2 V_{dc}/(2R_L)$.

References

[1] Krauss, H. L., Bostian, C. W. and Raab, F. H. *Solid State Radio Engineering*, New York: John Wiley, 1980.
[2] Raab, F. H., High efficiency amplification techniques, *IEEE Circuits and Systems*, Vol. 7, No. 10, pp. 3–11, December 1975.
[3] Raab. F. H., *Idealized operation of the class-E tuned power amplifier, IEEE Trans. Circuits Syst.*, Vol. CAS-25, pp. 725–735, Dec. 1977.
[4] Sokal, N. O. and Sokal, A. D., *Class-E – A new class of high-efficiency tuned single-ended switching power amplifiers, IEEE J. Solid-State Circuits*, vol. 10, no. 3, pp. 168–176, 1975.
[5] Eimac division of CPI, Inc., *Care and Feeding of Power Grid Tubes*, 5th edn, 2003, CPI, Inc., 301 Industrial Rd., San Carlos, CA. PDF: http://www.cpii.com/docs/related/22/C&F1Web.pdf

10 Transmission lines

We draw circuit diagrams with "lumped"components: ideal R's, C's, L's, transistors, etc., connected by lines that represent zero-length wires. But all real wires, if not much shorter than the shortest relevant wavelength, are themselves complicated circuit elements; the current is not the same everywhere along such a wire, nor is voltage uniform, even if the wire has no resistance. On the other hand, when interconnections are made with transmission lines, which are well-understood circuit elements, we can accurately predict circuit behavior. In this section we will consider two-conductor lines such as coaxial cables and open parallel wire lines. "Microstrip lines" (conducting metal traces on an insulation layer over a metal ground plane) behave essentially in the same way, but they have some subtle complications, which are mentioned in Appendix 10.1.

10.1 Characteristic impedance

The first thing one learns about transmission lines is that they have a parameter known as *characteristic impedance*, denoted Z_0. How "real" is characteristic impedance? If we connect an ordinary dc ohmmeter to the end of a 50-ohm cable will it indicate 50 ohms? Yes, if the cable is *very* long, so that a reflection from the far end does not arrive back at the meter before we finish the measurement. Otherwise, the meter will simply measure whatever is connected to the far end, which could be short, an open circuit, or a resistance. However, using a pulse generator and an oscilloscope, you can easily make an ohmmeter set-up that is fast enough that, even for a short cable, you can determine V_{in} and I_{in} and then calculate $V_{in}/I_{in} = Z_0$.

To make a theoretical determination of Z_0, we first model the transmission line as a ladder network made of shunt capacitors and series inductors, as shown in Figure 10.1.

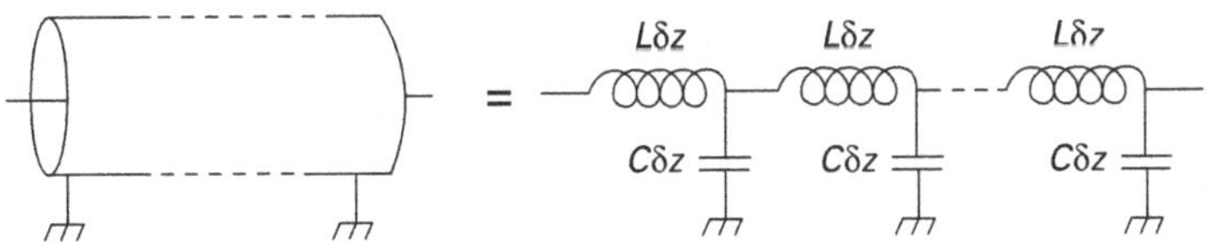

Figure 10.1. Transmission line model – a ladder network of infinitesimal LC sections.

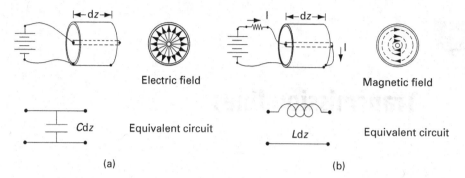

Figure 10.2. Capacitance and inductance per unit length.

To see that this model is reasonable, consider Figure 10.2(a), which shows the electric field lines in a length of coaxial cable connected to a voltage source. The field lines are radial and their number is obviously proportional to the length of the cable, so that capacitance per unit length is a constant. Likewise, a current through the cable (b) sets up a magnetic field, so another characteristic of the cable is its inductance per unit length. We will follow common convention and use the symbols C and L to denote capacitance and inductance per unit length. That convention is obvious when capacitors and inductors are labeled, respectively, $C\delta z$ and $L\delta z$, where δz is a short increment of length along the z-axis, i.e., parallel to the cable.

Every increment of a transmission line contributes series inductance and shunt capacitance; the ladder network shown in Figure 10.1 models a real transmission line in the limit that δz goes to zero. For some situations, e.g., baseband telephony and digital data transmission through long cables, the model must also include series and shunt resistance. At radio frequencies, however, the series reactance is usually much greater than the series resistance and the shunt reactance is usually much less than the shunt resistance so both resistances can be neglected. (See Problem 10.3.)

To see that $Z_0^2 = L/C$, consider the circuit of Figure 10.3, where we have added another infinitesimal LC section to the model transmission line, which is either infinitely long or terminated with a resistance equal to the characteristic impedance, so as to appear infinitely long. After adding the section, the line is still infinitely long and the impedance looking into it must still be Z_0. If the voltage and current at the input of the line were V and I, they will be modified to become $V+\delta V$ and $I+\delta I$ at the input to the new section. (This does not imply an increase in power; $V+\delta V$ and $I+\delta I$ are merely phase-shifted versions of V and I.)

Since the impedance looking into the line must stay the same, we have

$$\frac{V + \delta V}{I + \delta I} = \frac{V}{I},$$

(10.1)

Figure 10.3. Adding another infinitesimal section must leave Z_0 unchanged.

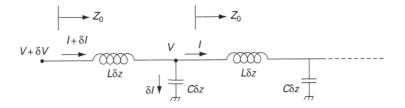

from which $\delta V/\delta I = V/I = Z_0$.

Using this, and substituting $\delta I = (C\delta z)\, dV/dt$ and $\delta V = (L\delta z)\, d/dt(I+\delta I)$ and ignoring the vanishingly small $\delta z\, \delta I$ term, we have

$$Z_0 = \frac{\delta V}{\delta I} = \frac{L\delta z\,(\mathrm{j}\omega I)}{C\delta z\,(\mathrm{j}\omega V)} = \frac{L}{C}\,Z_0^{-1}. \qquad (10.2)$$

Looking at the first and last terms of this equation, we see that $Z_0 = (L/C)^{1/2}$. Note: you can verify that, because δV and δI are infinitesimal, the above equations are the same if the network starts with a capacitor instead an inductor.

To evaluate Z_0, it is sufficient to know either L or C, since it follows from electrodynamics that they are related by $LC = \varepsilon_r/c^2$ where ε_r is the dielectric constant (relative to vacuum), and c is the speed of light. This relation between L and C holds for any two-conductor structure with translational symmetry such as an unlikely transmission line consisting of a square inner conductor inside a triangular outer conductor.

For a coaxial transmission line, $C = 2\pi\varepsilon_r\varepsilon_0 / \ln(b/a)$ farads/meter, where a and b are the inner and outer radii and ε_0, the "permittivity of free space," is equal to $(4\pi \times 10^{-7}c^2)^{-1}$. Using this, together with the relation $LC = \varepsilon_r/c^2$, gives us $Z_0 = (\varepsilon_r)^{-1/2}\, 60 \ln(b/a)$. Note that Z_0 depends on the ratio a/b, but not on the size of the cable.

10.2 Waves and reflected waves on transmission lines

We will use a simple ac analysis to show that an applied sinusoidal voltage causes a spatial voltage sine wave to propagate down the line: Let us apply a voltage $e^{\mathrm{j}\omega t}$ and find the voltage drop across an incremental length of line (see Figure 10.4).

Since we already know the input impedance is Z_0, the input current must be V/Z_0 and the voltage across the inductor can be written $\delta V = -(V/Z_0)\,(\mathrm{j}\omega L\delta z)$. But this is just the differential equation

$$\frac{dV}{dz} = -\mathrm{j}\omega\frac{L}{Z_0}V = -\mathrm{j}\omega\sqrt{LC}\,V. \qquad (10.3)$$

Figure 10.4. Finding the change in voltage, δV, over a distance δz.

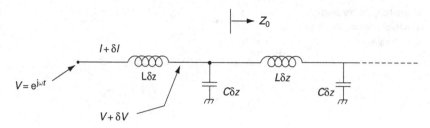

The solution to this familiar equation is

$$V = V_f e^{-j\omega\sqrt{LC}z} = V_f e^{-jkz} \quad \text{where} \quad k = \omega\sqrt{LC}, \tag{10.4}$$

where V_f is a constant, the amplitude. The constant k is known as the *propagation constant* and is the number of radians the wave progresses per unit length. The wave therefore repeats in a distance (the wavelength) given by $\lambda = 2\pi/k$. Since $V/I = Z_0$, the current along the line is also a wave: $I = (V_f/Z_0)e^{-jkz}$. If we include the otherwise implicit multiplicative time dependence factor $e^{j\omega t}$, the voltage is

$$V = V_f e^{j\omega t}e^{-jkz} = V_f e^{j(\omega t - kz)}. \tag{10.5}$$

This is just a sine wave running in the forward z-direction. The complex exponential now contains space as well as time but, as always, the physical voltage is the real part, i.e., Re[$V_f e^{j(\omega t - kx)}$] which is a weighted superposition of $\sin(\omega t - kz)$ and $\cos(\omega t - kz)$. For a point of constant phase, $\omega t - kz =$ constant, we have $\delta z/\delta t = \omega/k$. This velocity, $\omega/k = c/\sqrt{\varepsilon_r}$, is known as the *phase velocity*, v_{phase}. Figure 10.5 shows a forward-running wave on a coaxial cable. The electric and magnetic field lines are drawn only at the points where they reach their peak values. A graph shows the spatial distribution. Everything has the same phase, i.e., the voltage, current, and charge density all rise and fall together along the z-axis. Note that a wave of amplitude V transfers power at a rate $|V|^2/(2Z_0)$.

A transmission line can equally well support waves running in the negative z-direction. If we had assumed a current in the $(-z)$-direction, the phase would progress as $\omega t + kz$. A transmission line in a circuit operating at a frequency ω will, in general, have both a forward wave and a reverse wave. The waves have complex amplitudes, V_f and V_r, each containing magnitude and phase. Of course both waves have the same frequency and propagation constant. We regard current as positive when it is in the $(+z)$-direction, so the current of a forward wave is $I_f(z,t) = V_f(z,t)/Z_0$, but the current of a reverse wave is $I_r(z,t) = -V_r(z,t)/Z_0$, since the reverse wave is traveling in the $(-z)$-

Figure 10.5. Forward wave on a
transmission line.

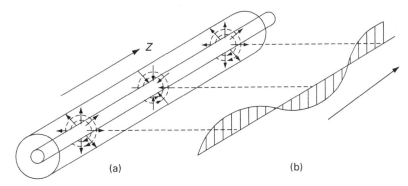

direction. Together, the forward and reverse waves are, in general, equivalent
to a stationary (*standing*) wave plus a single propagating wave.

Note also that the phase velocity is independent of ω; there is no
dispersion in this kind of lossless transmission line. Therefore, if we apply
an arbitrary voltage waveform, $V_{\text{arb}}(t)$, at the input to the line, this waveform,
considered as a Fourier superposition of sine waves, will propagate down the
line without distortion. At any point z, the voltage will be $V_{\text{arb}}(t - z/v_{\text{phase}})$,
a delayed but undistorted version of the input signal. For example, if, at
$t = 0$, we connect a dc voltage to the line, a step function propagates down
the line.

The *electrical length* of a line is the phase change imparted by the line. For
example, a "quarter wave line" imparts a 90° phase shift, $kl = \pi/2$, and therefore
$l = \pi/(2k) = \pi c/(2\omega\sqrt{\varepsilon_{\text{r}}}) = (c/f)/(4\sqrt{\varepsilon_{\text{r}}}) = 1/4(\lambda_0/\sqrt{\varepsilon_{\text{r}}})$, where λ_0 is the
wavelength in free space.

Standing waves

When both a forward and a reverse wave are present on a transmission line
the voltage along the line, which is the sum of the contributions from the
two waves, forms an interference pattern or standing wave. To see this, let
$V(z,t) = V_f e^{j(\omega t - kz)} + V_r e^{j(\omega t + kz)}$. The real parts of these two rotating phasors will
be in phase at points along the transmission line which are separated by $\lambda/2$. At
these points, the magnitude of the sum will be $|V_f| + |V_r|$. Halfway between
these points, the real parts of the phasors will be out of phase and the
magnitude of the sum will be $\||V_f| - |V_r|\|$. The ratio of these maximum and
minimum voltage magnitudes is called the *voltage standing wave ratio*:
$\text{VSWR} = (|V_f| + |V_r|)/\||V_f| - |V_r|\|$. If $|V_f| = |V_r|$ there is only a standing wave and
the VSWR is infinite. When $|V_f| \neq |V_r|$, the weaker one, along with an equal
portion of the stronger one, form a standing wave, leaving the remainder of the
stronger one as a travelling wave.

10.3 Modification of an impedance by a transmission line

From the discussion above, you can see that a transmission line terminated by a resistor of value Z_0 will always present an input impedance of Z_0. But a piece of transmission line that is terminated with an arbitrary impedance, $Z \neq Z_0$, as shown in Figure 10.6, will produce a modified ("transformed") impedance, Z'.

This figure shows a line of length l whose right-hand end ($z = 0$) is connected to some impedance Z_L (L denotes "load"). Assume that some constant ac source produces a constant incident wave traveling to the right, $V_f e^{-jkz}$ (we will not bother writing the always present factor $e^{j\omega t}$), and that Z_L causes a constant reflected wave, $\Gamma V_f e^{jkz}$, to travel to the left.[1] The factor Γ is known as the *reflection coefficient*. At any point, z, the voltage on the line is $V(z) = V_f e^{-jkz} + \Gamma V_f e^{jkz}$. The corresponding current is $I(z) = (V_f/Z_0)(e^{-jkz} - \Gamma e^{jkz})$. The minus sign occurs because the current in the reflected wave flows in the negative z-direction. At the right-hand end ($z = 0$), the load ensures that $V(0)/I(0) = Z_L$. This will give us Γ:

$$\frac{V(0)}{I(0)} = Z_L = \frac{(1 + \Gamma)}{(1 - \Gamma)/Z_0} \quad \text{so} \quad \frac{Z_L}{Z_0} = \frac{(1 + \Gamma)}{(1 - \Gamma)} \quad \text{and} \quad \Gamma = \frac{(Z_L - Z_0)}{(Z_L + Z_0)}.$$

$$(10.6)$$

Putting this expression in Equation (10.6) for Γ, together with the expressions for $V(z)$ and $I(z)$, we can immediately find $V(-l)/I(-l)$ which is what we are after, i.e., Z', the input impedance at a point l to the left of the load:

$$Z' = \frac{V(-l)}{I(-l)} = \frac{e^{-jk(-l)} + \Gamma e^{jk(-l)}}{e^{-jk(-l)}/Z_0 - \Gamma e^{jk(-l)}/Z_0}$$

$$= Z_0 \frac{(Z_L + Z_0)e^{jkl} + (Z_L - Z_0)e^{-jkl}}{(Z_L + Z_0)e^{jkl} - (Z_L - Z_0)e^{-jkl}}$$

or

$$Z' = Z_0 \frac{Z_L + jZ_0 \tan(kl)}{Z_0 + jZ_L \tan(kl)}.$$

$$(10.7)$$

Figure 10.6. An impedance is modified when seen through a transmission line.

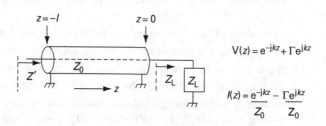

$$V(z) = e^{-jkz} + \Gamma e^{jkz}$$

$$I(z) = \frac{e^{-jkz}}{Z_0} - \frac{\Gamma e^{jkz}}{Z_0}$$

---

[1] Since everything is linear, superposition holds and the incident and reflected waves do not collide or interact in any way. They simply pass through one another unaltered. At any point, the current is the sum of their currents and the voltage is the sum of their voltages.

This important result, the modification of an impedance Z_L by a length l of transmission line, is not hard to remember; it has no minus signs and is symmetric. Just remember $(1+j\tan)/(1+j\tan)$. Once you have written this framework, you will remember how to put in the coefficients. Some important special cases are listed below:

- If $Z_L = Z_0$, then $Z' = Z_0$ for any length of line.
- If $Z_L = 0$ (a short) then $Z' = jZ_0\tan(kl)$, a pure reactance, which is inductive[2] for $kl < \pi/2$, then capacitive, etc.
- If $Z_L = $ infinity (an open circuit) then $Z' = Z_0/j\tan(kl)$ which is capacitive for $kl < \pi/2$, then inductive, etc.
- An impedance is left unchanged by a line of arbitrary Z_0 whose length is a half-wave ($kz = \pi$) or any integral multiple of a half-wave.
- A quarter-wave line ($kz = \pi/2$) or an odd multiple of a quarter-wave line, *inverts* an impedance: $Z' = Z_0^2/Z_L$. A short is transformed into an open and an open into a short, an inductor is transformed into a capacitor and vice versa, etc.

10.4 Transmission line attenuation

In a lossy transmission line, i.e., a line that causes attenuation of the signal, the e^{-jkz} or e^{jkz} spatial dependence of the wave is replaced by $e^{-jkz}e^{-\alpha z} = e^{-j(k-j\alpha)z}$ (forward wave) or $e^{jkz}e^{\alpha z} = e^{j(k-j\alpha)z}$ (reverse wave), where α is the attenuation constant. In a distance $1/\alpha$, the amplitude falls by a factor $1/e$ and the power falls by a factor $(1/e)^2$. Note that k for a lossless line is simply replaced by $k - j\alpha$, i.e., the propagation constant becomes complex. You can put this complex k into the "tan tan" formula to see how an impedance is modified by a lossy cable.

Transmission line attenuation is usually expressed in units of dB/meter. To find α for a line whose loss is A_L dB/m, note that, since the amplitude falls by a factor $e^{-\alpha \cdot 1}$ in 1 meter, we can write $-A_L = 10\log(e^{-\alpha \cdot 1})^2 = -20\alpha\log(e)$ from which $\alpha = A_L/(20\log(e))$.

10.5 Impedance specified by reflection coefficient

We have seen that an impedance Z produces a reflection coefficient given by $\Gamma = (Z - Z_0) / (Z + Z_0)$. This relation is easily inverted, $Z = Z_0(1+\Gamma)/(1-\Gamma)$, so there is a one-to-one mapping between Z and Γ. In antenna and microwave work, especially when using S-parameter analysis (Chapter 28), it is customary to think in terms of Γ, rather than Z.

One big advantage of working in the complex Γ-plane is that the modification of an impedance (represented by its equivalent Γ) is extremely simple. The

---

[2] Note that "inductive" does not mean equivalent to a lumped inductor since $Z_0\tan(kl) = Z_0\tan(\omega l/v_{phase})$ is not proportional to ω, except for small kl. Likewise, a short open-ended line is not equivalent to a lumped capacitor, except for small $\omega l/v_{phase}$.

reflection coefficient for the given impedance as seen through a length l of transmission line is just

$$\Gamma' = \Gamma e^{-j2kl}, \tag{10.8}$$

which means we simply rotate the point clockwise around the origin ($\Gamma = 0$) by an angle $2kl$ to give Γ', the modified reflection coefficient. This is easy to see: when we add a length of cable, the incident wave's phase is delayed by kl getting to the end of the cable and the reflected wave is delayed by the same kl getting back again. The effect of a cable is therefore to rotate the complex number Γ clockwise by an angle $2kl$.[3] (Since the time dependence is $e^{j\omega t}$, the round-trip time delay is a clockwise displacement.) Keep in mind that the Γ-plane is a complex plane but that it is *not* the $R + jX$ plane. Let us look at a few special points in the Γ-plane.

1. The center of the plane, $\Gamma = 0$, corresponds to a reflected wave of zero amplitude, so this point represents the impedance $Z_0 + j0$.
2. The magnitude of Γ (radius from the origin) must be less than or equal to unity for passive impedances. Otherwise the reflected wave would have more power than the incident wave.
3. The point $\Gamma = -1 + j0$ corresponds to $Z = 0$, a short circuit.
4. The point $\Gamma = 1 + j0$ corresponds to $Z = \infty$, an open circuit.
5. Points on the circle $|\Gamma| = 1$ correspond to pure reactances, $Z = 0 + jX$. All points inside this circle map to impedances with positive nonzero R.
6. The point $\Gamma = 0 + j1$ corresponds to an inductance, $Z = 0 + jZ_0$. All points in the top half of the Γ-plane are "inductive," i.e., $Z = R + j|X|$ or, equivalently, $Y = G - j|B|$.
7. The point $\Gamma = 0 - j1$ corresponds to a capacitance, $Z = 0 - jZ_0$. All points in the bottom half of the Γ-plane are "capacitive," i.e., $Z = R - j|X|$ or, equivalently, $Y = G + j|B|$.

These special cases of mapping of Z into Γ are shown in Figure 10.7.

In the Γ-plane, if you plot $\Gamma = R + jX$, where R is a constant and X varies, you will get a circle centered on the real axis and tangent to the line $\mathrm{Re}(\Gamma) = 1$. For every value of R there is one of these "resistance circles." The resistance circle for $R = 0$ is the unit circle in the Γ-plane. The resistance circle for $R = \infty$ is a circle of zero radius at the point $\Gamma = 1 + j0$. Likewise, if you plot $\Gamma(R + jX)$ where X is a constant and R varies, you will get "reactance circles" centered on the line $\mathrm{Re}(\Gamma) = 1$ and tangent to the line $\mathrm{Im}(\Gamma) = 0$. These circles are shown in Figure 10.8.

If you now trim the circles to leave only the portions within the $|\Gamma| = 1$ circle (corresponding to passive impedances, i.e., impedances whose real part is

---

[3] If the line is lossy, the magnitude of Γ decreases as it rotates around the origin, forming a spiral. For a long enough length of lossy line, Γ spirals all the way into the origin producing $Z = Z_0$, no matter what value of Z terminates the far end of the cable.

Figure 10.7. Impedances mapped into the reflection plane.

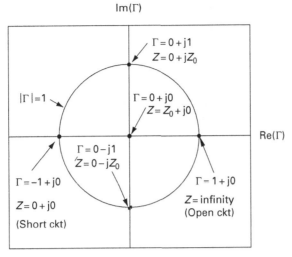

Reflection plane

Figure 10.8. Loci of constant resistance and of constant reactance – circles in the Γ-plane.

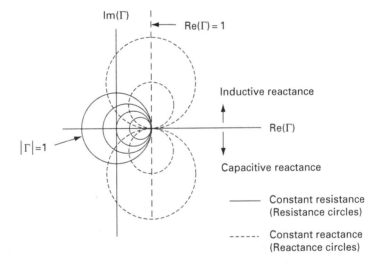

positive) you are left with a useful piece of graph paper, the famous *Smith chart*, shown in Figure 10.9.

The circular R and X "axes" on the Smith chart allow you to locate the Γ-point that corresponds to $Z=R+jX$. We have already seen that when we have located an impedance on the Γ-plane, we can find how that impedance is modified by a length of transmission line (whose Z_0 is the same as the Z_0 used to draw the chart) by rotating the point clockwise around the origin. We simply rotate the point clockwise around the origin by an angle equal to twice the electrical length of the line. The values of R and X corresponding to the rotated point can be read

Figure 10.9. The Smith chart – resistance and reactance circles on the Γ-plane.

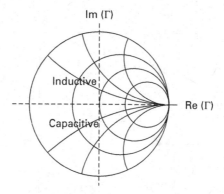

Figure 10.10. Conductance and susceptance circles.

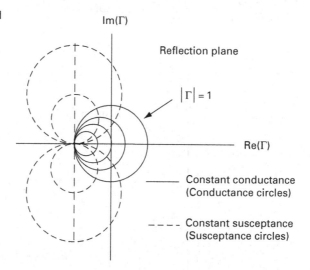

off the chart's R and X "axes." We can also use the chart to find how an impedance is modified by adding a series R or series X. In this operations, the Smith chart can be considered something of a calculator. Note that the Smith chart can also be made with G and B "axes". As you might guess, these produce "G circles" and "B circles" as shown in Figure 10.10.

Sometimes the Smith chart contains G and B circles as well as R and X circles. This full-blown chart, which can be quite dense, is shown in Figure 10.11. Again, remember the Smith chart is actually a rectangular graph of Γ; the x-axis is Re(Γ) and the y-axis is Im(Γ). Because only the area inside the circle $|Γ| = 1$, i.e., $x^2+y^2 = 1$, is used, the Smith chart resembles a polar graph. And, indeed, when we rotate a point around the origin to how a transmission line modifies an impedance, we are using it in a polar fashion. Sometimes the Smith chart is scaled for a specific Z_0 (usually 50 ohms or 75 ohms). Other charts are normalized; the $R = 1$ circle would be the 50-ohm circle if we are dealing with 50-ohm cable, etc.

Figure 10.11. Smith chart with R, X, G, and B circles.

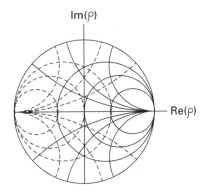

10.6 Transmission lines used to match impedances

Designing a matching network becomes an exercise in moving from a given Γ to a desired Γ' in the reflection plane. Working graphically, it is often easy to find a matching strategy. Let us use the Smith chart and revisit the 1000-ohm-to-50-ohm matching circuit example of Chapter 2.

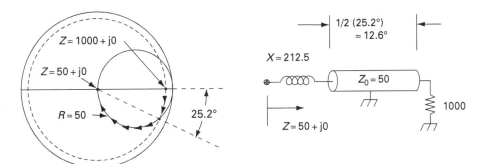

Figure 10.12. Conversion from 1000 ohms to 50 ohms – transmission line and inductor circuit.

I. The starting impedance, 1000 ohms, and the final target impedance, 50 ohms, are indicated on the chart in Figure 10.12. Also shown is the 50-ohm circle. We can use a (50-ohm) transmission line to move along the dashed circle until we reach the 50-ohm circle. Now we have $R = 50$ plus a capacitive reactance. A series inductor will cancel the capacitive reactance, taking us to $Z = 50 + j0$ (the center of the chart).

II. Another solution (Figure 10.13) would be to use a longer piece of cable to circle most of the chart, hitting the 50-ohm circle in the top half of the plane. At this point we have $Z = 50 + jX$ where X is positive (inductive). We can add a series capacitor to cancel this X and again arrive at $Z = 50 + j0$.

Figure 10.13. Transmission line and capacitor matching circuit.

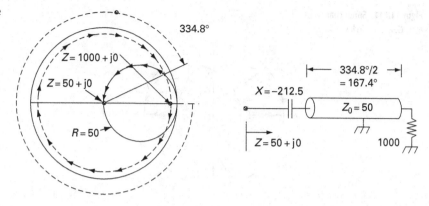

Figure 10.14. Series and shunt transmission line matching circuit.

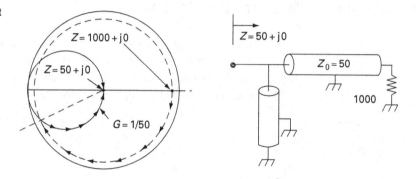

III. So far we have only used series elements. Let us now start by traveling around to the $G = 1/50$ circle. Then we can add a shunt element to reach the center of the chart. The first intersection of the $G = 1/50$ circle is in the lower half-plane (capacitive) so, to get from this point to the center, we need a shunt inductor. Instead of a lumped inductor we might use a shorted length of transmission, as shown in Figure 10.14, to make a matching circuit using only transmission line elements.

IV. Figure 10.15 shows a solution that uses no transmission line. We start on the $G = 1/1000$ circle, at $G = 0$. If we apply shunt reactance we can move along this circle. Let us pick shunt inductance which will move us upward along the G circle to the 50-ohm circle. We now have $R = 50$, but there is inductive reactance. As in the above example, we can now cancel the inductance reactance with a series capacitor. This is just the L-network found in Chapter 2.

V. If we had used shunt capacitance rather than shunt inductance, we would have moved downward to the 50-ohm circle, as shown in Figure 10.16. The remaining series capacitance can be cancelled with an inductor. This produces an L-network where the positions of the L and C are reversed.

Figure 10.15. *LC* matching network.

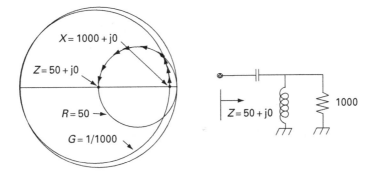

Figure 10.16. *CL* matching network.

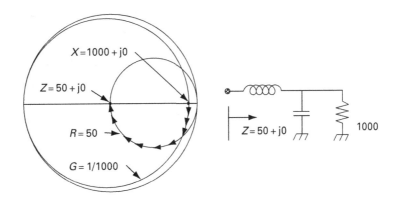

Figure 10.17. An impedance-admittance chart.

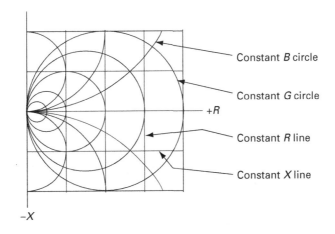

In these examples, our final impedance was at the center of the chart ($Z = 50 + j0$), but you can see that these techniques allow us to transform any point on the chart (i.e., any impedance) into any other point on the chart (any other impedance).

The Smith chart is a favorite because it handles networks that include transmission lines as well as inductors and capacitors. If we did not care about transmission lines, then any chart that maps R, X into G, B would do. For example, take the R, X plane (half-plane, since we will exclude negative R). Draw in the curves for G = constant and B = constant. The resulting chart, shown in Figure 10.17, can be used to design lumped element L, C,R ladder networks, such as the networks of Figures 10.15 and 10.16.

Appendix 10.1. Coaxial cable – Electromagnetic analysis

This chapter began with a derivation of Z_0 based on an equivalent lumped-element circuit model of a transmission line. That derivation required only elementary ac circuit theory, but is a rather indirect approach to what is really a problem in electromagnetics. Even then, some electromagnetic theory is needed to derive the expressions for capacitance and inductance per unit length.

An electromagnetic analysis of a coaxial transmission line is presented here for the reader who has some familiarity with Maxwell's equations. We make use of the fact that the propagation velocity of a *TEM wave*[4] is given by $v = (\mu\varepsilon)^{-1/2}$, where μ is the magnetic permeability and ε is the electrical permitivity of the material through which the fields propagate.[5]

To find the impedance of the coaxial line, we will first assume that the current on the inner conductor is given by $I = I_0 \cos(\omega t - kz)$, which is a wave traveling in the (+z)-direction. This is illustrated in Figure 10.18.

Figure 10.18. Transmission line element.

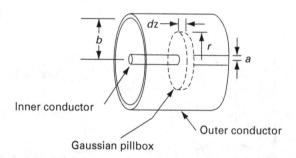

Inner conductor

Gaussian pillbox

Outer conductor

We will then proceed to find the charge density, the electric field, and then the voltage, which will have the form $V_0 \cos(\omega t - kz)$. Once we have the voltage, the

---

[4] In a TEM wave, by definition, both the electric field and the magnetic field are transverse, i.e., perpendicular to the direction along which the wave propagates. In most applications of coaxial cables and parallel-wire transmission lines, the wavelength is much greater than the transverse dimensions of the line and only TEM waves can propagate. Waves in free space are also TEM waves.

[5] To find the propagation velocity of a TEM wave: The variables t and z appear in E_x, E_y, B_x, and B_y only in the factor $e^{j(\omega t - kz)}$. Using the condition $E_z = 0$, the x-component of the Maxwell equation curl$(E) = -\partial B/\partial t$ gives us $B_x = -(k/\omega)E_y$. Likewise, using the condition $B_z = 0$, the y-component of the Maxwell equation curl$(B/\mu) = \partial(\varepsilon E)/\partial t$ gives us $B_x = -(\omega\mu\varepsilon/k)E_y$. Equating these two expressions for B_x gives $k^2 = \omega^2\mu\varepsilon$.

characteristic impedance is simply given by V_0/I_0. (Note that the current in the outer conductor is just the negative of the current in the inner conductor.)

Consider an incremental segment of the inner conductor from z to $z + \delta z$. The rate at which charge accumulates on this element is $\partial/\partial t\,(\rho_L)\delta z$, where ρ_L is the charge per unit length. But the rate at which charge accumulates in δz is nothing more than the difference between the current flowing into δz and the current flowing out of δz. Therefore, we can write

$$\frac{\partial \rho_L}{\partial t} = \frac{-\partial I}{\partial z} = -k\,I_0 \sin(\omega t - kz). \qquad (10.9)$$

Integrating $d\rho_L/dt$ with respect to time gives us the linear charge density (coulombs/meter):

$$\rho_l(z, t) = \frac{1}{\omega}(k I_0 \cos(\omega t - kz)) = \frac{k I_0}{\omega} \cos(\omega t - kz). \qquad (10.10)$$

Now that we know the charge density, we can find the electric field. The field is radial with field lines like spokes of a wheel. Imagining a Gaussian "pillbox" of radius r and height δz around the center conductor, we use Gauss's law: the integral of the E field over the sidewall surface must be equal to the enclosed charge divided by ε:

$$E(r, z, t)(2\pi r \delta z) = \frac{1}{\varepsilon}\rho_l(z, t)\delta z. \qquad (10.11)$$

Substituting for λ and solving for E, we have

$$E(r, z, t) = \frac{k I_0 \cos(\omega t - kz)}{2\pi r \varepsilon \omega}. \qquad (10.12)$$

Integrating this electric field from $r = a$ to $r = b$ gives us the voltage between the inner and outer conductors:

$$V(z, t) = \int_a^b E(r, z, t)dr = \frac{k I_0 \cos(\omega t - kz)\ln(b/a)}{2\pi \varepsilon \omega}. \qquad (10.13)$$

Finally, we divide $V(z,t)$ by $I(z,t)$ to get the characteristic impedance:

$$Z_0 = \frac{V(z, t)}{I(z, t)} = \frac{k \ln(b/a)}{2\pi \varepsilon \omega} = \frac{\ln(b/a)}{2\pi \varepsilon v} = \frac{1}{2\pi}\left(\frac{\mu}{\varepsilon}\right)^{1/2}\ln(b/a), \qquad (10.14)$$

which is the same as the result we obtained using the $\delta L\,\delta C$ ladder network equivalent circuit.

This derivation (as well as the LC derivation) for Z_0 is for TEM waves, where both E and H are perpendicular to z. For TEM solutions to exist, the line must be uniformly filled with homogenous dielectric material or vacuum. The dielectric can be lossy, but the metal conductors must, strictly speaking, have no resistance. In practice, these conditions are usually not satisfied perfectly, and the waves will be slightly different from the TEM waves corresponding to ideal conditions. In particular, the waves will have a small E_z or H_z field, or both. Microstrip lines are a case of nonuniform dielectric; some of the E-field lines arch through the air above the conductor, before plunging through the dielectric to the ground plane. The wave must have a unique phase velocity, but $(\mu\varepsilon)^{-1/2}$ has one value in the air and another value in the dielectric. The waves, therefore, cannot be TEM. They turn out to have both E_z and H_z components. Known as quasi-TEM waves, they show some frequency dependence in both Z_0 and v_{phase}, which can be important at millimeter-wave frequencies. Closed form expressions have not been derived for a microstrip; designers find Z_0 and v_{phase} vs. frequency by using graphs or approximate formulas based on numerical solutions of Maxwell's equations.

Problems

Problem 10.1. A common 50-ohm coaxial cable, RG214, has a shunt capacitance of 30.8 pF/ft. Calculate the series inductance per ft and the propagation velocity.

Problem 10.2. (a) Use the "tan tan" formula to show that a short length, δz, of transmission line, open-circuited at the far end, behaves as a capacitor, i.e., that it has a positive susceptance, directly proportional to frequency. Express the value of this capacitor in terms of the cable's capacitance per unit length. (Hint: $\tan(\theta) \approx \theta$ for small θ.)

(b) Show that a short length, δz, of transmission line, short-circuited at the far end, acts as an inductor, i.e., that it has a negative susceptance inversely proportional to frequency. Express the value of this inductor in terms of the cable's inductance/unit length.

Problem 10.3. (a) Find a formula for the characteristic impedance of a lossy cable where the loss can be due to a series resistance per unit length, R, as well as a parallel conductance per unit length, G. R represents the ohmic loss of the metal conductors while G represents dielectric loss.

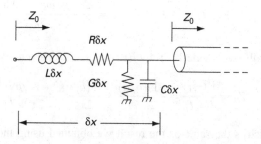

Hint: You can generalize the result for the lossless cable by simply replacing L by $L + R/(j\omega)$ and C by $C + G/(j\omega)$.

(b) Find the formula for the propagation constant k of this lossy cable. Hint: apply the substitutions given above to the formula $k = \omega\sqrt{LC}$. What distance (in wavelengths) is required to reduce by $1/e$ the power of a signal at frequency ω_1 if $R/(\omega_1) = 0.01L$?

Problem 10.4. If the (sinusoidal) voltage, V, and current, I, at the right-hand end of a transmission line are given, find the corresponding voltage, V', and current, I', at the left-hand end.

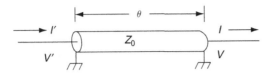

Hint: assume the (complex) voltage on the line is given by $V(\phi) = V_F e^{-j\phi} + V_R e^{j\phi}$. The corresponding current is given by $Z_0 I(\phi) = V_F e^{-j\phi} - V_R e^{j\phi}$. Let $\phi = 0$ at the right-hand end. Show that $V_F = (V + IZ_0)/2$ and $V_R = (V - IZ_0)/2$. Then show that, at the left-hand end, where $\phi = -\theta$, that $V' = V\cos\theta + IZ_0 j \sin\theta$ and $I' = I\cos\theta + j\sin\theta\, V/Z_0$.

Problem 10.5. Use the results of Problem 10.4 to upgrade your ladder network analysis program (Problem 1.3) to handle another type of element, a series lossless transmission line. Three parameters are necessary to specify the line. These could be the characteristic impedance, the physical length, and the velocity of propagation. For convenience in later problems, however, let the three parameters be the characteristic impedance (Z_0), the electrical length (θ_0) in degrees for a particular frequency, and that frequency (f_0). A 50-ohm cable that has an electrical length of 80° at 10 MHZ would appear in the circuit file as "TL, 50, 80, 10E6." For any frequency, f, the electrical length is then $\theta = \theta_0 f/f_0$.

Example answer: For the MATLAB program shown in Problem 1.3, insert the following lines of code in "elseif chain":

```
elseif strcmp(component,'TL')==1
ckt_index=ckt_index+1; Z0=ckt{ckt_index}; %characteristic impedance
ckt_index=ckt_index+1; refdegrees=ckt{ckt_index};%electrical length
ckt_index=ckt_index+1; reffreq =ckt{ckt_index};    %at ref. frequency
eleclength=pi/180*f(i)*(refdegrees/reffreq);
Iold=I; I=I*(cos(eleclength))+V*(1j/Z0*sin(eleclength));
V=V*(cos(eleclength))+1j*Z0*Iold*(sin(eleclength));
```

Problem 10.6. Use your program to analyze the circuit of Figure 10.13. Assume a design frequency, say 1 MHZ, in order to determine the value of the capacitor. Run the analysis from 0 to 2 MHz. Then make the transmission line 360° longer and repeat the analysis. What form will the response take if the transmission line is made very long?

Problem 10.7. A 50-ohm transmission line is connected in parallel with an equal length transmission line of 75 ohms, i.e., at each end the inner conductors are connected and the outer conductors are connected. The cables have equal phase velocities. Show that the characteristic impedance of this composite transmission line is given by $(50·75)/(50+75)$, i.e., the characteristic impedances add like parallel resistors.

Problem 10.8.   In the circuit shown below, the impedance, Z, is modified by a transmission line in parallel with a lumped impedance, Z_1, which could be an R, C, or L or a network.

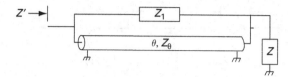

Show that the admittance looking in from the left, $Y' = 1/Z'$, is given by

$$\frac{1}{Z'} = Y' = Y_0 \frac{Y + jY_0 \tan\theta + \left(2Y_1 - \dfrac{2Y_1}{\cos\theta} + j\dfrac{YY_1}{Y_0}\tan\theta\right)}{Y_0 + jY\tan\theta + (jY_1\tan\theta)}.$$

Hint: extend the argument used in the text to find Z' for a cable without a bridging lumped element. Assume a forward and reverse wave in the cable with amplitudes 1 and Γ. The voltage on the cable is then $V(z) = e^{j\omega t}\,(e^{-jkz} + \Gamma e^{jkz})$ and the current is $I(z) = Z_0^{-1}e^{j\omega t}\,(e^{-jkz} - \Gamma e^{jkz})$. The current into Z is the sum of the current from the cable and the current from Z_1 while the current into the circuit is the sum of the current into the cable and the current into Z_1.

Problem 10.9.   Using a 50-ohm network analyzer, it is found that a certain device, when tested at 1 GHz, has a (complex) reflection coefficient of 0.6 at an angle of $-22°$ (standard polar coordinates: the positive x-axis is at $0°$ and angles increase in the counterclockwise direction).
(a) Calculate the impedance, $R+jX$.
(b) Find the component values for both the equivalent series R_sC_s circuit and the equivalent parallel R_pC_p circuit that, at 1 GHz, represent the device.

Problem 10.10.   The circuit below matches a 1000-ohm load to a 50-ohm source at a frequency of 10 MHz. The characteristic impedance of the cable is 50 ohms.

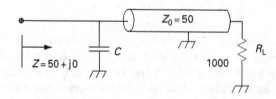

(a) Make a Smith chart sketch that shows the derivation of this circuit.
(b) Find the length of the (shortest) cable and the value of the capacitor. Specify the length in degrees and the capacitance in picofarads. Calculate these values rather than reading them from an accurately drawn Smith chart.
(c) Use your ladder network analysis program (Problems 1.3 and 10.5) to find the transmission from 9 MHz to 11 MHz in steps of 0.1 MHz.

Problem 10.11.   Find a transmission line element to replace the capacitor in the circuit of Problem 10.9.

Problem 10.12.   Suppose that a transmission line has small shunt susceptance (capacitive or inductive) at a point z. By itself, this will cause a small reflection. If an identical shunt reactance is placed one quarter-wave from the first, its reflection will compensate the first and the cable will have essentially perfect transmission. Show that this is the case (a) analytically, using the "tan tan" formula for Z' and B', and (b) graphically, using the Smith chart (the area around the center of the chart).

Problem 10.13.   Find the size and position of the constant resistance circles on the normalized Smith chart. Use the following procedure:

We have $z(x) = r + jx$ where x is a variable and r is a constant. This vertical line in the z-plane maps into the ρ-plane via the equation $\rho(x) = [z(x) - 1]/[z(x) + 1]$. We want to show that the locus of points in the ρ-plane is a circle with radius $1/(r+1)$ centered at $[r/(r+1), 0]$.

Assume that the locus will be a circle centered on the real axis at $[a,0]$. Write the equation $|\rho(x) - a| = radius$. This equation has the form

$$|[N_{\text{Re}}(x) + jN_{\text{Im}}(x)]/[D_{\text{Re}}(x) + jD_{\text{Im}}(x)]| = radius, \tag{1}$$

where $N_{\text{Re}}(x)$ and $N_{\text{Im}}(x)$ are the real and imaginary parts of the numerator and $D_{\text{Re}}(x)$ and $D_{\text{Im}}(x)$ are the real and imaginary parts of the denominator. If every point on the circle is to have the same value of r, the radius of the circle must be independent of x.

$$|\rho(x) - a|^2 = [(N_{\text{Re}}(x))^2 + (N_{\text{Im}}(x))^2]/[(D_{\text{Re}}(x))^2 + (D_{\text{Im}}(x))^2] = radius^2 \tag{2}$$
$$= \text{function only of } r.$$

In this case, the way to satisfy Equation (2) is to set $N_{\text{Re}}(x)/D_{\text{Re}}(x) = -N_{\text{Im}}(x) / D_{\text{Im}}(x)$.

This will let us find a and *radius*. Other ways to make the radius constant will produce circles on which both r and x vary.

11 Oscillators

Oscillators are autonomous dc-to-ac converters. They are used as the frequency-determining elements of transmitters and receivers and as master clocks in computers, frequency synthesizers, wristwatches, etc. Their function is to divide time into regular intervals. The invention of mechanical oscillators (clocks) made it possible to divide time into intervals much smaller than the Earth's rotation period and much more regular than a human pulse rate. Electronic oscillators are analogs of mechanical clocks.

11.1 Negative feedback (relaxation) oscillators

The earliest clocks used a "verge and foliot" mechanism which resembled a torsional pendulum but was not a pendulum at all. These clocks operated as follows: torque derived from a weight or a wound spring was applied to a pivoted mass. The mass accelerated according to Torque $= I\, d^2\theta/dt^2$ (the angular version of $F = ma$). When θ reached a threshold, θ_0, the mechanism reversed the torque, causing the mass to accelerate in the opposite direction. When it reached $-\theta_0$ the torque reversed again, and so on. The period was a function of the moment of inertia of the mass, the magnitude of the torque, and the threshold setting. These clocks employed negative feedback; when the controlled variable had gone too far in either direction, the action was reversed. Most home heating systems are negative feedback oscillators; the temperature cycles between the turn on and turn off points of the thermostat. Negative feedback electronic oscillators are called "relaxation oscillators." Most of these circuits operate by charging a capacitor until its voltage reaches an upper threshold and then discharging it until the voltage reaches a lower threshold voltage. In Figure 11.1(a), when the voltage on the capacitor builds up to about 85 V, the neon bulb fires. The capacitor then discharges quickly through the ionized gas (relaxes) until the voltage decays to about 40 V. The bulb then extinguishes and the cycle begins anew.

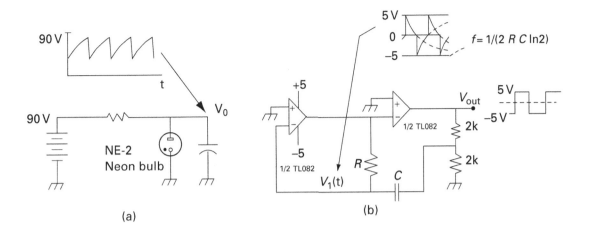

$f = 1/(2\,R\,C\,\ln 2)$

Figure 11.1. Relaxation (negative feedback) oscillators.

The circuit of Figure 11.1(b) alternately charges the capacitor, C, until its voltage reaches 2.5 V, and then discharges it until the voltage has fallen to -2.5 V. ($V_1(t)$ decays alternately toward $+5$ or -5 volts. When it reaches zero volts, the left-hand op-amp abruptly saturates in the opposite direction, kicking $V_1(t)$ to the voltage it had been approaching. The voltage then begins to decay in the opposite direction, and so forth.) Voltage-to-frequency converters are usually relaxation oscillators in which the control voltage determines the slope, and hence the oscillation period, of a fixed-amplitude sawtooth wave. Relaxation oscillators typically contain waveforms that are ramps or exponential decays. In the verge and foliot clock, the angle $\theta(t)$ consists of a sequence of parabolic arcs. Note that relaxation oscillators are nonlinear circuits which switch alternately between a charge mode and a discharge mode. Positive feedback oscillators, the main subject of this chapter, are nominally linear circuits. They generate sine waves.

11.2 Positive feedback oscillators

Clock makers improved frequency stability dramatically by using a true pendulum, a moving mass with a restoring force supplied by a hair spring or gravity.[1] As first observed by Galileo, a pendulum has its own natural frequency, independent of amplitude. It moves sinusoidally in simple harmonic motion. A pendulum clock uses *positive* feedback to push the pendulum in the direction of its motion, just as one pushes a swing to restore energy lost to friction.

---

[1] The Salisbury Cathedral clock, when installed around 1386, used a verge and foliot mechanism. Some 300 years later, after Christian Huygens invented the pendulum clock based on Galileo's observations of pendulum behavior, the Salisbury clock was converted into a positive feedback pendulum clock, its present form.

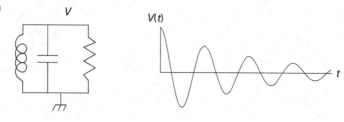

Figure 11.2. Damped oscillation in a parallel *LCR* circuit.

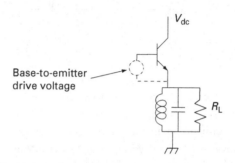

Figure 11.3. Transistor and dc supply replace energy lost to damping.

Electronic versions of the pendulum clock are usually based on resonators such as parallel or series LC circuits or electromechanical resonators such as quartz crystals. They use positive feedback to maintain the oscillation. A resonator with some initial energy (inductor current, capacitor charge, or mechanical kinetic energy) will oscillate sinusoidally with an exponentially decaying amplitude as shown in Figure 11.2. The decay is due to energy loss in the load and in the internal loss of the finite-Q resonator. In Figure 11.2 the resonator is a parallel LCR circuit.

To counteract the exponential decay, a circuit pumps current into the resonator when its voltage is positive and/or pulls current out when its voltage is negative. Figure 11.3 shows how a transistor and a dc supply can provide this energy. In this example circuit, the transistor is shown in the emitter-follower configuration simply because it is so easy to analyze; the emitter voltage tracks the base voltage and the base draws negligible current. The single transistor cannot supply negative current but we can set it up with a dc bias as a class-A amplifier so that current values less than the bias current are equivalent to negative current.

All that remains to complete this oscillator circuit is to provide the transistor drive, i.e., the base-to-emitter voltage. We want to increase the transistor's conduction when the output voltage (emitter voltage) increases, and we see that the emitter voltage has the correct polarity to be the drive signal. Since an emitter follower's voltage gain is slightly less than unity, the base needs a drive

Figure 11.4. Feedback loop details define (a) Armstrong; (b) Hartley; (c) Colpitts oscillators.

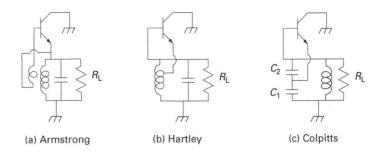

(a) Armstrong   (b) Hartley   (c) Colpitts

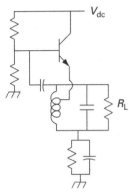

Figure 11.5. Hartley oscillator circuit including bias circuitry.

signal with slightly more amplitude than the sine wave on the emitter. Figure 11.4 shows three methods to provide this drive signal. The Armstrong oscillator adds a small secondary winding to the inductor. The voltage induced in the secondary adds to the emitter voltage. The Hartley oscillator accomplishes the same thing by connecting the emitter to a tap slightly below the top of the inductor. This is just an autotransformer version of the Armstrong oscillator if the magnetic flux links all the turns of the inductor. But the top and bottom portions of the inductor do not really have to be magnetically coupled at all; most of the current in the inductor(s) is from energy stored in the high-Q resonant circuit. This current is common to the two inductors so they essentially form a voltage divider. (Note, though, that the ratio of voltages on the top and bottom portions of the inductor ranges from the turns ratio, when they are fully coupled, to the square of the turns ratio, when they have no coupling.) If we consider the totally decoupled Hartley oscillator – no mutual inductance – and then replace the inductors by capacitors of equal (but opposite) reactance and replace the capacitor by an inductor, we get the Colpitts oscillator. Note that each oscillator in Figure 11.4 is an amplifier with a positive feedback loop. No power supply or biasing circuitry is shown in these figures; they simply indicate the ac signal paths.

Using the Hartley circuit as an example, Figure 11.5 shows a practical circuit. It includes the standard biasing arrangement to set the transistor's operating point. (A resistor voltage divider determines the base voltage and an emitter resistor then determines the emitter current, since V_{be} will be very close to 0.7 V.) A blocking capacitor allows the base to be dc biased with respect to the emitter. A bypass capacitor puts the bottom of the resonant circuit at RF ground.

In practice, one usually finds oscillators in grounded emitter circuits, as shown in Figure 11.6. The amplitude of the base drive signal must be much smaller than the sine wave on the resonant circuit. Moreover, the polarity of the base drive signal must be inverted with respect to the sine wave on the collector. You can inspect these circuits to see that they do satisfy these conditions. But, on closer inspection, you can note the circuits are *identical* to the circuits of Figure 11.4, except that the ground point has been moved from the collector to

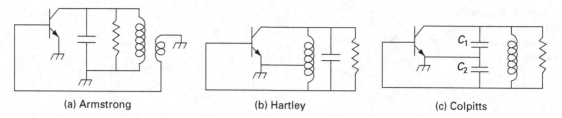

(a) Armstrong   (b) Hartley   (c) Colpitts

Figure 11.6. Grounded-emitter oscillator circuits.

the emitter. In these oscillators, an amplifier is enclosed in a positive feedback loop. But, because there is no input signal to have a terminal in common with the output signal, oscillators, unlike amplifiers, do not have common-emitter, common-collector, and common-base versions.

The Colpitts oscillator, needing no tap or secondary winding on the inductor, is the most commonly used circuit. Sometimes the transistor's parasitic collector-to-emitter capacitance is, by itself, the top capacitor, C_1, so this capacitor may appear to be missing in a circuit diagram. A practical design example for the Colpitts circuit of Figure 11.6(c) is presented later in this chapter.

11.2.1 Unintentional oscillators

In RF work it is common for a casually designed amplifier to break into oscillation. One way this happens is shown in Figure 11.7. The circuit is a basic common-emitter amplifier with parallel resonant circuits on the input and output (as bandpass filters and/or to cancel the input and output capacitances of the transistor). When the transistor's parasitic base-to-collector capacitance is included, the circuit has the topology of the decoupled Hartley oscillator. If the feedback is sufficient, it will oscillate. The frequency will be somewhat lower than that of the input and output circuits so that they look inductive as shown in the center figure. This circuit known as a TPTG oscillator, form Tuned-Plate Tuned-Grid, in the days of the vacuum tube.

Figure 11.7. Tuned amplifier as an oscillator

Tuned amplifier ⎯⎯⎯⎯⎯⎯⎯⎯⎯⎯⎯→ Hartley oscillator

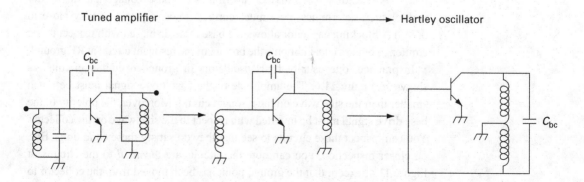

With luck, the loop gain of any amplifier will be less than unity at any frequency for which the total loop phase shift is 360° and an amplifier will be stable. If not, it can be *neutralized* to avoid oscillation. Two methods of neutralization are shown in Figure 11.8.

Figure 11.8. Amplifier neutralization.

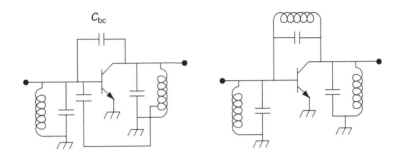

In Figure 11.8(a), a secondary winding is added to provide an out-of-phase voltage which is capacitor-coupled to the base to cancel the in-phase voltage coupled through C_{bc}. In Figure 11.8(b), an inductor from collector to base resonates C_{bc} to effectively remove it (a dc blocking capacitor would be placed in series with this inductor). In grounded-base transistor amplifiers and grounded-grid vacuum tube amplifiers the input circuit is shielded from the output circuit. These are stable without neutralization (but provide less power gain than their common-emitter and common-cathode-counterparts).

11.2.2 Series resonant oscillators

The oscillators discussed above were all derived from the parallel resonant circuit shown in Figure 11.2. We could just as well have started with a series LCR circuit. Like the open parallel circuit, a shorted series LCR circuit executes an exponentially damped oscillation unless we can replenish the dissipated energy. In this case we need to put a voltage source in the loop which will be positive when the current is positive and negative when the current is negative, as shown in Figure 11.9.

Figure 11.9. Series-mode oscillator operation.

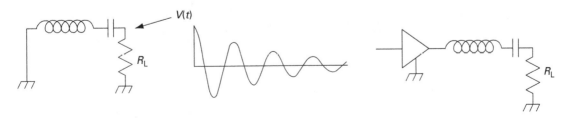

While a bare transistor with base-to-emitter voltage drive makes a good current source for a parallel-mode oscillator, a low-impedance voltage source is needed for a series-mode oscillator. In the series-mode oscillator shown in Figure 11.10, an op-amp with feedback is such a voltage source.

Figure 11.10. An op-amp series-mode oscillator.

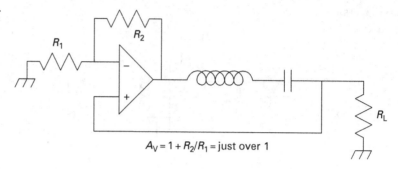

$$A_V = 1 + R_2/R_1 = \text{just over } 1$$

Since no phase inversion is provided by the tank circuit, the amplifier is connected to be noninverting. An emitter-follower has a low output impedance and can be used in a series-mode oscillator (see Problem 11.4). When the series *LC* circuit is replaced by a multisection RC network, the resulting oscillator is commonly known as a *phase-shift oscillator* (even though every feedback oscillator oscillates at the frequency at which the overall loop phase shift is 360°). An *RC* phase-shift oscillator circuit is shown in Figure 11.11. Op-amp voltages followers make the circuit easy to analyze.

For the three cascaded *RC* units, the transfer function is given by $V_2(t)/V_1(t) = 1/(\omega RC+1)^3$. The inverting amplifier at the left provides a voltage gain of $-16/2 = -8$, so $V_1(t)/V_2(t) = -8$. Combining these two equations yields a cubic equation with three roots: $\omega RC = 3j$, $\sqrt{3}$, and $-\sqrt{3}$. The first root corresponds to an exponential decay of any initial charges on the capacitors while the two imaginary roots indicate that the circuit will produce a steady sine-wave oscillation whose frequency is given by $\omega RC = \sqrt{3}$. In practice, the 16k resistor would be increased to perhaps twice that value to ensure oscillation. Note that

Figure 11.11. An *RC* phase-shift oscillator.

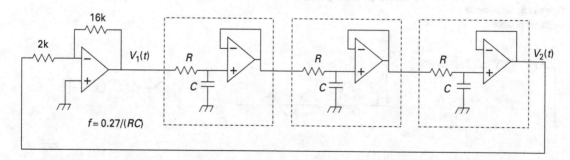

this circuit is a positive-feedback sine-wave oscillator even though it does not contain a resonator. When the 16k resistor value is increased, the loop gain for the original frequency becomes greater than unity, but for the new gain, there will be a nearby complex frequency, $\omega - j\alpha$, for which the loop gain is unity. The time dependence therefore becomes $e^{j(\omega - j\alpha)t} = e^{j\omega t}e^{\alpha t}$, showing that the oscillation amplitude grows as $e^{\alpha t}$. This circuit illustrates how any linear circuit with feedback will produce sine-wave oscillations if there is a (complex) frequency for which the overall loop gain is unity and the overall phase shift is 360°. (Of course α must be positive, or the oscillation dies out exponentially.)

11.2.3 Negative-resistance oscillators

In the circuits described above, a transistor provides current to an *RLC* circuit when the voltage on this circuit is positive, i.e., the transistor behaves as a negative resistance. But the transistor is a three-terminal device and the third terminal is provided with a drive signal derived from the *LCR* tank. Figure 11.12 shows how two transistors can be used to make a two-terminal negative resistance that is simply paralleled with the *LCR* tank to make a linear sine wave oscillator that has no feedback loop.

The two transistors form an emitter-coupled differential amplifier in which the resistor to $-V_{ee}$ acts as a constant current source, supplying a bias current, I_0. The input to the amplifier is the base voltage of the right-hand transistor. The output is the collector current of the left-hand transistor. The ratio of input to output is $-4V_T/I_0$, where V_T is the thermal voltage, 26 mV. This ratio is just the negative resistance, since the input and output are tied together. This negative-resistance oscillator uses a parallel-resonant circuit, but a series-resonant version is certainly possible as well.

Any circuit element or device that has a negative slope on at least some portion of its *I–V* curve can, in principle, be used as a negative resistance. Tunnel diodes can be used to build oscillators up into the microwave frequency range. At microwave frequencies, single-transistor negative-resistance oscillators are common. A plasma discharge exhibits negative resistance and provided a pre-vacuum tube method to generate coherent sine waves. High-efficiency

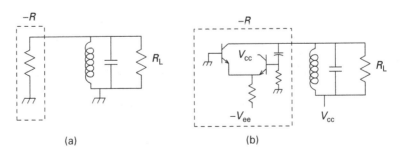

Figure 11.12. A negative-resistance oscillator.

(a)   (b)

Poulsen arc transmitters, circa World War I, provided low-frequency RF power exceeding 100 kW.

11.3 Oscillator dynamics

These resonant oscillators are basically linear amplifiers with positive feedback. At turn-on they can get started by virtue of their own noise if they run class A. The tiny amount of noise power at the oscillation frequency will grow exponentially into the full-power sine wave. Once running, the signal level is ultimately limited by some nonlinearity. This could be a small-signal non-linearity in the transistor characteristics. Otherwise, the finite voltage of the dc power provides a severe large-signal nonlinearity, and the operation will shift toward class-C conditions. The fact that amplitude cannot increase indefinitely shows that some nonlinearity is operative in every real oscillator. Any non-linearity causes the transistor's low- frequency $1/f$ noise to mix with the RF signal, producing more noise close to the carrier than would exist for linear operation. An obvious way to mitigate large-signal nonlinearity is to detect the oscillator's output power and use the detector voltage in a negative feedback arrangement to control the gain. This can maintain an amplitude considerably lower than the power supply voltage. Alternatively, if the oscillator uses a device (transistor or op-amp circuit) with a soft saturation characteristic, the amplitude will reach a limit while the operation is still nearly linear. For example, the amplifier in the oscillator of Figure 11.10 might have a small cubic term, i.e., $V_{OUT} = AV_{IN} - BV_{IN}^3$, where B/A is very small (see Problem 11.5).

11.4 Frequency stability

Long-term (seconds to years) frequency fluctuations are due to component aging and changes in ambient temperature and are called *drift*. Short-term fluctuations, known as oscillator noise, are caused by the noise produced in the active device, the finite loaded Q of the resonant circuit, and nonlinearity in the operating cycle. The higher the Q, the faster the loop phase-shift changes with frequency. Any disturbances (transistor fluctuations, power supply variations changing the transistor's parasitic capacitances, etc.) that tend to change the phase shift will cause the frequency to move slightly to reestablish the overall 360° shift. The higher the resonator Q, the smaller the frequency shift. Note that this is the *loaded Q*, so the most stable oscillators, besides having the highest Q resonators, are loaded as lightly as possible. In LC oscillators, losses in the inductor almost always determine the resonator Q. A shorted piece of transmission line is sometimes used as a high-Q inductor. Chapter 24 treats oscillator noise in detail.

11.5 Colpitts oscillator theory

Let us look in some detail at the operation of the Colpitts oscillator. Figure 11.13 shows the Colpitts oscillator of Figure 11.6(c) redrawn as a small-signal equivalent circuit (compare the figures). The still-to-be-biased transistor is represented as a voltage-controlled current source. The resistor r_{be} represents the small-signal base-to-emitter resistance of the transistor.

The parallel combination of L and the load resistor, R, is denoted as Z, i.e., $Z = j\omega LR/(j\omega L+R) = j\omega L_S + R_S$, where L_S and R_S are the component values for the equivalent series network. Likewise, it is convenient to denote r_{be}^{-1} as g. The voltage V_{be}, a phasor, is produced by the current I (a phasor) from the current source. This is a linear circuit, so V_{be} can be written as $V_{be} = IZ_T$, where Z_T is a function of ω. We will calculate this "transfer impedance" using standard circuit analysis. Since the current I is proportional to V_{be}, we can write an equation expressing that, in going around the loop, the voltage V_{be} exactly reproduces itself:

$$-g_m V_{be} Z_T = V_{be} \quad \text{or} \quad \frac{1}{Z_T} = -g_m. \tag{11.1}$$

This equation will let us find the component values needed for the circuit to oscillate at the desired frequency, i.e., the values that will make the loop gain equal to unity and the phase shift equal 360°.

We can arbitrarily select L, choosing an inductor whose Q is high at the desired frequency. Equation (11.1), really two equations (real and imaginary parts), will then provide values for C_1 and C_2. To derive an expression for Z_T, we will assume that $V_{be} = 1$ and work backward to find the corresponding value of I. With this assumption, inspection of Figure 11.13 shows that the current I_1 is given by

$$I_1 = j\omega C_2 + g. \tag{11.2}$$

Now the voltage V_c is just the 1 volt assumed for V_{be} plus $I_1 Z$, the voltage developed across Z:

$$V_c = 1 + (j\omega C_2 + g)Z. \tag{11.3}$$

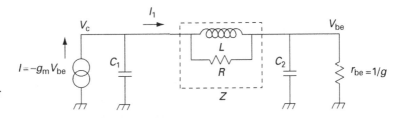

Figure 11.13. Colpitts oscillator small-signal equivalent circuit.

Finally, the current I is just the sum of I_1 plus $V_c\, j\omega C_1$, the current going into C_1:

$$I = (j\omega C_2 + g) + [1 + (j\omega C_2 + g)Z]\, j\omega C_1. \tag{11.4}$$

Since we had assumed that $V_{be} = 1$, we have $Z_T = 1/I$ or

$$\frac{1}{Z_T} = j\omega C_2 + g + [1 + (j\omega C_2 + g)Z]\, j\omega C_1. \tag{11.5}$$

Using this, the condition for oscillation, Equation (11.1) becomes

$$g_m + j\omega C_2 + g + [1 + (j\omega C_2 + g)Z]\, j\omega C_1 = 0. \tag{11.6}$$

The job now is to solve Equation (11.6) for C_1 and C_2. If we assume that ω is real i.e., that the oscillation neither grows nor decays, we find from the imaginary part of this equation, that

$$\frac{C_2 + C_1 + gR_S C_1}{L_S C_1 C_2} = \omega^2 \tag{11.7}$$

and, from the real part, that

$$\omega^2 C_1 C_2 R_S + g(\omega^2 L_S C_1 - 1) = g_m. \tag{11.8}$$

Solving Equations (11.7) and (11.8) simultaneously for C_2 and C_1 produces

$$C_2 = \frac{g_m L_S}{2R_S}\left(1 + \sqrt{\frac{4R_S(1 + gR_S)(g_m + g)}{g_m^2 L_S \omega^2}}\right) \tag{11.9}$$

and

$$C_1 = \frac{C_2}{\omega^2 L_S C_2 - 1 - gR_S}. \tag{11.10}$$

Normally C_2 will have a much larger value than C_1 and $\omega \approx 1/\sqrt{LC_1}$. Moreover, the second term in the square root of Equation (11.9) is usually much less than unity so $C_2 \approx g_m\, L_S/R_S$.

11.5.1 Colpitts oscillator design example

Let us design a practical grounded-emitter Colpitts oscillator. Suppose this oscillator is to supply 1 mW at 5 MHz and that it will be powered by a 6 V dc supply. Assuming full swing, the peak output sine wave voltage will be 6 V. The output power is given by $0.001\,\text{W} = (6\,\text{V})^2/(2R_L)$ so the value of the load resistor, R_L, will be 18 k ohms. Assuming class-A operation, the bias current in the transistor is made equal to the peak current in the load: $I = I_{pk} = 6\,\text{V}/18\,\text{k} = 0.33\,\text{mA}$. If we let the emitter biasing resistor be 1.5 k, the emitter bias voltage will be $1500 \times 0.33\,\text{mA} = 0.5\,\text{V}$. Assuming the typical 0.7 V offset between the base and emitter, the base voltage needs to be 1.2 V. A voltage divider using a 40 k resistor and a 10 k resistor will produce 1.2 V from the 6 V

Figure 11.14. Colpitts oscillator:
5 MHz, 1 mW.

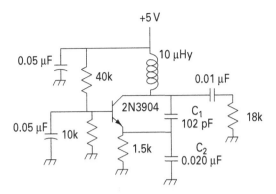

supply. These bias components are shown in the schematic diagram of Figure 11.14.

A 0.05 µF bypass capacitor pins the base to ac ground and another bypass capacitor ensures that the dc input is held at a firm RF ground. Note that the 1.5 k emitter bias resistor provides an unwanted signal path to ground. This path could be eliminated by putting an inductor in series with the bias resistor as an RF choke, but this is not really necessary; the 1.5k resistor is in parallel with C_2, which will have such a low reactance that the resistor will divert almost no current from it.

With the biasing out of the way, we now deal with the signal components. The transconductance of the transistor is found by dividing the bias current, I_0, by 26 mV, the so-called thermal voltage,[2] i.e., $g_m = 0.33$ mA$/26$ mV $= 0.013$ mhos. The small-signal base-to-emitter resistance, r, is given by $r = \beta V_{thermal}/I_0$. For a typical small-signal transistor, such as a 2N3904, β is about 100, so $r_{in} = 100 \cdot 0.026$ V$/0.33$ mA $= 8000$ ohms.

Using Equation (11.9) and (11.10), the values of C_1 and C_2 are 102 pF and 0.023 µF, respectively. These are the values for which the oscillator theoretically will maintain a constant amplitude. In practice, we increase the feedback by decreasing the value of C_2 to ensure oscillation. This produces a waveform that grows exponentially until it reaches a limit imposed by circuit nonlinearity. The frequency becomes complex, i.e., ω becomes $\omega - j\alpha$ and the time dependence therefore becomes $e^{j(\omega - j\alpha)t} = e^{j\omega t}e^{\alpha t}$. Suppose we want α to be, say 10^5, which will cause oscillation to grow by a factor e every 10 µsec. (Fast growth would be important if, for example, the oscillator is to be rapidly pulsed on and off.) How do we find the value of C_2 to produce the desired α? To avoid doing more analysis, it is convenient to use a standard computer program such as Mathcad to find the root(s) of Equation (11.6) for trial values of C_2. In this example, if we decrease C_2 to 0.020 µF, we obtain the desired α.

---

[2] The thermal voltage is given by $V_{thermal} = 0.026$V $= kT/e$, where k is Boltzman's constant, T is the absolute ambient temperature, and e is the charge of an electron.

Problems

Problem 11.1. Draw a schematic diagram (without component values) for a bipolar transistor Colpitts oscillator with the collector at ground for both dc RF. Include the biasing circuit. The oscillator is to run from a positive dc supply.

Problem 11.2. Design (without specifying component values) a single-transistor series-mode oscillator based on the emitter follower circuit.

Problem 11.3. A simple computer simulation can illustrate how an oscillator builds up to an amplitude determined by the nonlinearity of its active element. The program shown below models the negative-resistance oscillator of Figure 11.12(a). The LC resonant frequency is 1 Hz. This network is in parallel with a negative-resistance element whose voltage vs. current relation is given by $I = -(1/R_n)*(V - \varepsilon V^3)$, to model the circuit of Figure 11.12. The small-signal (negative) resistance is just $-R_n$. The term $-\varepsilon V^3$ makes the resistance become less negative for large signals. The program integrates the second-order differential equation for $V(t)$ and plots the voltage versus time from an arbitrary initial condition, $V = 1$ volt.

Run this or an equivalent program. Change the value of the load resistor R. Find the minimum value of R for sustained oscillation. Experiment with the values of R and R_n. You will find that when the loaded Q of the RLC circuit is high, the oscillation will be sinusoidal even when the value of the negative resistor is only a fraction of R. When Q is low (as it is for $R = 1$), a low value of R_n such as $R_n = 0.2$ will produce a distinctly distorted waveform.

```
'QBasic simulation of negative-resistance oscillator of Figure 11.12a.
SCREEN 2
R = 1 : L = 1 / 6.2832 : C = L 'the parallel RLC circuit: 1 'ohms, 1/2pi henries, 1/2pi farads
RN = .9 'run program also with RN=.2 to see non-'sinusoidal waveform
E = .01 'negative resistance: I= (1/RN)*(V-EV^3)
V = 1 : U = 0 'initial conditions, V is voltage, U is 'dV/dt
DT = .005 'step size in seconds
FOR I = 1 TO 3000
T = T + DT 'increment the time
VNEW = V + U * DT
U = U + (DT / C) * ((1 / RN) * (U - 3 * E * V * V * U) - V / L - U / R)
V = VNEW
PSET (40 * T, 100 + 5 * V) 'plot the point
NEXT I
```

Problem 11.4. In the oscillator shown below, the voltage gain of the amplifier decreases with amplitude. The voltage transfer function is $V_{out} = 2 V_{in} - 0.5 V_{in}^3$. This characteristic will limit the amplitude of the oscillation.

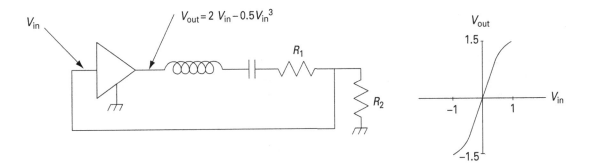

Find the ratio R_2/R_1 in order that the peak value of the sine wave V_{in} will be one volt. Hint: assume $V_{in} = \sin(\omega t)$. The amplifier output is then $2\sin(\omega t) - 0.5\sin^3(\omega t)$. The second term resembles the sine wave but is more peaked. The LC filter will pass the fundamental Fourier component of this second term. Find this term and add it to $2\sin(\omega t)$. Then calculate the ratio R_2/R_1 so that the voltage divider output is $\sin(\omega t)$.

12 Phase lock loops and synthesizers

Oscillators whose frequencies are derived from a stable reference source are used in transmitters and receivers as L.O.'s for accurate digital tuning. A VCO in a loop that tracks the frequency excursions of an incoming FM signal is a commonly used FM detector. The first widespread use of phase lock loops was in television receivers where they provide noise-resistant locking of the horizontal sweep frequency to the synchronization pulses in the signal.

12.1 Phase locking

A phase lock loop (PLL) circuit forces the phase of a voltage-controlled oscillator (VCO) to follow the phase of a reference signal. Once lock is achieved, i.e., once the phases stay close to each other, the frequency of the VCO will be equal to the frequency of the reference. In one class of applications, the PLL is used to generate a stable signal whose frequency is determined by an unstable (noisy) reference signal. Here the PLL is, in effect, a narrow bandpass filter that passes a carrier, while rejecting its noise sidebands. Examples include telemetry receivers that lock onto weak pilot signals from spacecraft and various "clock smoother" circuits. It is often necessary to lock an oscillator to the suppressed carrier of a modulated signal, an operation known as carrier recovery. In yet another class of applications the PLL is designed to detect all the phase fluctuations of the reference. An example is the PLL-based FM demodulator where the VCO reproduces the input signal, which is usually in the IF band. The voltage applied by the loop to the VCO is proportional to the instantaneous frequency (in as much as the VCO has a linear voltage-to-frequency characteristic), and this voltage is the audio output.

12.1.1 Phase adjustment by means of frequency control

In a PLL, the VCO frequency, rather than phase, is determined by the control voltage, but it is easy to see how frequency control provides phase adjustment.

Figure 12.1. Phase lock loop
concept.

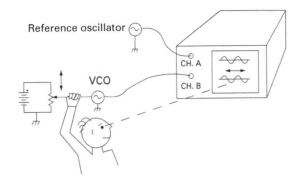

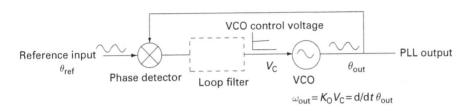

$$\omega_{out} = K_O V_C = d/dt\, \theta_{out}$$

Figure 12.2. Phase lock loop
block diagram.

Suppose you have two mechanical clocks. You want to make Clock B agree
with Clock A. You notice that Clock B is consistently five minutes behind Clock
A. You could exert direct phase control, using the time adjustment knob to set
Clock B ahead 5 min to agree with Clock A. Or you could use frequency control,
regulating the speed of Clock B to run somewhat faster. Once Clock B catches
up with Clock A, you reset the frequency control to its original value. You may
have done an electronic version of this in the lab (Figure 12.1). Suppose you
have a two-channel oscilloscope. The first channel displays a sine wave from a
fixed-frequency reference oscillator. The scope is synchronized to this reference
oscillator so Channel A displays a motionless sine wave. Channel B is connected
to a variable frequency oscillator which might be a laboratory instrument with a
frequency knob or a VCO fitted with a potentiometer. Suppose the frequencies
are the same. Then the sine wave seen on Channel B is also motionless. If you
lower the VCO frequency slightly, the Channel B trace will drift to the right.
And if you raise the VCO frequency, the trace will drift to the left. To align the
two traces, you can shift the frequency slightly to let the Channel B trace drift
into position, and then return the VCO frequency to the reference frequency
value to stop the drift. The operator keeps the traces aligned and is therefore an
element of this phase lock loop.

To automate the loop we use an electronic phase detector, i.e., an element that
produces a voltage proportional to the phase difference between the reference
and the VCO. Figure 12.2 shows the block diagram for a PLL.

If the VCO phase gets behind (lags) the reference phase, the phase detector
will produce a positive "error voltage" which will speed up the VCO. As the
phase error then decreases, the error voltage also decreases and the VCO slows

back down. The exact way in which equilibrium is established, i.e., the mathematical form of the VCO phase, $\theta_{out}(t)$, depends on the yet-unspecified loop filter or "compensation network," shown as a dotted box in Figure 12.2. We will examine phase detectors and loop filters below.

12.1.2 Mechanical analog of a PLL

The PLL is a feedback control system – a servo. The electromechanical system shown in Figure 12.3 is a positioning servo. If an operator adjusts the input crank angle, the output shaft automatically turns so that the output angle, θ_{out}, tracks the input shaft angle, θ_{ref}. The gear teeth are in constant mesh but the top gear can slide axially. A dc motor provides torque to turn the output shaft. The input shaft might be connected to the steering wheel of a ship while the output shaft drives the rudder.

Consider the operation of this servo system. Assume for now that the voltage, V_{bias}, is set to zero, and that the system is at rest. When the input and output angles are equal, the slider on the potentiometer is centered and produces zero error voltage. But suppose the reference phase gets slightly ahead of the output phase – maybe the input crank is abruptly moved clockwise. During this motion, the threads on the input shaft have caused the top gear to slide to the left, producing a positive voltage at the output of the potentiometer. This voltage is passed on by the loop filter to the power amplifier and drives the motor clockwise, increasing θ_{out}. This rotates the top gear clockwise. The input shaft is now stationary, so the top gear moves along the threads to the right, reducing the potentiometer output voltage back to zero. The output angle (output phase) is again in agreement with the input or "reference" angle (input phase). In this system the gear/screw thread mechanism is the phase detector. Note that the

Figure 12.3. Positioning system with feedback control – a mechanical phase lock loop.

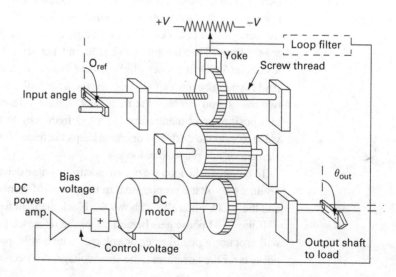

range of this phase detector can be many turns – the number of threads on the shaft. The negative feedback error signal provides the drive to force the (high-power) output to agree with the (low-power) input command.

We see how this mechanical system operates as a power-steering servo. To see that it can also operate as a PLL, suppose that an operator is cranking the input shaft at a constant or approximately constant rate (input frequency). We will now set the bias voltage so that, with the bias voltage alone, the motor turns at the nominal input frequency. (A VCO can be regarded as having an internal bias that determines its nominal frequency, i.e., its frequency when the control voltage is zero.) At equilibrium, the top gear is being turned by the middle gear but the threaded shaft is turning just as fast. The top gear therefore does not move along the threads, but stays at an equilibrium position. The output frequency is locked to the input frequency. How about the phases; will they also agree exactly? They did in the power steering model,[1] but now, with continuous rotation, they may not. The key is the loop filter. Suppose the loop filter is just a piece of wire (no filter at all). If the input frequency changes from its nominal value, there will have to be a constant phase error in order to keep the potentiometer off-center and produce a control voltage which will add to or subtract from the bias voltage. How much phase error (*tracking* error) is necessary depends on the gain of the power amplifier. Increasing the gain of the amplifier/motor and/or the gain (sensitivity) of the phase detector will reduce the error. But, if the loop filter contains an integrator, the control voltage will be the integral of the error voltage history. The steady-state error voltage can be zero. In fact, it *must* be zero, on average, or the integrator output, and hence the output frequency, would increase indefinitely.

Note: a servo's *type* is given by the number of integrators contained in its loop. Even without an integrator in the loop filter block, a PLL contains one integrator, the VCO, since phase is the time integral of frequency. In the mechanical analog, rotation angle is the integral of angular velocity.

12.1.3 Loop dynamics

Let us now look at how a PLL responds to disturbances in the input phase. We have already described qualitatively the response to a step function; the loop catches up with the reference. The way in which it catches up, i.e., quickly, slowly, with overshoot, or with no overshoot, depends on the loop filter and the characteristics of the phase detector, amplifier and motor. An exact analysis is straightforward if the entire system is *linear*, i.e., if the system can be described with linear differential equations. In this case it can be analyzed with the standard techniques applied to linear electronic circuits, i.e., complex numbers, Fourier and Laplace transforms, superposition, etc. In the PLL, the loop filter is linear since it consists of passive components: resistors, capacitors, and

---

[1] We will assume the motor speed is proportional to the applied voltage, independent of load.

op-amps. The VCO is linear if its frequency is strictly proportional to the control voltage. Of course, over a small operating region, any smooth voltage–frequency characteristic is approximately linear. The most commonly used phase detector is a simple multiplier (mixer). You can show[2] that it produces an error voltage proportional to $\sin(\theta_{out}(t) - \theta_{ref}(t) + \pi/2)$. For small x, $\sin(x) \approx x$, so a multiplier is a fairly linear phase detector over a restricted region around $\pi/2$. The phase detector in our mechanical analog is linear over a range limited only by the length of the screw thread.

To describe the loop dynamics, we will find its response to an input phase disturbance, $\theta_{dist\_in} = e^{j\omega t}$. The complete input function is therefore $\theta_{ref}(t) = \omega_0 t + e^{j\omega t}$. Since the system is linear, we invoke superposition to note that the output will consist of a response to the $\omega_0 t$ term plus an output response to the $e^{j\omega t}$ disturbance term. The input term, $\omega_0 t$, results in an output contribution $\omega_0 t + \Theta$, where the constant Θ might be 90°, if the phase detector is a mixer, and might include another constant, if the nominal frequency of the VCO is not quite ω_0 and the loop filter includes no integrator. Since the input disturbance function is sinusoidal, the output response to it will also be sinusoidal, $A(\omega)e^{j\omega t}$, where the complex amplitude $A(\omega)$ is known as the input-to-output transfer function. Before we can calculate the transfer function, we must specify the loop filter.

12.1.4 Loop filter

A standard filter for type-II loops uses an op-amp to produce a weighted sum of the phase detector output plus the integral of the phase detector output. Figure 12.4 shows the circuit commonly used.

Note that if we let C go to infinity and set $R_2 = R_1$, the response of this filter is just unity (except for a minus sign) so this circuit will serve to analyze the type-I PLL as well as the type-II PLL. Remembering that the negative input of the op-amp is a virtual ground, we can write the transfer function of this filter.

$$\frac{-V_2}{V_1} = \frac{R_2 + 1/j\omega C}{R_1} = \frac{R_2}{R_1} + \frac{1}{j\omega C R_1} = \frac{\tau_2}{\tau_1} - \frac{j}{\omega \tau_1}. \tag{12.1}$$

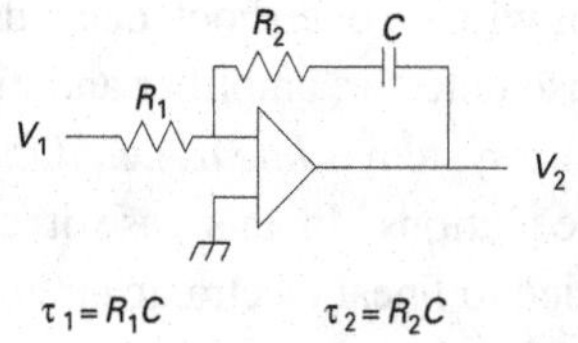

Figure 12.4. Loop filter circuit. $\tau_1 = R_1 C$ $\tau_2 = R_2 C$

---

[2] Just expand the product $\cos(\omega_0 t + \theta_{out}(t)) \cos(\omega_0 t + \theta_{ref}(t))$ and ignore (filter off) components at $2\omega_0$.

In the time domain, the relation between V_2 and V_1 is given by

$$\frac{dV_2}{dt} = \frac{-R_2}{R_1}\frac{dV_1}{dt} - \frac{V_1}{R_1 C}. \tag{12.2}$$

12.1.5 Linear analysis of the PLL

Figure 12.5 is a block diagram of the loop, including the loop filter of Figure 12.4.

Figure 12.5. Type-II, second-order PLL block diagram.

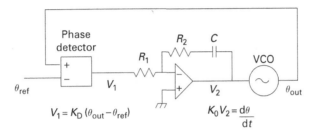

$$V_1 = K_D\,(\theta_{out} - \theta_{ref})$$

$$K_O V_2 = \frac{d\theta}{dt}$$

For this linear system, let us find the frequency response of the output phase to a sinusoidal disturbance of the reference phase, $\theta_{dist}e^{j\omega t}$, where ω is the disturbing frequency. From inspection of Figure 12.5 we can write

$$-K_D(\theta_{out} - \theta_{dist})\left(\frac{\tau_2}{\tau_1} - \frac{j}{\omega\tau_1}\right)K_O = j\omega\theta_{out}. \tag{12.3}$$

Solving for θ, we have the frequency response

$$\frac{\theta_{out}(\omega)}{\theta_{dist}(\omega)} = \frac{1}{1 - [K_D K_O/(\omega^2\tau_1) + jK_D K_O\tau_2/(\omega\tau_1)]^{-1}}. \tag{12.4}$$

12.1.5.1 Frequency response of the type-I loop

We will look at the frequency response for the type-I loop by letting C go to infinity and letting $R_1 = R_2$, effectively eliminating the loop filter. The frequency response, Equation (12.4), becomes

$$\frac{\theta_{out}(\omega)}{\theta_{dist}(\omega)} \rightarrow \frac{1}{1 + j\omega/K}, \tag{12.5}$$

where $K = K_D K_O$, and is called the *loop gain*. The frequency response is identical to that of a simple RC lowpass filter with a time constant $RC = 1/K$. We saw earlier that a type-I PLL (a PLL with no integrator in the loop filter) needs high loop gain to have good tracking accuracy (the ability to follow a changing reference frequency without incurring a large phase error). Here we see that high gain implies a large bandwidth; the type-I PLL cannot have both high gain (for good tracking) and a narrow bandwidth (to filter out high-frequency reference phase noise).

12.1.5.2 Frequency response of the type-II loop

To deal with the type-II loop (the most common PLL) it is standard to define two constants, the *natural frequency*, ω_n, and the *damping coefficient*, ζ, as follows:

$$\omega_n = \sqrt{\frac{K_O K_D}{\tau_1}} \quad \zeta = \frac{\tau_2 \omega_n}{2}. \tag{12.6}$$

We will see soon that these names relate to the transient response of the loop but, for now, we will substitute them into Equation (12.4), the expression for the frequency response, to get

$$\frac{\theta_{out}(\omega)}{\theta_{dist}(\omega)} = \frac{\omega_n^2 + 2j\omega\zeta\omega_n}{\omega_n^2 + 2j\omega\zeta\omega_n - \omega^2}. \tag{12.7}$$

This transfer function is plotted vs. ω/ω_n in Figure 12.6 for several different damping coefficients.

Figure 12.6. Frequency response vs. ω/ω_n for a second-order type-II loop with various damping coefficients.

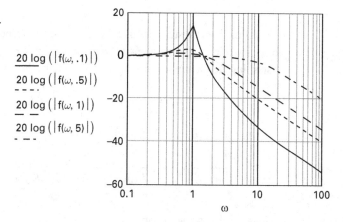

$$20 \log\left(\,|\,f(\omega, .1)\,|\,\right)$$

$$20 \log\left(\,|\,f(\omega, .5)\,|\,\right)$$

$$20 \log\left(\,|\,f(\omega, 1)\,|\,\right)$$

$$20 \log\left(\,|\,f(\omega, 5)\,|\,\right)$$

12.1.5.3 Transient response

The transient response can be determined from the frequency response by using Fourier or Laplace transforms, but this loop is simple enough that we can work directly in the time domain. Equation (12.2) gives the time response for the loop filter. We assume the loop has been disturbed but the reference is now constant. By inspection of Figure 12.5 we can write

$$\frac{dV_1}{dt} = K_D K_O V_2 = KV_2. \tag{12.8}$$

Combining this with Equation (12.2) we get

$$\frac{d^2 V_1}{dt^2} + K\frac{\tau_2}{\tau_1}\frac{dV_1}{dt} + \frac{K}{\tau_1}V_1 = 0 \quad \text{or} \quad \frac{d^2 V_1}{dt^2} + 2\omega_n\zeta\frac{dV_1}{dt} + \omega_n^2 V_1 = 0. \tag{12.9}$$

Figure 12.7. Phase error for a
step of -1 radian in the
reference phase at $t=0$.

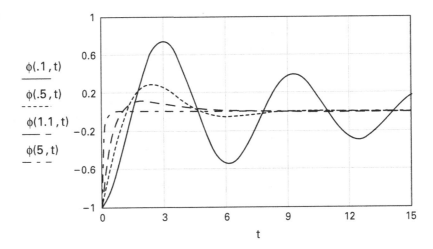

Assuming a solution of the form $e^{j\omega t}$, we find an equation for ω:

$$-\omega^2 + 2j\omega_n\zeta\omega + \omega_n^2 = 0. \tag{12.10}$$

The roots of this equation are

$$\omega = -\omega_n\zeta \pm \omega_n\sqrt{\zeta^2 - 1}. \tag{12.11}$$

When ζ is less than unity, the two solutions for ω are complex conjugates, giving sine and cosine oscillations with damped exponential envelopes. This is the underdamped situation. When ζ is greater than unity, the two solutions for ω are both damped exponentials, but with different time constants. If ζ is exactly unity, the two solutions are $e^{-\omega t}$ and $te^{-\omega t}$. In every case, a linear combination of the two appropriate solutions can match any given set of initial conditions, i.e., the $t=0$ values of V_1 and d/dt V_1. Figure 12.7 shows the transient phase response to a step function in the reference phase of one radian at $t=0$. This initial condition, $V_1(0) = -1$, determines the other initial condition, $dV_1/dt = 2\zeta\omega_n$, through the action of the loop filter circuit. Curves are shown for the recovery from this transient for damping coefficients of 0.1, 0.5, 1.1 and 5. In each case, $\omega_n = 1$. The underdamped cases show the trademark oscillatory behavior. When the damping is 5, the fast exponential recovery is essentially that of a wideband type-I loop.

Note: Equation (12.10) (which is the same as the denominator of the transfer function) is known as the *characteristic equation* of the system. Here it is a second-order equation and this type-II loop is therefore also a second-order loop. Equivalently, the order of a given system is the order of the differential equation describing its transient response.

12.1.6 A multiplier as a phase detector

A multiplier (mixer) is the most commonly used phase detector, especially at higher frequencies. When the reference sine wave is multiplied by the VCO sine wave, the usual sum and difference frequencies are generated. The desired baseband output is the difference component. If the VCO and reference frequencies are equal, this baseband component will be a dc voltage, zero when the VCO and reference phases differ by 90°, and linearly proportional to the phase difference in the vicinity of 90°. Therefore, when a multiplier is used as a phase detector, the loop will lock with the output phase 90° away from the reference phase. If this is a problem (in most applications it is not), a 90° phase shift network can be put at one of the multiplier inputs or in the output line. Note that a multiplier phase detector puts out zero volts when the phase shifts are different by 90° in either direction. One of these is a point of metastable equilibrium. As soon as it "falls off," the feedback will be positive, rather than negative, and the loop will rush to the stable equilibrium point.

As a phase detector, a mixer has a linear response of about $\pm \pi/4$, where $\sin(x) \approx x$. For somewhat larger phase differences, the sine curve flattens out and K_D begins to decrease. The system becomes nonlinear and the techniques of linear analysis fail. What is more important, if $\theta_{out} - \theta_{dist}$ needs to increase beyond $\pm \pi/2$, the $\sin(\theta_{out} - \theta_{dist})$ response of the mixer produces alternating positive and negative error signals, which will push the VCO frequency one way and then the other, and the loop will become unlocked. This is in contrast to the mechanical analog of Figure 12.3, whose phase detector range is 2π times the number of threads on the phase detector shaft. Electronic *phase/frequency detectors* (PFDs) are digital circuits that operate as linear phase detectors while the phase difference remains small, but then produce an output voltage proportional to the frequency difference, driving the loop in the correct direction to achieve lock. How a simple mixer-based PLL ever manages to lock without assistance is discussed below.

12.1.7 Frequency range and stability

The type-II loop can operate over the full range of the VCO since its integrator can build up as much bias as needed. The type-I loop, if $K_D K_O$ is small and if the phase detector has a limited range, may not be able to track over the full range of the VCO. Both the loops discussed above are unconditionally stable since the transient responses are decaying exponentials for any combination of the parameters K_D, K_O, R_1, R_2, and C. When you build one and it oscillates it is usually because you have high-frequency poles you did not consider (maybe in your op-amp or circuit parasitics) and you have actually built a higher-order loop.

12.1.8 Acquisition time

A high-gain type-I loop can achieve lock very quickly. A type-II loop with a small bandwidth can be very slow. Acquisition depends on some of the beat note from the phase detector getting through the filter to FM modulate the VCO. That beat note puts a pair of small sidebands on the VCO output. One of these sidebands will be at the reference frequency and will mix with the reference to produce dc with the correct polarity to push the VCO toward the reference frequency. The integrator gradually builds up this dc until the beat frequency comes within the loop bandwidth. From that point the acquisition is very fast. This reasoning, applied to the type-II loop, predicts an acquisition time of about $4f^2/B^3$ seconds where f is the initial frequency error and B is the bandwidth of the loop (see Problem 12.6). If the bandwidth is 10 Hz and the initial frequency error is 1 kHz, the predicted time is about one hour. (In practice, unavoidable offsets in the op-amp would make the integrator drift to one of the power supply rails and never come loose.) Obviously in such a case some assist is needed for a lockup. One common method is to add search capability, i.e., circuitry that causes the VCO voltage to sweep up and down until some significant dc component comes out of the phase detector. At that point the search circuit turns off and the loop locks itself up. The integrator can form part of the sweep circuit. Another way to aid acquisition is to use a frequency/phase detector instead of a simple phase detector. This is a digital phase detector which, when the input frequencies are different, puts out dc whose polarity indicates whether the VCO is above or below the reference. This dc quickly pumps the integrator up or down to the correct voltage for lock to occur. You will see these circuits described in Motorola literature. In digitally controlled PLLs, it is common for a microprocessor with a lookup table in ROM to pretune the VCO to the commanded frequency. This pretuning allows the loop to acquire lock quickly.

12.1.9 PLL receiver

There are many inventive circuits and applications in the literature on PLLs. Here is an example: deep space probes usually provide very weak telemetry signals. The receiver bandwidth must be made very small to reject noise that is outside the narrow band containing the modulated signal. A PLL can be used to do the narrowband filtering and to do the detection as well. One simple modulation scheme uses slow-frequency shift keying (FSK) where the signal is at one frequency for a "0" and then slides over to a second frequency for a "1." The loop bandwidth of the PLL is made just wide enough to follow the keying. The control voltage on the VCO is used as the detected signal output. (With a wide loop bandwidth, such a circuit can demodulate ordinary FM audio broadcast signals.) In this narrowband example, the PLL circuit has the advantage that it will track the signal automatically when the transmitter frequency drifts or is Doppler shifted. Another modulation scheme uses phase shift keying (PSK). Suppose the transmitter phase is shifted 180° to distinguish a "1" from a "0." This would confuse the PLL receiver because, for random data, the average output from the phase detector would be zero and the PLL would be unable to lock. A simple cure is to use the "doubling loop" shown in Figure 12.8. The incoming signal is first sent through a doubler to produce an output that looks like the output of a full-wave rectifier. (The waveform loops are positive no

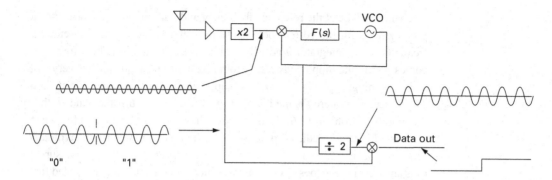

Figure 12.8. Doubling loop for detection of binary phase-shift modulation.

matter what the modulation does.) The VCO output frequency is then divided by two to produce a phase-locked reference. This reference is used as one input of a mixer. The other input of the mixer is the raw signal and the output is the recovered modulation.

12.2 Frequency synthesizers

By *frequency synthesizer* we usually mean a signal generator which can be switched to put out any one of a discrete set of frequencies and whose frequency stability is derived from a standard oscillator, either a built-in crystal oscillator or an external "station standard." Most laboratory synthesizers generate sine waves but some low-frequency synthesized function generators also generate square and triangular waves. General-purpose synthesizers have high resolution; the step between frequencies is usually less than 1 Hz and may be millihertz or even microhertz. Many television receivers and communications receivers have synthesized local oscillators. Special-purpose synthesizers may generate only a single frequency.

At least three general techniques are used for frequency synthesis. *Direct synthesizers* use frequency multipliers, frequency dividers, and mixers. *Indirect synthesizers* use phase lock loops. *Direct digital synthesizers* use a digital accumulator to produce a staircase sawtooth. A lookup table then changes the sawtooth to a staircase sinusoid, and a D-to-A converter provides the analog output. Some designs combine these three techniques.

12.2.1 Direct synthesis

The building blocks for direct synthesis are already familiar. Frequency multiplication can be done with almost any nonlinear element. A limiting amplifier (limiter) or a diode clipper circuit will convert sine waves into square waves, which include all the odd harmonics of the fundamental frequency. A delta function pulse train (in practice, a train of narrow pulses) contains all harmonics. When frequencies in only a narrow range are to be multiplied, class-C amplifiers can be used. (A child's swing does not need a push on each and every

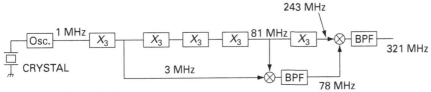

Figure 12.9. A direct synthesizer that produces 321 MHz from a 1-MHz reference.

cycle.) Frequency division, used in all three types of synthesizers, is almost always done digitally; the input frequency is used as the clock for a digital counter made of flip-flops and logic gates. Frequency translation can be done with any of the standard mixer circuits.

Single-frequency synthesizers are usually ad hoc designs; the arrangement of mixers, multipliers, and dividers depends on the ratio of the desired frequency and the reference frequency. As an example of direct synthesis, the circuit of Figure 12.9 generates 321 MHz from a 1-MHz standard. A prime factor of 321 is 107. It would be difficult to build a times-107 multiplier. This design uses only triplers and mixers. Laboratory instruments that cover an entire range of frequencies must, of course, use some general scheme of operation.

12.2.2 Mix and divide direct synthesis

Some laboratory synthesizers use an interesting mix and divide module. An n-digit synthesizer would use n identical modules. An example is shown in Figure 12.10. Each module has access to a 16-MHz source and ten other

Figure 12.10. A mix-and-divide direct synthesizer.

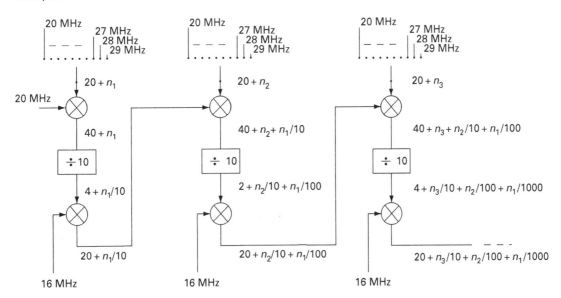

reference sources from 20 to 29 MHz. (These references are derived from an internal or external standard, often at 5 MHz). In this kind of design the internal frequencies must be chosen carefully so that after each mixer the undesired sideband can be filtered out easily.

12.2.3 Indirect synthesis

The phase lock loop circuit of Figure 12.11 is an indirect synthesizer to generate the frequency Nf_{std}/M where N and M are integers. If the $\div N$ and/or the $\div M$ blocks are variable modulus counters, the synthesizer frequency is adjustable.

Figure 12.11. Basic indirect (PLL) synthesizer.

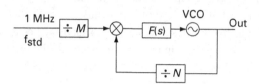

The 321-MHz synthesizer of Figure 12.9 could be built as the indirect synthesizer of Figure 12.11 by using a divide-by-321 counter (most likely a divide-by-3 counter followed by a divide-by-107 counter). Circuits like that of Figure 12.11 are used as local oscillators for digitally tuned radio and television receivers. For an AM broadcast band receiver, the frequency steps would be 10 kHz (the spacing between assigned frequencies). This requires.that the phase detector reference frequency, f_{std}/M, be 10 kHz so only the modulus N would be adjustable. What about a synthesized local oscillator for a short-wave radio? We would probably want a tuning resolution of, say, 10 Hz. The reference frequency in the simple circuit of Figure 12.11 would have to be 10 Hz and the loop bandwidth would have to be about 2 Hz. This low bandwidth would make fast switching difficult. Moreover, with such a narrow loop bandwidth, the close-in noise of the VCO would not be cleaned up by the loop and the performance of the radio would suffer. For this application a more sophisticated circuit is needed. One method is to synthesize a VHF or UHF frequency with steps of 1 or 10 kHz, divide it to produce the necessary smaller steps, and then mix it to a higher frequency. Other circuits use complicated multiple loops. The newest receivers use L.O. synthesizers based on the principle of direct digital synthesis.

12.2.4 Direct digital synthesis (DDS)

This technique, illustrated in Figure 12.12, uses an adder with two n-bit inputs, A and B, together with an n-bit register to accumulate phase. The output of the adder is latched into the register on every cycle of a high-frequency clock. The inputs to the adder are the current register contents (the phase) and an adjustable addend (the phase increment). On every clock cycle, the accumulated phase increases by an amount given by the addend. Since the accumulator has a finite

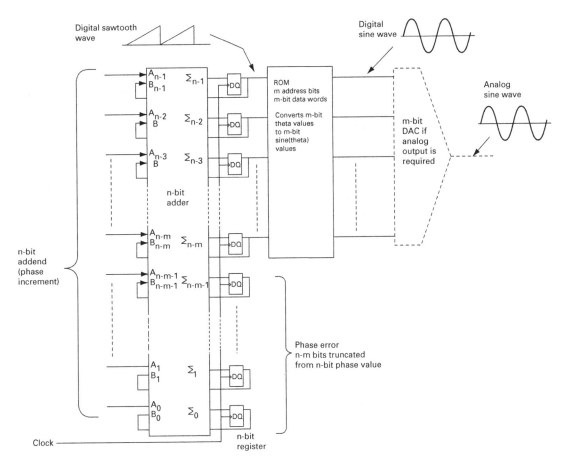

Figure 12.12. Direct digital
synthesizer (DDS).

length, it rolls over, like an automobile odometer, and the sequence of its output
values forms a sawtooth wave. The maximum addend value is 2^{n-1}, where n is the
number of bits in the adder/accumulator. Therefore, the output frequency of the
DDS extends up to one-half the clock frequency. At this maximum frequency,
the MSB toggles on every clock pulse and the lower bits remain constant. The
output of the accumulator is used to address a ROM (read-only memory) that
converts the stairstep digital sawtooth into a stairstep digital sine wave.

Let us consider the operation of the DDS in some detail. Suppose the selected
addend is the integer A. Let θ_i denote the number held in the register after the i-th
clock pulse. Because the accumulator rolls over when the output value would
have reached or exceeded 2^n, we can write $\theta_i = (\theta_{i-1}+A) \bmod 2^n$, a formula
which lets us easily simulate the DDS. On average, it will take $2^n/A$ clock pulses
to fill the accumulator,[3] so the average period of the sawtooth wave will be f_{clk}^{-1}
$2^n/A$ and the output frequency will be $f_{clk} A/2^n$. This frequency can be changed

[3] Consider that, after every $2^n A$ clock pulses, the accumulator must return to the same value. During
that period, the value that would be accumulated without rollover is $A(2^n A)$. Thus the number
of rollovers would be A^2. The average time per rollover is therefore $f_{clk}^{-1} (2^n A) /A^2 = f_{clk}^{-1} 2^n/A$.

in very fine increments if the number of bits in the accumulator is large. With a 32-bit accumulator, for example, the frequency resolution will be $f_{clock}/2^{32}$. For a clock frequency of 100 MHz the resolution will be $10^8/2^{32} = 0.023$ Hz.

One might think that, since the most significant bit (MSB) of the accumulator toggles at the desired output frequency, we could use it alone, simply filtering away its harmonics to obtain a sine wave at the desired frequency. However, while the MSB has the right *average* frequency it can be quite irregular. If the addend is a power of 2, the MSB will toggle at uniform time intervals but otherwise the MSB will have a jitter which depends on the value of the addend. A sine wave made from just the MSB would, therefore, contain too much phase noise (FM noise) for most RF applications.

The solution is to use more than just the top bit. If we use the top m bits, then the phase error is never greater than $360/2^m$ degrees. A table-lookup ROM uses these m-bit phase values, θm_i, as address bits to produce an output sequence, $\sin(\theta m_i)$, which has the desired wave shape (except for the roundoff error), together with low phase noise. An analog output can be provided by adding an m-bit D-to-A converter but, in digital radio applications, the digital sine wave is used directly. The more bits used to form the output sine wave, the lower the phase noise. Note that the DDS can incorporate the technique of digital pipe-lining in the adder/accumulator to achieve simultaneously high output frequencies and fine frequency resolution. The pipe delay is of no consequence for most system applications. Note also that the DDS can easily be modified to produce a chirped frequency by adding another accumulator to produce a sawtooth addend value. A precise FM modulator consists of a DDS whose addend is the digitized audio signal. Clock rates in DDS ICs have reached several GHz.

Output spectrum of the DDS

The exact spectrum of the DDS output can be calculated directly using Fourier analysis. For any value of the addend, the sequence of accumulator values will always repeat after at most 2^n clock pulses. Therefore, the sequence of digital values produced by the lookup ROM also repeats after at most 2^n cycles. This repetitive waveform can be represented as a Fourier series whose components are spaced in frequency by at least $f_{clk}/2^n$. The squares of the Fourier coefficients at each frequency are proportional to power. A simpler and more instructive approach is to note that the number contained in truncated accumulator bits (the bits below $n - m$) is just the phase error, $\delta\theta_i = \theta_i - \theta m_i$. The output sequence formed by the m-bit numbers can therefore be written as

$$\sin(\theta m_i) = \sin(\theta_i - \delta\theta_i) \approx \sin(\theta_i) - \delta\theta_i \cos(\theta_i), \qquad (12.12)$$

where we have assumed $\delta\theta_i$ is small. The term $\sin(\theta_i)$ is the desired output, while the term $\delta\theta_i \cos(\theta_i)$ is the noise. The error sequence, $\delta\theta_i \cos(\theta_i)$, will repeat after, at most, 2^{n-m} clock pulses, so the noise forms a Fourier series with components separated in frequency by at least $f_{clk}/2^{n-m}$. The nature of the noise will depend on the value of the bottom part of the addend, i.e., the lowest $n - m$ bits of the addend. If that number is a power of 2, the noise components will fall at harmonics of the output frequency, distorting the output waveform slightly, but contributing no phase noise. If the bottom part of the addend contains no factors of 2, the noise may be a grass of components at all multiples of $f_{clk}/2^{n-m}$. The more factors of 2 in the bottom part of the addend, the wider the spacing between noise components. Thus, the nature of the noise can change dramatically from one output frequency to the next.

12.2.5 Switching speed and phase continuity

Indirect synthesizers cannot change frequency faster than the time needed for their phase lock loops to capture and settle. Direct synthesizers and direct digital synthesizers can switch almost instantly. Sometimes it is necessary to switch frequencies without losing phase continuity. The DDS is perfect for this since the addend is changed and the phase rate changes but there is no sudden phase jump. Other times, when the synthesizer is retuned to a previously selected frequency, the phase must take on the value it would have had if the frequency had never been changed. This second kind of continuity might be called phase memory. A frequency synthesizer that provides the first kind of phase continuity clearly will not provide the second and vice versa. Continuity of the second kind can be obtained with a direct synthesizer that uses only mixers and multipliers (no dividers – which can begin in an arbitrary state).

12.2.6 Phase noise from multipliers and dividers

It is important to see how any noise on the reference signal of a synthesizer determines the noise on its output signal. Let us examine how noise is affected by the operations of frequency multiplication and division. We will assume the input noise sidebands are much weaker than the carrier. Suppose the input signal has a discrete sideband at 60 Hz. (Such a sideband would normally be one of a pair but for this argument we can consider them one at a time.) Let this sideband have a level of -40 dBc, i.e., its power is 40 dB below the carrier power. If this signal drives a times-N frequency multiplier, it turns out that the output signal will also have a sideband at 60 Hz but its level will be $-40 + 20\log(N)$ dBc. The relative sideband power increases by the *square* of the multiplication factor.

Let us verify this, at least for a specific case – a particular frequency tripler. Let the input signal be $\cos(\omega t) + \alpha\cos([\omega+\delta\omega]t)$, i.e., a carrier at ω having a sideband at $\omega+\delta\omega$ with relative power of α^2 or $20\log(\alpha)$ dBc. We assume that $\alpha \ll 1$. Here the tripler will be a circuit whose output voltage is the cube of the input voltage. Expanding the output and keeping only terms of order α or higher whose frequencies are at or near $3f$ we have

$$[\cos(\omega t) + \alpha\cos[(\omega + \delta\omega)t]]^3 \rightarrow \cos^3(\omega t) + 3\alpha\cos^2(\omega t)\cos[(\omega + \delta\omega)t]$$

$$(12.13)$$

$$\to 1/2 \, \cos(\omega t)\cos(2\omega t) + 3/2 \, \alpha \cos(2\omega t) \cos[(\omega + \delta\omega)t]$$

$$\to 1/4 \, \cos(3\omega t) + 3/4 \, \alpha\cos[(3\omega + \delta\omega)t]$$

$$= 1/4 \, [\cos(3\omega t) + 3\alpha\cos[(3\omega + \delta\omega)t]]. \tag{12.14}$$

Note that the carrier-to-sideband spacing is still $\delta\omega$ but the relative amplitude of the sideband has gone up by 3, the multiplication factor. The relative power of the sideband has therefore increased by 3^2, the square of the multiplication factor. A continuous distribution of phase noise $S(\delta\omega)$ is like a continuous set of discrete sidebands so, if the noise spectrum of a multiplier input is $S(\delta\omega)$, the noise spectrum of the output will be $n^2 S(\delta\omega)$. Sideband noise enhancement is a direct consequence of multiplication. If the multiplier circuit itself is noisy, the output phase noise will increase by more than n^2. Fortunately, most multipliers contribute negligible additive noise.

Division, the inverse of multiplication, reduces the phase noise power by the square of the division factor. Mixers just translate the spectrum of signals; they do not have a fundamental effect on noise. Additive noise from mixers is usually negligible. If a direct synthesizer is built with ideal components, the relation between the output phase noise and the phase noise of the standard will be as if the synthesizer were just a multiplier or divider, no matter what internal operations are used. The phase noise produced by indirect synthesizers depends on the quality of the internal VCOs and the bandwidths of the loop filters.

Problems

*These more difficult problems could be used as projects.

Problem 12.1. How would you modify the gear train in Figure 12.3 so that the output shaft would turn N/M times faster than the input shaft?

Problem 12.2. Show that Equation (12.2) describes the time domain input/output relation for the loop filter circuit of Figure 12.4.

Problem 12.3. Suppose the VCO in the block diagram of Figure 12.5 is noisy and that its noise can be represented as an equivalent noise voltage added at the control input of the VCO. Redraw the block diagram showing a summing block just in front of the VCO. Find the frequency transfer function for this noise input. Show that the loop is able to "clean up" the low-frequency noise of the VCO.

Problem 12.4. * Write a computer program to numerically simulate the PLL shown in Figure 12.5. Use a multiplier type phase detector and investigate the process of lock-up. Let the phase detector output be proportional to $\cos(\theta - \theta_{\mathrm{ref}})$. Use numerical integration on the simultaneous first-order differential equations for $V_1(t)$ and $V_2(t)$.

Problem 12.5. * Invent a phase detector circuit that would have a range of many multiples of 2π rather than the restricted range of the multiplier phase detector.

Problem 12.6.[*] Derive the formula given for lock-up time, $\tau_{\text{LOCKUP}} \approx 4(\Delta\omega)^2/B^3$, where $\Delta\omega$ is the initial frequency error and B is the loop bandwidth (in radians/sec). Consider the type-II loop of Figure 12.5. Find the ac component on the VCO control input. Assume that $\Delta\omega$ is high enough that the gain of the loop filter for this ac voltage is just $-R_2/R_1$. Find the amplitude of the sidebands caused at the output of the VCO from this ac control voltage component. Use this amplitude to find the dc component at the output of the phase detector caused by product of the reference signal and the VCO sideband at the reference frequency. This dc component will be integrated by the loop filter. Find dV/dt at the output of the loop filter. Find an expression for τ_{LOCKUP} by estimating the time for the dc control voltage to change by the amount necessary to eliminate the initial error.

Problem 12.7. Design a direct synthesizer (ad hoc combination of mixers, dividers, multipliers) to produce a frequency of 105.3 MHz from a 10-MHz reference. Avoid multipliers higher than ×5 and do not let the two inputs of any mixer have a frequency ratio higher than 5:1 (or lower than 1:5).

Problem 12.8. Design a direct digital synthesizer with 10-kHz steps for use as the tunable local oscillator in a middle-wave broadcast band (530–1700 kHz) AM receiver with a 455 kHz IF frequency. Assume a high-side L.O., i.e., the synthesizer frequencies range from $530 + 455$ to $17800 + 455$ An accurate reference frequency is available at 10.24 MHz.

Problem 12.9. Design a synthesizer with the range of 1–2 MHz that has phase memory, i.e., when the synthesizer is reset to an earlier frequency, its phase will be the same as if it had been left set to the earlier frequency. The required step size is 50 kHz and the available frequency reference is 5 MHz. Hint: one approach is to generate a frequency comb with a spacing of 50 kHz and then phase-lock a tunable oscillator to the desired tooth of the comb.

Problem 12.10. Explain why a direct synthesizer that includes one or more dividers will not have phase memory.

Problem 12.11. Draw a block diagram for an FM transmitter in which a phase locked loop keeps the average frequency of the VCO equal to a stable reference frequency.

Problem 12.12. Draw block diagrams for PM and FM generators based on the direct digital synthesizer (DDS) principle.

References

[1] Crawford, James A. *"Frequency Synthesizer Design Handbook,"* Boston: Artech House, 1994.

[2] Gardner, Floyd M., *Phaselock Techniques*, 2nd edn. New York: John Wiley, 1979.

[3] Kuo, Benjamin, *Automatic Control Systems*, 5th edn. Englewood Cliffs: Prentice Hall, 1987.

[4] Manassewitsch, V., *Frequency Synthesizers Theory and Design*, 2nd edn. New York: John Wiley, 1980.

13 Coupled-resonator bandpass filters

We saw in Chapter 4 that the straightforward transformation of a prototype lowpass filter to a bandpass filter yields a circuit with alternating parallel resonant circuits and series resonant circuits as shown in Figure 13.1.

Figure 13.1. Conversion of a lowpass filter to a canonical bandpass filter.

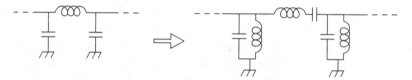

If the prototype lowpass filter has a response $F_{LP}(\omega)$, the corresponding bandpass filter will have the response $F_{BP}(\omega) = F_{LP}(|\omega - \omega_C|)$, where ω_C is the center frequency. These canonical bandpass filters work perfectly – when simulated with a network analysis program. But usually they call for impractical component values. The inductors in the shunt branches must be smaller than the inductors in the series branches by a factor on the order of the square of the fractional bandwidth. For a 5% bandwidth filter, the ratio of the inductor values would be of the order of 1:400. For a given center frequency we might be lucky to find a high-Q inductor of any value, let alone high-Q inductors with such different values. Low-Q (resistive) inductors make a filter lossy and change its nominal passband shape. The series and shunt capacitor values have the same ratio. Generally Q is not a problem with capacitors, but very small values are impractical when they become comparable to the stray wiring capacitances.

13.1 Impedance inverters

This component value problem can be solved by transforming canonical bandpass filters into *coupled-resonator bandpass filters*, which can be built with identical or almost identical LC resonant circuits. The coupled-resonator filters have the same filter shapes, based on prototype lowpass designs, such as Butterworth or Chebyshev. Figure 13.2 shows some coupled-resonator fiter designs.

These filters are based on *impedance inverters*. Three examples of impedance inverters are shown in Figure 13.3, a 90° length of transmission line and two lumped *LC* circuits.

Figure 13.2. Three examples of coupled-resonator bandpass filter circuits.

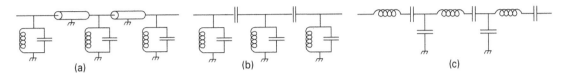

(a)   (b)   (c)

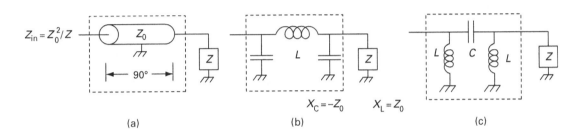

(a)   (b)   (c)

Figure 13.3. Three impedance inverter circuits

In every case, an impedance Z, when seen through the inverter, becomes Z_0^2/Z where Z_0 can be called the characteristic impedance of the inverter. For the transmission line inverter, a 90° length of line, Z_0 is just the characteristic impedance of the line. For the *LC* inverters, both the inductor's reactance, X_L, and the capacitor's reactance, X_C, must be equal to the desired Z_0. Like the 90° cable, the lumped element circuits are perfect inverters only at one frequency but, in practice, they are adequate over a considerable range. An inverted capacitor is an inductor. An inverted inductor is a capacitor. Figure 13.4 shows an inverter (in this example, a 90° transmission line) used to invert a parallel circuit, making an equivalent series circuit.

The mathematics of this inversion is just

$$Z_{in} = Z_0^2 Y = Z_0^2 \left(\frac{1}{j\omega L_p} + j\omega C_p + \frac{1}{R_p} \right) = \frac{1}{j\omega(L_p/Z_0^2)} + j\omega(Z_0^2 C_p) + \frac{Z_0^2}{R_p}.$$

(13.1)

Let us look at four inverters which include inductors or capacitors with negative values. For these inverters, shown in Figure 13.5, $X_C = Z_0$ or $X_L = Z_0$.

Figure 13.6 verifies the inverter action of the all-capacitor T-section inverter. You can use this kind of reasoning to verify the inverter action of the other circuits:

Figure 13.4. Impedance inverter makes a parallel circuit appear as a series circuit.

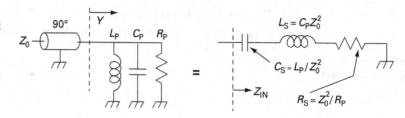

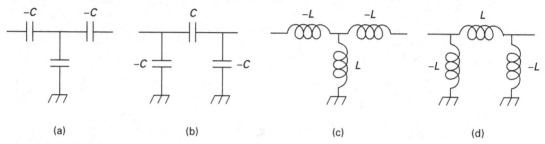

(a)   (b)   (c)   (d)

Figure 13.5. Impedance inverters based on negative value components.

Figure 13.6. Operation of the T-network negative capacitor inverter.

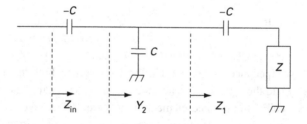

$$Z_1 = \frac{1}{j\omega(-C)} + Z \tag{13.2}$$

$$Y_2 = j\omega C + \frac{1}{Z_1} = \frac{j\omega C Z}{Z - 1/j\omega C} \tag{13.3}$$

$$Z_{in} = \frac{1}{j\omega(-C)} + \frac{1}{Y_2} = \frac{1}{\omega^2 C^2 Z} = \frac{Z_0^2}{Z}. \tag{13.4}$$

Because they contain negative capacitances or negative inductances, the four inverters in Figure 13.5 might seem to be only mathematical curiosities. Not at

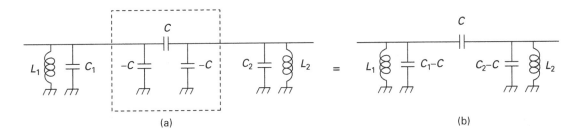

Figure 13.7. Negative capacitors absorbed into adjacent positive capacitors.

all; the negative elements can be absorbed by positive elements in the adjacent circuitry as shown in Figure 13.7, where a π-section capacitor inverter is placed between two parallel LC "tanks."

13.2 Conversion of series resonators to parallel resonators and vice versa

Ladder network filters have alternating series and shunt branches. Let us see how inverter *pairs* are used in ladder filters. Suppose we embed a series capacitor between a pair of inverters at some point along a ladder network.

Figure 13.8. A series capacitor between inverters is equivalent to a shunt inductor.

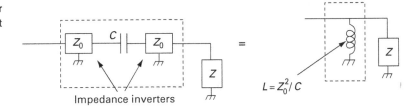

The combination of the capacitor and the inverter pair is equivalent to a shunt inductor, as shown in Figure 13.8.

You can show just as easily that any series impedance, Z_s, together with a pair of bracketing inverters of characteristic impedance Z_0 is equivalent to a shunt admittance $Y_p = Z_0^{-2} Z_s$. Likewise, the combination of any shunt admittance Y and a pair of bracketing inverters is equivalent to a series impedance $Z = Z_0^2 Y$. Figure 13.9 illustrates this, showing how a series resonant series branch in an ordinary bandpass filter can be replaced by a parallel resonant shunt branch imbedded between a pair of inverters.

Likewise, a parallel resonant shunt branch can be realized as a series resonant series branch imbedded between a pair of inverters, as shown in Figure 13.10.

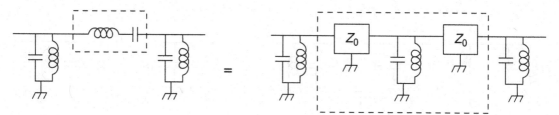

Figure 13.9. A shunt resonator between inverters is equivalent to a series resonator.

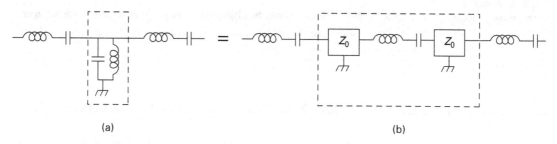

(a)　　　　　　　　　(b)

Figure 13.10. A series resonator between inverters is equivalent to a shunt resonator.

13.3 Worked example: a 1% fractional bandwidth filter

Consider a 50-ohm, 1-dB Chebyshev filter with a 10-MHz center frequency and a bandwidth of 100 KHz between the 1-dB points. The filter, which results from the straightforward lowpass to bandpass transformation (Chapter 4) is shown in Figure 13.11 and its response is shown in Figure 13.12.

We might find 86-μH inductors with high Q at 10 MHz but the 3.728 nH and 2.645 nH inductors would be tiny single turns of wire with very poor Q. To get around these component limitations, we will convert this filter into a coupled-resonator filter. Suppose we have in hand some adjustable 0.3 to 0.5 microhenry

Figure 13.11. A straightforward (but impractical) bandpass filter. The calculated response of this filter is shown in Figure 13.12.

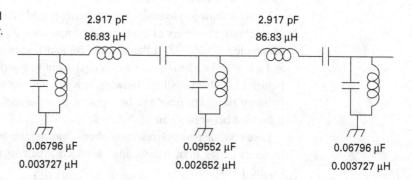

Figure 13.12. Calculated response for filter of Figure 13.11.

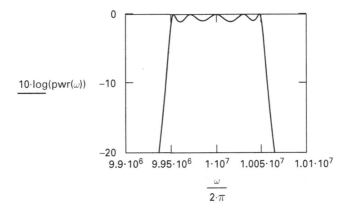

Figure 13.13. Filter of Figure 13.11, scaled from 50 to 6705 ohms.

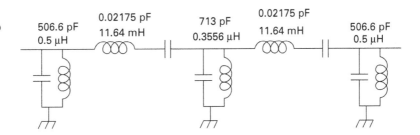

inductors which, at 10 MHz, have very high Q (we will see later just how much Q is required). Let us first change the working impedance of the filter so that the parallel resonators at the end will use 0.5 μH, which is 134.1 times the original end inductors and implies that the filter will be scaled to $50 \times 134.1 = 6705$ ohms. We multiply the other inductors by 134.1 and divide the capacitors by 134.1 to get the circuit of Figure 13.13.

The parallel resonators now use the desired inductors but the series resonators call for inductors of 11.6 mH, a very large value for which we surely will not find high Q components. Moreover, the series capacitors are only 0.02 pF, a value far too small to be practical. We can solve this problem by using impedance inverters to convert the series resonators into parallel resonators. Let us use the all-capacitor π-section inverters of Figure 13.5(b) and the same parallel resonators we used for the end sections. Figure 13.14 shows how two inverters and the parallel resonator replace each series resonator.

We can calculate the inverter's characteristic impedance, Z_0, as follows:

$$Z_0^2 Y = Z; \quad Z_0^2(j\omega C_p + 1/j\omega L_p) = j\omega L_S + 1/j\omega C_S \tag{13.5}$$

$$Z_0^2 = L_p/C_S = 0.5 \times 10^{-6}/0.02175 \times 10^{-12} = 4796^2. \tag{13.6}$$

Figure 13.14. Inverters transform a 0.5-μH shunt inductor into a 11.644-mH series inductor.

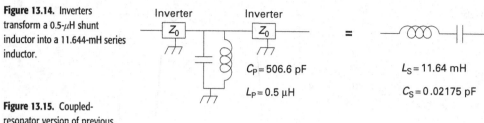

Figure 13.15. Coupled-resonator version of previous bandpass filter.

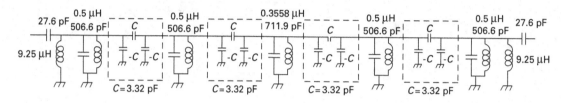

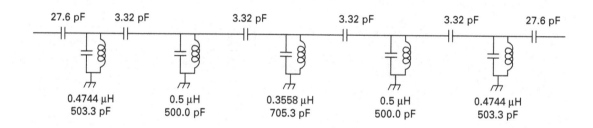

Figure 13.16. Finished coupled-resonator filter.

For this type of inverter, we had seen that $Z_0 = X_C$, so $C = 3.32$ pF. We now have our coupled-resonator filter but since it works at 6705 ohms we will add L-section matching networks at each end to convert it back to 50 ohms. The filter, at this point, is shown in Figure 13.15. All the resonators are now parallel resonators. (In other situations we might use inverters to convert series resonators into equivalent parallel resonators to make an all-series-resonator filter – see Figure 13.1.)

The final clean-up step is to absorb the -3.32 pF capacitors into the resonator capacitors and combine the matching section inductors with the end-section resonator inductors as shown in Figure 13.16.

The response of the finished filter is shown in Figure 13.17 and is almost identical to the response of the prototype filter of Figure 13.11. The difference, a fraction of a dB, occurs because the inverters work perfectly only at the center frequency.

13.4 Tubular bandpass filters

A popular bandpass filter design, the "tubular filter" is produced by many filter manufacturers. Figure 13.18 shows the construction of a three-resonator tubular filter.

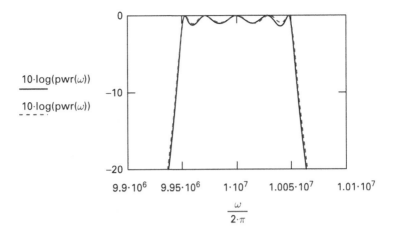

The only standard electronic components are the coaxial connectors at the ends. There are also (in this example) three inductors (wire coils), four metal cylinders, two dielectric spacers, two (or one long) dielectric sleeves, and a tubular metal body. Figure 13.19 shows how a coupled-resonator filter design, of the type we have discussed, is transformed into the tubular filter design. You can verify that Figure 13.19(d) is the circuit diagram of the tubular filter. The three-capacitor π-sections are formed by the capacitance between the adjacent faces of the metal cylinders and the capacitors are formed between the outside surfaces of the cylinders and the tubular body.

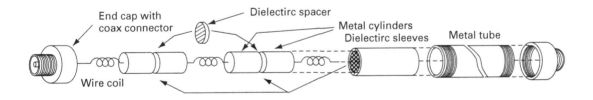

Figure 13.18. Tubular bandpass filter.

Beginning with Figure 13.19(a), we have a standard coupled-resonator bandpass filter using series resonators. In the canonical prototype for this filter, the middle section is a parallel resonator, but this has been replaced by a series resonator sandwiched between two impedance inverters. In (b), the center capacitor has been replaced by two capacitors (each of twice the value of the original capacitor so that, in series, the total series capacitance is the same). The capacitors have been shifted slightly in (c) to identify a T-section capacitor network at each side of the central inductor. Finally, in going from (c) to (d), these T-networks are replaced by equivalent π-networks, to arrive at the circuit of the tubular filter. Any

Figure 13.19. Tubular bandpass filter evolution.

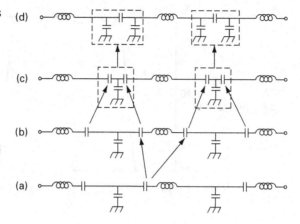

Figure 13.20. Equivalent π-section and T-section networks.

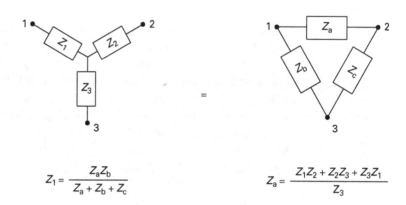

$$Z_1 = \frac{Z_a Z_b}{Z_a + Z_b + Z_c}$$

$$Z_a = \frac{Z_1 Z_2 + Z_2 Z_3 + Z_3 Z_1}{Z_3}$$

T-network has an equivalent π-network and vice versa (Problem 13.5). These transformations, also known as T–π and π–T are shown in Figure 13.20. Formulas are given for one element in each network; the others follow from symmetry.

13.5 Effects of finite Q

These calculated filter responses assume components of infinitely high Q. We can calculate the effects of finite Q by paralleling the (lossless) inductors in our model with resistors equal to Q times the inductor reactances at the center frequency. If, for example, the Q is 500 (quite a high value for a coil), we would parallel the inductors in the filter of Figure 13.15 with resistors of about 15 000 ohms. Reanalyzing the circuit response, we would find that the filter will have a midband insertion loss of 7 dB and that the flat (within 1 dB) passband response becomes rounded. The effect will be somewhat less for a filter with more gradual skirts, e.g., a 0.01 dB Chebyshev or a Butterworth filter. But the real problem is still the small fractional bandwidth. For a filter with small fractional bandwidth to have the ideal shape of Figure 13.17, the resonators must be quartz or ceramic or other resonators with Qs in the thousands. An approximate analysis predicts that the midband loss per section in a bandpass filter will be on the order of

$$\frac{\text{power transmitted}}{\text{power incident}} = \left(1 - \frac{L_0/2}{Q \cdot \text{fractional bandwidth}}\right) \qquad (13.7)$$

where L_0 represents the inductor value in the normalized lowpass prototype filter. For our five-section filter we can take L_0 to be about 1.5 henrys. If the inductor Q is 500, the predicted transmission of the five-section filter is 5×10 $\log[1-(1.5\ /2)/(500\cdot(1/100))] = -10$ dB, which is roughly equal to the actual value of -7 dB.

13.6 Tuning procedures

Filters with small fractional bandwidths and sharp skirts are extremely sensitive to component values. In the filter of Figure 13.16, for example, the resonators must be tuned very precisely or the shape will be distorted and the overall transmission will be lowered. (The values of the small coupling capacitors – all that remains of the impedance inverters – are not as critical.) Usually each resonator is adjustable by means of a variable capacitor or variable inductor. All the adjustments interact and, if the filter is totally out of tune, it may be hard to detect any transmission at all. A standard tuning procedure is to monitor the input impedance of the filter while tuning the resonators, one-by-one, beginning at input end. While resonator N is being adjusted, resonator $N+1$ is short circuited. The tuning of one resonator is done to produce a maximum input impedance while the tuning of the next is done to produce a minimum input impedance. The procedure must sometimes be customized to account for matching sections at the ends.

13.7 Other filter types

The coupled-resonator technique is used from HF through microwaves. Not all RF bandpass filters, however, use the coupled-resonator technique. The IF bandpass shape in television receivers is usually determined by a SAW (surface acoustic wave) bandpass filter. SAW filters are FIR (finite impulse response) filters, whereas all the LC filters we have discussed are IIR (infinite impulse response) networks. This classification is made according to the behavior of the output voltage following a delta function (infinitely sharp impulse) excitation. Digital filters can be designed to be either FIR or IIR filters.

Problems

Problem 13.1. Use your network analysis program to verify that the filter of Figure 13.16 does indeed give the response shown in Figure 13.17.

Problem 13.2. Verify that the two *LC* circuits in Figure 13.3 are impedance inverters.

Problem 13.3. The filter shown below was developed in Chapter 4 as an example of the straightforward conversion from a prototype lowpass filter to a bandpass filter. This Butterworth (maximally flat) filter has a bandwidth of 10 kHz and a center frequency of 500 kHz. Suppose you have available some 30 μH inductors with a *Q* of 100 at 500 kHz. Convert the filter into a coupled-resonator filter that uses these inductors. Use your ladder network analysis program to verify the performance of your filter.

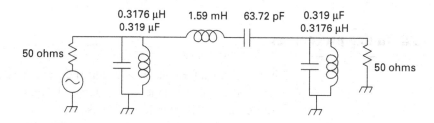

Problem 13.4. A bandpass filter is to have the following specifications:

Center frequency: 10 MHz; shape: three-section 1-dB Chebyshev; bandwidth: 3 KHz (between outermost 1-dB points); source and load Impedances: 50 ohms. Since the loaded *Q* of this filter is very high, $10^6/3000 = 333$, it is important to use very high-*Q* resonators. Suppose you have located some resonators (cavities, crystals, or whatever) with adequate *Q*. These resonators are all identical. At 10 MHz they exhibit a parallel resonance, equivalent to a parallel *LC* circuit. At 10 MHz, they have a susceptance slope of 10^{-6} (1 mho/MHz).

(a) Find the *LC* equivalent circuit for these resonators (in the vicinity of 10 MHz).

(b) Design the filter shown below around these resonators.

(c) Use your ladder network analysis program to verify the frequency response of your design.

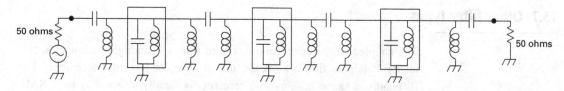

Problem 13.5. Derive expressions for Z_a, Z_b, and Z_c in terms of Z_1, Z_2, and Z_3 for the equivalent *T* and *π* networks shown in Figure 13.19. Hint: consider the connections

shown below. The sketched-in wires show that $Y_A + Y_B = (Z_3 + Z_1 \| Z_2)^{-1}$. If you write the corresponding $Y_B + Y_C$ and $Y_C + Y_A$ equations, then add the first two and subtract the third, you will have the formula for Y_B. A similar technique yields the expressions for Z_1, Z_2, and Z_3.

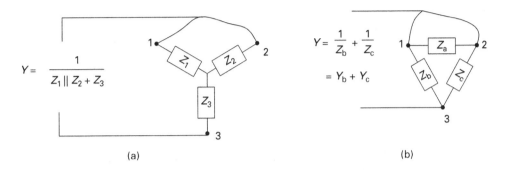

(a)   (b)

Problem 13.6. The bridge circuit shown below in (a) is the simplest network whose resistance cannot be found immediately by series and parallel reduction. Rather than resorting to loop or node equations, note that the circuit contains two πs and two Ts. Replace a π by its equivalent T or a T by its equivalent π. Now find the resistance of the network by simple reduction. The circuit at the right shows how one of the π's can be replaced by a T.

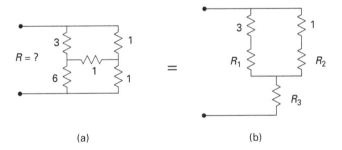

(a)   (b)

References

[1] Christiansen, D., Alexander, C., Jurgen, R. K. *Standard Handbook of Electronic Engineering*, 5th edn, New York: McGraw-Hill, 2004.

[2] Matthaei, G., Young, L. and Jones, E. M. T. *Microwave Filters, Impedance-Matching Networks, and Coupling Structures*, New York: McGraw Hill, 1964, reprinted, Boston: Artech House, Inc. 1980.

14 Transformers and baluns

Transformers are harder to understand than resistors, capacitors, and single inductors. First, transformers have two terminal pairs rather than one, so we must deal with two voltages and two currents. Second, we can be misled by the deceptive simplicity of the simplest mathematical model, the "ideal transformer." In this chapter we discuss the conventional transformers used in power supplies, switching power supplies, amplifiers, and RF matching networks. We will then examine transmission line transformers, which work to higher frequencies, and *baluns*, which are devices used to connect *bal*anced circuits to *un*balanced circuits.

Figure 14.1(a) shows two inductors (here, wire coils) with arbitrary placement. The region does not have to be otherwise empty; it can contain any distribution of clumped and/or continuous magnetic materials.

The inductance values ("self-inductances") of these coils are L_1 and L_2, each measured with the other coil open circuited, so that it carries no current. We will refer to these coils as L_1 and L_2. Two representative magnetic flux lines are shown, corresponding to a current in L_1. Note that one of these flux lines is encircled by three turns of L_2. Therefore, when L_1 carries an ac current, by Faraday's law there will be an ac voltage induced in L_2, proportional to the time derivative of the encircled magnetic flux. We can write Faraday's law for this situation as

$$V_2 = j\omega t M I_1 + j\omega t L_2 I_2, \tag{14.1}$$

where the constant M is known as the "mutual inductance." The $j\omega$ factor represents the time derivative since we are using standard ac circuit analysis, where the time dependence is contained in an implicit factor $e^{j\omega t}$. The second term, containing the self-inductance, L_2, is a voltage induced by the current, if any, flowing in L_2 itself. Note that we have assumed that the wires have negligible resistance – no IR voltage drop.[1] The directions of the currents are

---

[1] To take account of the winding resistances, we would add a term $I_2 R_2$ to the right-hand side of Equation (14.1) and a term $I_1 R_1$ to the right-hand side of Equation (14.2). The effect of the resistance distributed in the windings is the same as if the resistance were consolidated into two external resistors, R_1 and R_2, in series with the windings.

Figure 14.1. (a) Coupled inductors; (b) prototype transformer.

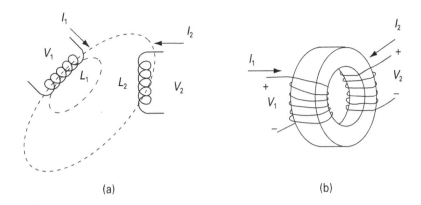

(a)   (b)

defined by arrows in Figure 14.1 as entering the positive end of each coil. This symmetric assignment produces a symmetric equation for V_1:

$$V_1 = j\omega t L_1 I_1 + j\omega t M I_2. \tag{14.2}$$

Note that both equations contain the same constant M. There is no need to write M_{12} and M_{21}, since the mutual inductances are always equal. This can be seen quite easily for the arrangement of coupled coils shown in Figure 14.1(b). This a conventional transformer, in which a toroid of iron or other magnetic material effectively contains all the field lines, forcing them to thread through every turn of both L_1 and L_2. An ac current in L_1 produces a flux proportional to N_1, the number of turns in L_1. The ac voltage induced in L_2 is proportional to N_2 times the flux. Hence M_{21} is proportional to the product $N_1 N_2$. Likewise the ac current in L_2 produces an ac voltage in L_1 with the same proportionality to $N_1 N_2$, so M_{12} and M_{21} are equal.[2]

If all the flux lines thread both windings, the transformer is said to be *perfectly coupled*. A *coupling coefficient*, k, is defined by $M = k(L_1 L_2)^{1/2}$. The value of k ranges from zero, for inductors with no coupling, to unity, for perfect coupling. In the transformer of Figure 14.1(b), perfect coupling is approached by using a core material of extreme magnetic permeability. For such a transformer, it makes no difference whether the windings are side-by-side, as shown, or wound one on top of the other. Nor is it necessary that they be wound tightly around the core; loose windings can be used to allow circulation in an oil-cooled power transformer.

14.1 The "ideal transformer"

If a transformer is perfectly coupled and the windings have negligible resistance, then the ratio of primary-to-secondary[3] voltages is equal to the turns ratio

---

[2] The equality of M_{12} and M_{21} is a general reciprocity relation that holds true for any passive two-port network, e.g., any network made from resistors, capacitors, inductors, and transformers.

[3] The names "primary" and "secondary" are arbitrary and refer only to the way the transformer is used; power usually flows into the primary and out of the secondary.

and these voltages have the same phase. This strict proportionality of voltages follows directly from Faraday's law, since the time derivative of the magnetic flux is equal in both windings. The situation with the currents is not as simple. The primary and secondary currents are not strictly proportional. This follows from the transformer's ability to store magnetic energy. If the transformer could *not* store energy, the instantaneous net power into the transformer would have to be zero. The proportionality of primary and secondary voltages would demand that the currents be proportional, i.e., setting $V_p I_p = V_s I_s$, we would have $I_s/I_p = V_p/V_s =$ constant $= n_P/n_S$, the effective turns ratio. This hypothetical transformer, which stores no energy, is known as the *ideal transformer* and is a useful abstraction. We will discuss below a circuit with two inductors and an ideal transformer which together, are equivalent to a real transformer. But first let us emphasize how the ideal transformer, by itself, is an unrealistic model.

Consider an ideal transformer with an effective turns ratio n_1/n_2. If an impedance Z_{load} is connected to the secondary, the ratio of primary voltage to primary current will be $(n_1/n_2) V_2/[(n_2/n_1)I_s] = (n_1/n_2)^2 V_s/I_s = (n_1/n_2)^2 Z_{\text{load}}$ times the secondary current. The primary will therefore present an impedance of $(n_1/n_2)^2 Z_{\text{load}}$, a simple impedance multiplication. If Z_{load} is infinite (an open circuit), the impedance looking into the primary of the ideal transformer is also infinite. But inspection of Equation (14.2) shows that, in this case, the impedance looking into a real transformer is $j\omega L_1$. Another "unreal" feature of an ideal transformer is that it contains no magnetic field! The field (or flux) from the primary winding is exactly cancelled by the field from the secondary winding. And with no magnetic field there would be no $d\Phi/dt$ and therefore no voltage across either winding. A real transformer approaches the ideal transformer model only when the number of turns approaches infinity and the magnetic coupling approaches 100%. Transformers used in practice are usually far from ideal. This is in contrast to resistors and capacitors which, at least at low frequencies, are almost ideal components obeying the relations $Z_R = R$, $Z_C = 1/j\omega C$, and $Z_L = j\omega L$.

14.2 Transformer equivalent circuit

An equivalent circuit for a real transformer is shown in Figure 14.2(b).

This model circuit consists of two inductors plus an ideal transformer having an equivalent turns ratio n_1/n_2. The shunt inductor at the left is known as the *magnetizing inductance* and its value is L_1, the inductance of the primary. Note that, even if the secondary is left open, a voltage V applied to the primary will produce a primary current $I = V/(j\omega L_1)$. The series inductor at the right is known as the *leakage inductance*. Its value is $L_2 (1- k^2)$, so a perfectly coupled transformer ($k = 1$) has no leakage inductance. In practice, maximum coupling is limited to maybe 98% at low frequencies and less at RF frequencies. The useful frequency range of a transformer is determined by these two inductances. Suppose we put a transformer between a resistive load and a signal generator.

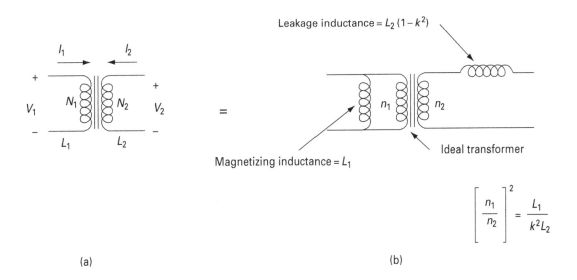

(a)   (b)

Figure 14.2. (a.) Transformer symbol with voltage and current assignments; (b.) an equivalent circuit.

Below the useful frequency range, the magnetizing inductance becomes a short circuit across the generator. Above the useful range, the leakage inductance becomes a high impedance in series with the load. In both extremes, the power delivered to the load becomes negligible.

Let us demonstrate that the model circuit of Figure 14.2(b) does, indeed, agree with the fundamental equations (14.1) and (14.2), i.e., that it really is an equivalent circuit. We have already seen that the value of the magnetizing inductance must be L_1, the inductance value of the primary winding. Leaving the secondary open, so that $I_2 = 0$, the fundamental equations (14.1) and (14.2) produce the relation $V_2/V_1 = M/L_1 = k\sqrt{L_1 L_2}/L_1 = k\sqrt{L_2/L_1}$. In this situation, the model circuit gives $V_2/V_1 = n_2/n_1$. For the model to agree with the fundamental equations, $n_2/n_1 = k\sqrt{L_2/L_1}$. Finally, consider the situation in which the primary is shorted, so that $V_1 = 0$. The secondary current given by Equations (14.1) and (14.2) must be the same as the current predicted by the model. Equation (14.1) produces $I_1 = -I_2 M/L_1$. Putting this into Equation (14.2) gives $V_2/I_2 = j\omega\ (L_2/M - M^2/L_1) = j\omega L_2(1-k^2)$. Looking at the model, the impedance at the secondary, with the primary shorted, is just $j\omega$ time the leakage inductance, so the leakage inductance must be assigned the value $L_2(1-k^2)$. With these assignments for the values of the magnetization and leakage inductances, the model correctly reproduces Equations (14.1) and (14.2).

This is not the only possible equivalent circuit. We could just as well have constructed this equivalent circuit with the magnetizing inductance on the right side and the leakage inductance on the left side. Or we could "push" either the magnetizing inductance or the leakage inductance through the transformer, correcting the inductance by a factor $(n_1/n_2)^2$ or $(n_2/n_1)^2$, so that they are both on the same side. Another equivalent circuit, is shown in Figure 14.3. Using arguments like those presented above, you can show that this circuit also

Figure 14.3. An all-inductor equivalent circuit.

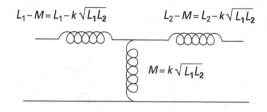

$$L_1 - M = L_1 - k\sqrt{L_1L_2} \qquad\qquad L_2 - M = L_2 - k\sqrt{L_1L_2}$$

$$M = k\sqrt{L_1L_2}$$

satisfies Equations (14.1) and (14.2). This circuit contains no unphysical ideal transformer (and therefore has no dc isolation between primary and secondary, making it not quite as equivalent). But you can see from the labels in the figure that, in general, one of the inductors must have an unphysical negative inductance. However, note that if $k < L_1/L_2$ and $k < L_2/L_1$ all the inductors are positive. As the turns ratio becomes close to unity, the inductors all remain positive as the coupling is increased. In the case of a 1:1 transformer with perfect coupling, the values of the series inductors go to zero and the equivalent circuit is just a single shunt inductor, the magnetizing inductance. (You could put a 1:1 ideal transformer on each side of this inductor to produce a symmetric equivalent circuit that preserves dc isolation.) And, of course, the circuit of Figure 14.3 could be converted from the T configuration to an equivalent pi configuration.

To approach the ideal, a transformer must have a very high magnetizing inductance and a very small leakage inductance. You can increase the magnetizing inductance by increasing the number of turns (keeping the turns ratio constant) but, in practice, this only increases the leakage inductance and the ohmic resistance of the windings. You can decrease the leakage inductance by using fewer turns, but this lowers the magnetizing inductance. A compromise is generally needed. However, we will see that there are applications in which the leakage and/or magnetizing inductances become useful circuit components.

14.3 Power transformer operation

Power transformers are usually iron-core transformers with high coupling. The best power transformer would be an ideal transformer; its stored energy, excitation current and leakage inductance would all be negligibly small. The primary is connected to the ac line, which can be regarded as a perfect voltage source with negligible source impedance. If the coupling is high enough to make the leakage inductance negligible and the winding resistances are low, the secondary voltage will be constant, $V_2 = (n_2/n_1)V_1$, independent of the load. The magnetizing inductance will draw a constant "magnetizing" current from the line, $I_M = V/(j\omega L_1)$. Since this current is 90° out of phase with respect to the primary voltage, it consumes no average power, but does cause "excitation" energy to slosh in and out of the transformer. When a resistive load is connected to the secondary, additional "working" currents flow in both the primary and the secondary. These currents have the ratio n_2/n_1. They are in phase with the primary voltage and transfer power from the source to the load. Sometimes a capacitor is placed across the primary to resonate with the magnetizing inductance. The excitation energy will then slosh back and forth between the capacitor and the magnetizing inductance. This corrects the power factor; the power line now only has to supply the component of current that is in phase with the voltage. In practice, the magnetizing current may be comparable to the maximum working current. In a power transformer, the magnetic core is used close to saturation. When the magnetizing current is at its maximum, the inductance of the core is reduced. This nonlinear behavior of the core distorts the otherwise sinusoidal waveform of the magnetizing current. Nevertheless, the voltages on the primary and secondary remain proportional and sinusoidal, because of the low source impedance of the power line, low IR drops in the windings, and negligible leakage inductance.

14.4 Mechanical analogue of a perfectly coupled transformer

A transformer transfers ac power, usually with a step-up or step-down in voltage. Figure 14.4 shows how a lever could be used to step down the velocity of a sinusoidally reciprocating arm. The resistive load on the right-hand side is a dashpot (damper), which produces a reaction force proportional to velocity. A voltage step-down transformer increases current. This lever steps down velocity (and amplitude) and provides increased force. For an ideal transformer (infinite magnetizing inductance) or an ideal lever (zero mass) the input power (primary voltage times primary current or primary velocity times primary force) is equal to the output power at every instant. But, for a real transformer with finite magnetizing inductance, there is also the "excitation" current, lagging the voltage by 90°, pumping energy in and out of the core. Likewise, for a real lever, with nonzero mass, there is an additional component of input force, leading the velocity by 90°, that pumps mechanical kinetic energy in and out of the lever. For both the transformer and the lever the average reactive power is zero but the excitation current or force can be considerable.

Figure 14.4. Mechanical analogue of a transformer.

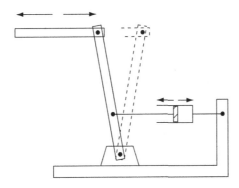

14.5 Magnetizing inductance used in a transformer-coupled amplifier

In Chapter 3 we saw a circuit whose operation cannot be explained if its transformer is modeled as an ideal transformer. That circuit, a transformer-coupled class-A amplifier, is shown in Figure 14.5. Since the transformer windings have almost no dc resistance, the average voltage at the collector must be V_{cc}. Under maximum signal conditions the collector voltage swings between 0 and $2V_{dc}$, applying a peak-to-peak voltage of $2V_{dc}$ to the transformer primary.

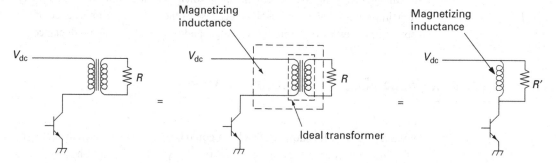

Figure 14.5. Transformer-coupled amplifier.

If we do not include the magnetizing inductance, the transformed load is a pure resistance. We would mistakenly conclude that the quiescent collector voltage must be $V_{cc}/2$ rather than V_{cc} and that the largest peak-to-peak collector signal would be V_{cc} rather than $2V_{cc}$. We would also conclude incorrectly that the frequency response would be unlimited, rather than being limited at low frequencies by the magnetizing inductance and limited at high frequencies by leakage inductance.

14.6 Double-tuned transformer: making use of magnetization and leakage inductances

Leakage inductance and magnetizing inductance limit the performance of transformers used in audio and other baseband applications. But in RF work these parasitic inductances can be tuned out with capacitors. Sometimes the leakage and magnetizing inductance can be intentionally used as in the band-pass filter of Figure 14.6(a).

To see how this circuit works, consider Figure 14.6(b), where the transformer has been replaced by its equivalent circuit. Only the leakage and magnetizing inductances are shown; the ideal transformer in the equivalent circuit of the transformer is either one-to-one or the resistor and capacitor on the right-hand side have been multiplied by its ratio. This equivalent circuit, with its vertical

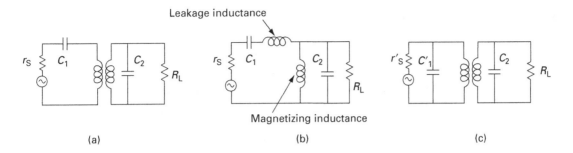

Figure 14.6. (a) Bandpass filter made with a loosely-coupled transformer; (b) equivalent circuit; (c) alternate circuit with two shunt capacitors.

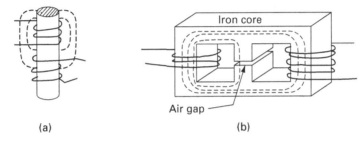

Figure 14.7. Loosly coupled transformers: (a) with powdered iron core; (b) with iron core for low frequencies.

parallel resonator and its horizontal series resonator, is a canonical two-section bandpass filter, as discussed in Chapter 4. The transformer, to have enough intentional leakage inductance, may be air-wound or may be wound on a permeable rod as shown in Figure 14.7(a). The alternate circuit of Figure 14.6(c) uses a parallel capacitor on each side of the transformer. The values of r'_S and C'_1 can be determined from the values of r_S and C_1 by noting that the equivalent Thévenin generator containing r'_S and C'_1 must have an impedance equal to the impedance of $r_S + jC_1$. (Note: whenever you see capacitors across both windings of a transformer, you can guess that the coupling is less than unity – otherwise the two capacitors would be effectively in parallel, and a single capacitor could be used.)

Power transformers are sometimes designed to have intentional leakage inductance to provide short-circuit protection (the leakage inductance limits the current). One way to build an iron core transformer with leakage inductance is shown in Figure 14.7(b). The magnetic path containing the air gap effectively shunts some of the flux generated by one winding from reaching the other winding. This design also provides magnetic shielding, in that the fringing fields are contained within the body of the transformer.

14.7 Loss in transformers

Large power distribution transformers are designed to have efficiencies around 99%. Inexpensive transformers are designed with enough efficiency to avoid premature burn-out. (Plug-in "wall-wart" transformers can be hot to the touch even without a load.) High-frequency transformers have efficiency limits set by core materials and the "skin effect" that excludes high-frequency currents from the interior of conductors, thus increasing the effective resistance of the windings.

Resistance of the windings ("copper loss") is an obvious loss mechanism. We have seen that this can be included easily in a transformer equivalent circuit by simply putting a resistor in series with the primary and another in series with the secondary.

A magnetic core made of a conductive material such as iron will dissipate energy as ordinary I^2R loss since closed paths in the iron core will act as shorted turns around magnetic flux lines. To minimize these *eddy current* losses, any such closed paths are kept short by making the core a stack of thin sheet iron laminations separated by insulating varnish or oxide. The core of a toroidal transformer can be a bundle of insulated iron wire rings, a stack of varnished sheet metal toroids, or a toroid wound of varnished thin metal tape. High-frequency transformers use cores made of magnetic particles, held together in an inert binder material. Eddy current losses can be included in the transformer equivalent circuit as a resistor in parallel with the ideal transformer.

The final loss mechanism comes from magnetic hysteresis in the core. Ideally, the magnetic flux density is proportional to the magnetic force, $B = \mu H$, where μ is the permeability. But B often lags H, as magnetic domains exhibit a kind of static friction before they break loose and reverse their direction. As a result, B vs. H through the ac cycle forms a closed curve whose included area is the energy loss per unit volume per cycle. Hysteresis loss is associated with the magnetizing current, since it is the magnetizing current that produces H, which induces B. We can therefore include hysteresis loss in the equivalent circuit as a resistor in parallel with the magnetizing inductance. This is satisfactory for a power transformer, where the magnetizing current is constant and the frequency is constant. But note that the hysteresis loss is proportional to frequency, since the B-H loop is traversed once per cycle. It also has a nonlinear amplitude dependence. Thus a simple resistor is not adequate to model hysteresis loss in a wideband transformer or a transformer operating over a range of input voltages.

14.8 Design of iron-core transformers

A transformer designer usually strives to find the smallest, lightest, and least expensive (usually synonymous) transformer that conforms to a set of electrical specifications. To see the issues involved, let us consider the design of a 60-Hz

Figure 14.8. Iron-core transformer geometry.

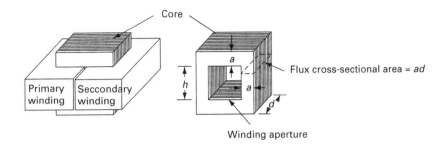

power transformer. Suppose the primary voltage is 220 volts rms and the power delivered to the load is 500 watts. The efficiency is to be 96% and the magnetizing current must be no greater than the "working" (in-phase) current.

We will pick a silicon-steel core material for which the maximum flux density before saturation, B_{max}, is 1.5 webers/m^2. The core, a square toroid, is shown in Figure 14.8. For minimum copper loss (neglecting the excitation current) the primary and secondary windings will have equal loss and will each occupy half of the winding aperture.

The transformer will be specified by four parameters: the number of turns on the primary, N, and the three linear dimensions of the core, a, h, and d. To determine these four parameters we must write equations for the maximum B field, the loss, and the inductance of the primary winding. Faraday's law of induction gives us the maximum B field:

$$V_{max} = N \frac{d\phi}{dt} = N\omega a d B_{max}. \tag{14.3}$$

Since the rms primary voltage is 220, $V_{max} = 220\sqrt{2}$ and we have

$$B_{max} = \frac{220\sqrt{2}}{\omega N a d} < 1.5 \text{ webers/m}^2. \tag{14.4}$$

The copper loss in the primary winding will be $I^2 R_p$ where I is the rms current in the primary and R_p is the resistance of the winding. The number of turns on the primary, N, is given by

$$N = \frac{(h^2/2)}{\sigma} \tag{14.5}$$

where $h^2/2$ is the winding area for the primary and σ is the cross-sectional area of the wire. The mean length per turn is given by $2(a+d+h)$ so the primary resistance is found to be

$$R_p = \rho \frac{\text{length}}{\sigma} = \frac{4\rho N^2(a+d+h)}{h^2}. \tag{14.6}$$

As for core losses, the 60-Hz loss for the selected core material (when B_{max}= 1.5) is 0.6 watts/lb = 11 000 watts/m^3. The overall loss is the sum of the winding losses and the core loss. Since this loss is to be $500(1-0.96) = 20$ watts, we can write

$$\text{Loss}_{\text{watts}} = 11\,000(4ad(a+h)) + \left(\frac{500}{220}\right)^2 \frac{4\rho N^2(a+d+h)}{h^2} = 20. \quad (14.7)$$

Finally, the specification on the magnetizing current is equivalent to specifying that the reactance of the primary, ωL, is greater than the equivalent input load resistance, or

$$\omega L \geq \frac{220^2}{500}. \quad (14.8)$$

The inductance of the primary, L, can be written as

$$L = \frac{\mu N^2(\text{flux area})}{\text{mean flux path length}}. \quad (14.9)$$

The mean flux path length, from Figure 14.8, is $4(h+a)$, so

$$L = \frac{\mu N^2 ad}{4(h+a)}. \quad (14.10)$$

We must use Equations 14.4, 14.7, and 14.8 to find transformer parameters, a, d, h, and N that will satisfy the given specifications and minimize the size of the transformer. This is not quite as simple as solving four equations in four unknowns. The equations are really inequalities and, in general, there will not be a solution that simultaneously produces the maximum allowable flux density, the maximum allowable loss, and the minimum allowable inductance. Instead, this problem in linear programming is most often solved by cut-and-try iterative methods, conveniently done using a spreadsheet program. In this particular example, such a procedure led to the set of parameters: $d = 5$ cm, $a = 2$ cm, $h = 5$ cm and $N = 580$ turns. These dimensions give a core weight of 5.1 lbs, a loss of 19.3 watts and B_{max} of 1.42. The reactance of the primary, assuming a relative permeability of 1000, is 5.9 times the input load resistance – five times more than the minimum required reactance. Note: in the equations presented above, no consideration was made for the space occupied by wire insulation and lamination stacking but these can be accounted for by simply increasing the value of the winding wire resistivity, ρ, and decreasing the permeability.

14.8.1 Maximum temperature and transformer size

The heat generated by a transformer makes its way to the outside surface to be radiated or conducted away. The interior temperature buildup must not damage the insulation or reach the Curie temperature where the ferromagnetism quits (a consideration with high-frequency ferrite cores). "Class-A" insulation materials

Figure 14.9. Weight vs. power rating for 16 commercial 60-Hz power transformers. The data points fit the solid line, for which (weight in lbs) = 0.18 (power in watts)$^{3/4}$.

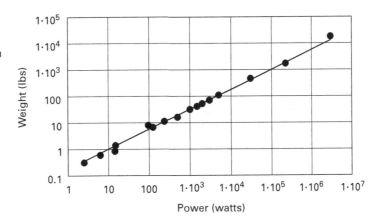

(cotton, silk, paper, phenolics, varnishes) are limited to a maximum temperature of 105 °C. If reliable theoretical or empirically determined equations are available to predict internal temperatures, they can be included in the iterative design procedure described above. Rules of thumb can be used, at least as a starting point to determine core sizes for conventional transformers. One such rule for transformers up to, say, 1 kW is that the flux cross-sectional area of the core ($a \times d$ in Figure 14.8) in square inches should be about $0.25\sqrt{\text{Power in watts}}$. Transformer weight vs. power rating (from catalog specifications) is plotted in Figure 14.9 for sixteen 60-Hz power transformers, ranging from 2.5 watts to 3 megawatts.

The solid line, which fits the data, shows that the weight is proportional to the power raised to the exponent 3/4. Transformer manufacturers seem to use the rule-of-thumb that makes core area proportional to power$^{1/2}$ since this results in the volume and weight being proportional to power$^{3/4}$.

14.9 Transmission line transformers

Leakage inductance and distributed capacitance eventually determine the high-frequency limit of conventional transformers. For wideband applications we cannot simply resonate away these parasitics. Wideband *transmission line transformers* [3] are built like ordinary core-type transformers except that the windings are made with transmission line – either a coaxial cable, as shown in Figure 14.10, or a bifilar winding. The core must have high permeability but modest loss is acceptable (cores in chokes and transformers store little energy so high Q is not necessary). The effect of the core is to choke off any common-mode current in the transmission line, leaving only differential currents. When the transformer is wound with a piece of coaxial cable, as shown in Figure 14.10, the core suppresses current flowing on the outside of the shield,

Figure 14.10. Transmission line reversing transformer.

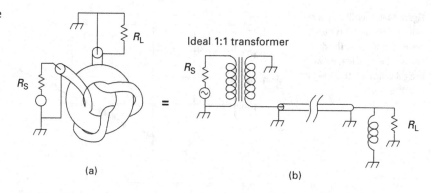

(a)   (b)

leaving only the equal and opposite currents on the inside of the shield and the inner conductor.

This circuit is a reversing transformer, i.e., $V_{out} = -V_{in}$. The polarity flip is achieved by reversing the transmission line connections at the load end where the center conductor is grounded. Normally this would simply short the generator, but the inductance provided by the magnetic core chokes off the otherwise short-circuit current. An equivalent circuit model is shown in (b). Here the reversal is done with an ideal transformer. The length of the coax is the same, to duplicate the additional phase shift between the generator and the load. The inductor in parallel with the load represents the inductance of the cable winding around the core. At the lowest frequencies, this inductor diverts current from the load, just as the magnetization inductance limits the low-frequency response of a conventional transformer. At high frequencies, however, the circuit becomes just a piece of transmission line and its response does not fall off. There is effectively no leakage inductance nor stray capacitances. The time lag through the transmission line, however, will shift the phase from the nominal 180° as the frequency increases. Nevertheless, if an application calls for a pair of signals, identical except for polarity, the "reference" signal can be provided by using an identical piece of transmission line to provide an identical delay.

Transmission line transformers extend the range of ordinary transformers by two octaves or more. In addition to this reversing transformer, many other transformers can be made with the transmission line technique [3, 4]. Commercial hybrids good from 0.1 MHz to 1000 MHz use transmission line transformers. Miniature transmission line transformers are commercially available as standard components.

14.10 Baluns

A balun is any device that converts a *bal*anced (double-ended symmetric) signal into an *un*balanced (single-ended) signal. Baluns are commonly used to feed

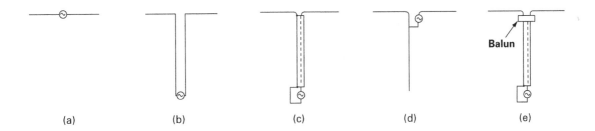

(a)   (b)   (c)   (d)   (e)

Figure 14.11. A symmetric dipole antenna fed (a) at the antenna feed point, (b) with a balanced feedline, (c) with an unbalanced coaxial feedline, (d) equivalent circuit for (c) showing how the antenna is modified by current on the feedline, (e) Balun provides symmetric feed.

symmetric antennas (e.g., dipoles) from unbalanced coaxial feed lines. Figure 14.11 shows what happens if we feed a dipole directly with a coaxial transmission line. In (a), the generator is at the feed point of the dipole so there is no question of balance or imbalance. In (b), a balanced feedline is used. Everything is still symmetric. At any point along the feedline, the current in one side is equal and opposite to the current in the other side. The spacing between the conductors is very small compared to the wavelength, so "cancellation" assures there is negligible radiation from the line. In (c) a coaxial line feeds the dipole improperly; the shield of the coax tied to the left-hand element of the dipole. An equivalent circuit (d) shows how the outer conductor of the coax becomes part of the left-hand dipole element. The antenna now has one straight element and one L-shaped element. The radiation pattern will not be the intended dipole pattern and there will be RF current flowing on the outside of the feedline. In (e) a balun at the end of the coaxial feedline provides equal and opposite voltages to each side of the dipole and eliminates any current from the outside of the feedline.

Figure 14.12 illustrates the requirement for a balun; with equal Z_1 and Z_2, i.e., a load structure symmetric with respect to ground, we want V_1 and V_2 to be equal and opposite with respect to ground.

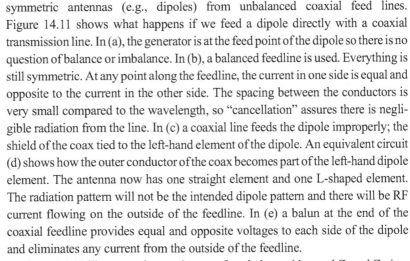

Figure 14.12. Balun operation: unbalanced-to-balanced.

The dotted ground symbol indicates that this point of symmetry will have zero voltage (when $Z_1 = Z_2$) and can be grounded if necessary or desirable. Figure 14.13 shows the equivalent situation with the load at the unbalanced side. When V_1 and V_2 are in phase (common mode) there must be no excitation of Z. But when V_1 and V_2 are 180° apart (differential mode) the load, Z, is fully excited. Baluns are normally reciprocal devices so the name "Unbal" is not needed.

From a transmitting standpoint, the balun eliminates common mode current on the feedline which otherwise would radiate and affect the pattern of the antenna. Figure 14.14(a) shows a reversing transformer used as a balun for this application. Note that this balun is also a 4-to-1 impedance transformer since the voltage across the dipole is $2V$. In Figure 14.14(b), the reversing transformer is replaced by a half-wave length of transmission line. The phase shift through this piece of line transforms V into $-V$ just as the transformer did, but the reversal is only correct over a narrow frequency band. (Note that the half-wave line can have any value for Z_0.)

The simple reversing transformer in Figure 14.14(a) can be replaced by the wideband transmission line transformer of Figure 14.10. The phase shift this transformer picks up at increasing frequencies is compensated by using a

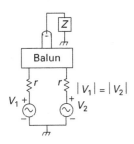

Figure 14.13. Balun operation: balanced-to-unbalanced.

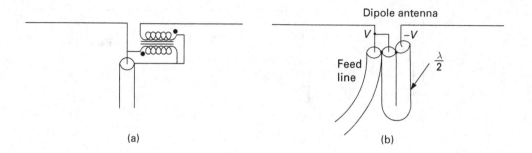

(a)   (b)

Figure 14.14. 4:1 baluns.

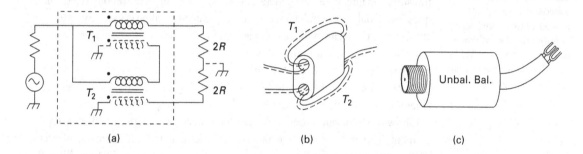

(a)   (b)   (c)

Figure 14.15. Wideband 4:1 balun made from two 1:1 transmission line transformers.

second transformer, identical except with no reversal, to provide an equal frequency-dependent phase shift. The combination of these transformers, connected in parallel at one end and in series at the other end, as shown in Figure 14.15(a), makes a very wideband 4-to-1 balun.

This is the circuit most often found in the television balun of Figure 14.15(c). The transformers are often wound on a "binocular core" (Figure 14.15b). This core operates as two separate cores, i.e., there is nominally no magnetic coupling between the two transformers, T_1 and T_2. For clarity, the figure shows the transformers wound with only two turns; in practice several turns are used.

Problems

Problem 14.1. Use Equations (14.1) and (14.2) to show that when the primary and secondary windings of a transformer are connected in series the total inductance is given by $L = L_1 + L_2 \pm 2M$ where M, the mutual inductance, is given by $M = k\sqrt{L_1 L_2}$ and the $\pm$ changes when one of the windings is reversed. (This is a standard method for measuring mutual inductance.)

Problem 14.2. Consider the following transformer: $L_1 = 0.81\,\text{H}$ (inductance of the primary winding), $L_2 = 1\,\text{H}$ (inductance of the secondary winding), $k=0.9$ (coupling coefficient).

(a) If the secondary is open circuited and 1 volt (ac, of course) is applied to primary, show that the secondary voltage is 1 V. (b) If the primary is open circuited and 1 volt is applied to the secondary, show that the primary voltage is 0.81 V.

Problem 14.3.   Calculate the low-frequency cutoff (half-power frequency) for a resistive load coupled by a particular transformer to a generator. The transformer has perfect coupling and a 1:1 turns ratio. The source and load impedances are both 100 ohms and the reactance of the transformer primary is 50 ohms at 20 Hz.

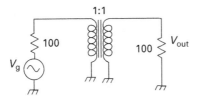

Problem 14.4.   Upgrade your ladder network analysis program (Problem 1.3) to handle conventional transformers. Let the transformer be specified by its primary inductance, secondary inductance, and coupling coefficient.

Example answer: For the MATLAB example solution given in Problem 1.3, add the element, "XFRMR" by inserting the following sequence of statements in the "elseif chain":

```
elseif strcmp(component,'XFRMR')==1
ckt_index=ckt_index+1; Lpri=ckt{ckt_index}; %primary inductance
ckt_index=ckt_index+1; Lsec=ckt{ckt_index};% secondary inductance
ckt_index=ckt_index+1; k =ckt{ckt_index}; %coupling coefficient
V=V+I*(1j*w*Lsec*(1-k^2));
ratio= sqrt(Lpri/(k^2*Lsec));
V=V*ratio; I=I/ratio;
I=I+V/(1j*w*Lpri);
```

Problem 14.5.   A lossless transformer is placed between a 50-ohm signal generator and a 4.5-ohm load. (a) Use your ladder network analysis program (or an equivalent program) to plot the relative power at the load vs. frequency. Use the following parameters: primary inductance=100 μH, secondary inductance=10 μH, coupling coefficient k=0.9. (b) Find the values of a capacitor to be shunted across the primary (i.e., in parallel with the magnetizing inductance) and another capacitor to be placed in series with the secondary (i.e., in series with the leakage inductance) so that the magnetizing and leakage inductances will be cancelled (resonated out) at 0.5 MHz. Plot the resulting frequency response to verify that the transmission is now perfect at 0.5 MHz.

Problem 14.6.   When a power transformer is first turned on, i.e., connected to the line, there is sometimes an initial inrush of current strong enough to dim lights on the same circuit and produce an audible "grunt" from the transformer itself. Decide whether this effect is strongest when the circuit is closed at a zero crossing of the line voltage or at a maximum of the line voltage. (This involves the magnetizing inductance of the transformer so simply analyze the transient when an inductor is connected to an ac line.)

Problem 14.7. (a) Suppose you have a power transformer designed to be fed from 220 V, 60 Hz but you want to use it in a country where the power line supplies 220 V, 50 Hz. Why is the transformer likely to overheat when fed with 50 Hz power? Consider the magnetizing current, copper losses, and core losses. (b) Consider the reverse situation. Would there be any mechanism to cause extra power dissipation if a 50 Hz transformer is used on a 60 Hz line?

Problem 14.8. Two identical perfectly coupled 1:1 transformers are connected in series, i.e., the secondary of the first is connected to the secondary. The primary and secondary inductances of each transformer are L. Show that this combination is equivalent to a single transformer and find its magnetizing inductance L'. If you enjoy algebra, assume the transformers are not perfectly coupled and find L' and k'.

Problem 14.9. Consider a lossless transformer with primary and secondary inductances L_1 and L_2. Suppose the coupling coefficient has a value that results in a 1:1 ideal transformer in the transformer's equivalent circuit. Find the values of the magnetizing inductance and the leakage inductance.

Problem 14.10. The transformer in the figure has a turns ratio of 1:1. The primary and secondary inductances are both L. The amplitude of the sine wave from the generator is V_0. Assume the transformer has no leakage inductance. Find an expression for the current in the resistor. Hint: use the equivalent circuit for the transformer: an inductor together with an ideal transformer.

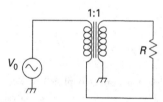

References

[1] Flanagan, W. M. *Handbook of Transformer Design & Applications*, 2nd edn. New York: McGraw-Hill, 1993.
[2] McLyman, Col. W. T. *Transformer and Inductor Design Handbook*, 3rd edn, Bora Rotan: CRC Press, 2004.
[3] Ruthroff, C. L., Some broadband transformers, *Proceedings of the IRE*, August 1968, pp 1357–1342.
[4] Sevick, J., *Transmission Line Transformers*, Newington CT: American Radio Relay League, 1987.

15 Hybrid couplers

Hybrid couplers, also known as hybrid junctions or simply "hybrids," are lossless passive four-port devices used to make interconnections between circuit elements. Hybrids are used as power dividers ("signal splitters") and combiners. They are also used in mixers and TR (transmit/receive) switches. A useful schematic representation for a hybrid, Figure 15.1, shows the four connection points (ports).

RF hybrids usually have unbalanced ports designed for coaxial transmission lines, so all four ports share a common ground, indicated in Figure 15.1 by a dotted ground symbol (usually not shown). Each port has a characteristic impedance. Most packaged RF hybrids with coaxial ports are made so that the characteristic impedance of all four ports is 50 or 75 ohms. The symbol in Figure 15.1 shows signal flow paths; power incident on Port 1 splits and exits through Ports 2 and 3. If both of these ports are properly terminated there will be no reflections and the impedance seen looking into Port 1 will be equal to the characteristic impedance of that port. In this case no power will reach Port 4 as opposite ports are isolated. But if Port 2 and/or Port 3 are not terminated in their own characteristic impedances, the power exiting these ports will be partially or completely reflected back into the hybrid. The reflection, which depends on the mismatch, is calculated exactly as if the power had exited from a transmission line whose impedance is equal to that of the respective port. Any power reflected back into the hybrid splits and follows the signal paths, just as if it had come from an external source. You can see that, with arbitrary terminations and arbitrary signals, the situation could become complicated. But usually we deal with continuous wave (cw) sinusoidal signals so, rather than analyze multiple reflections in the time domain, we only have to solve for the forward and reverse wave amplitudes on each of the four internal paths. In most applications, things are even simpler; hybrids usually have proper terminations and the signal flows are simple and can be determined by inspection of the signal flow diagram.

15.1 Directional coupling

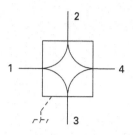

Figure 15.1. Schematic symbol for a hybrid coupler.

From inspection of the signal paths in Figure 15.1, we see that with the ports matched and with power flowing from Port 1 to Port 2 there will also be power flowing out of Port 3 but none out of Port 4. If the power is now reversed, to flow from Port 2 to Port 1, there will be power flowing out of Port 4 but none from Port 3. Port 3 is therefore coupled to power flowing from 1 to 2. Likewise, Port 4 is coupled to power flowing from 2 to 1. Therefore, a hybrid is a *directional coupler*, and can be used to determine how much power is flowing in each direction on a transmission line. Here we will use the term *hybrid* only for 3-dB directional couplers, i.e., directional couplers that split the incident power in half.

15.2 Transformer hybrid

The name *hybrid transformer* was first applied around 1920[1] to the simple center-tapped transformer shown in Figure 15.2. The primary winding has N turns while each half of the secondary winding has $N/\sqrt{2}$ turns. For this transformer hybrid (hybrid composed of a transformer), the characteristic impedances of Ports 1, 2, and 3 are equal and are twice the impedance of Port 4. (Here the port impedances are R, R, R, and $R/2$, where the value of R is arbitrary, as long as the transformer behaves as an ideal transformer, i.e., its magnetizing inductance has a reactance substantially larger than R.)

Let us confirm that this circuit has the power splitting and isolation characteristics of a hybrid. First consider a signal connected to Port 1. If Ports 2 and 3 have identical terminations, they will have equal and opposite voltages. The voltage at Port 4, since it is midway between the voltages at Ports 2 and 3, must be zero and Port 4 is indeed isolated from Port 1. Next note that a signal applied to Port 4 will appear unchanged at Ports 2 and 3 but will not appear at Port 1. (The currents to Ports 2 and 3 are in opposite directions so there is no net flux in the transformer to provide a voltage at Port 1 or produce an IX_L drop.) You can verify that Ports 2 and 3 are also isolated from each other (see Problem 15.1). Figure 15.2 also shows the symbol appropriate

Figure 15.2. Transformer hybrid.

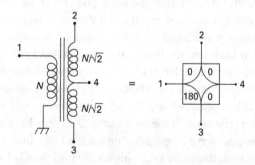

---

[1] The origin of the term seems to be lost – a hybrid of what? – but it came from the telephone industry, where the terms *hybrid transformer* and *hybrid coil* were both common.

Figure 15.3. Hybrids allow full-duplex communication over a single line.

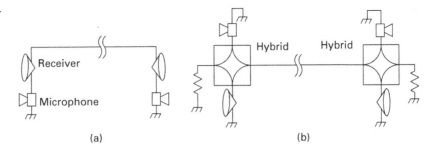

(a)   (b)

for this hybrid. The labels 0, 0, 0, and 180 indicate the phase shifts through the respective branches. A signal incident on Port 1, for example, appears at Port 2 with the phase unchanged (shifted 0°) and at Port 3 with its polarity inverted (shifted 180°). Any hybrid with these four phase shifts is called a *180° hybrid*.

15.2.1 Applications of the transformer hybrid

In telephony or other wired communication, hybrids allow a transmission line to carry independent signals in each direction. Figure 15.3 shows two telephone circuits. In the simple series circuit of Figure 15.3(a), each user hears his own voice as well as the voice from the other end. In the circuit of Figure 15.3(b), hybrids isolate each receiver from its own microphone.

If we are using the transformer hybrid of Figure 15.2, we would terminate Port 4 with a resistor of value $Z_0/2$, where Z_0 is the characteristic impedance of the phone line. The microphones and receivers must have impedances equal to Z_0. This arrangement provides two-way signaling , "full duplex," over a single cable.[2]

The circuit of Figure 15.4 uses two hybrids and two amplifiers to make a bidirectional repeater for a long (lossy) line. Here the hybrids let the signals in each direction be independently amplified without feedback and consequent oscillation.

It is convenient to make the characteristic impedances be the same for all four ports of a general-purpose hybrid. The transformer hybrid, fixed up to have equal impedances, is shown in Figure 15.5. This is the kind of circuit found inside an off-the-shelf wideband 3 dB hybrid. Transformer hybrids made with toroidal cores (ferrite beads) can work over large bandwidths, e.g., 10 KHz to 20 MHz and 1 MHz to 500 MHz.

[2] In telephony, the circuit is deliberately unbalanced – just enough so that the users, hearing their own voices or ambient noise, will sense that the call is connected, but not enough that the users hold the receiver (and hence the microphone) away from their heads. Of course cancellation should be as great as possible when this kind of full-duplex circuit carries two-way digital data.

Figure 15.4. Two-way telephone repeater for long lines.

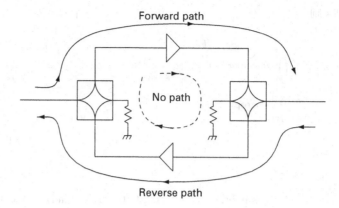

Figure 15.5. Two-transformers make a hybrid with the same impedance at all four ports.

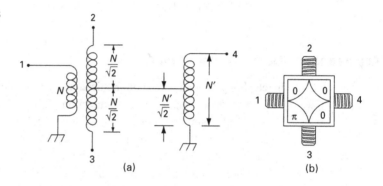

Hybrids are often used in circuits like those illustrated above, as well as signal splitting and combining, where one port is terminated in its characteristic impedance. An easy way to terminate Port 1 of the hybrid of Figure 15.5 is to put a resistor of value $2Z_0$ between Ports 2 and 3. When this is done, the Port 1 winding on the hybrid transformer can be eliminated. The resulting circuit, shown in Figure 15.6a, is

Figure 15.6. Internally terminated hybrid is a two-way splitter/combiner.

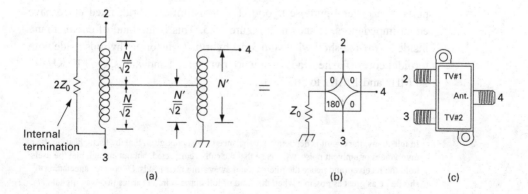

commonly found in the signal splitters used to connect two receivers to a single antenna (c) and in other packaged 2:1 splitter/combiners. When it is acceptable for the impedance of Port 4 to be $Z_0/2$, the right-hand transformer can be omitted, and the hybrid consists only of the center-tapped *hybrid coil.*

15.3 Quadrature hybrids

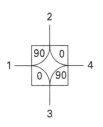

Figure 15.7. Symbol for a 90° hybrid.

The transformer hybrid is naturally a 180° hybrid. Other circuits are natural 90° hybrids, the symbol for which is shown in Figure 15.7.

Let us look at an interesting application of the 90° hybrid (often called a *quadrature hybrid*). Here the internal phase paths are zero and 90°. A 90° path means a phase shift equal to that produced by a quarter-wave length of cable. For circuit analysis, one deals with hybrids in terms of voltages. We will consider the hybrid to be connected to transmission lines (of the same impedance as the hybrid) in order to describe the signals in terms of incident and reflected waves. A signal incident at Port 1 will be split equally into signals exiting Ports 2 and 3. Since the power division is equal, the magnitudes of the voltages of the signals exiting Ports 2 and 3 will be $1/\sqrt{2}$ times the magnitude of the incident voltage. The phases of the exiting signals will be delayed as indicated on the symbol for the hybrid. For the hybrid of Figure 15.7, the signal exiting Port 3 has no additional phase shift but the signal exiting Port 2 is multiplied by $e^{-j\pi/2}$. Suppose a signal is also incident at Port 4. It will also split into signals exiting from Ports 2 and 3. The total voltage of the waves exiting Ports 2 and 3 is just the superposition of the waves originating from Ports 1 and 4.

15.3.1 Balanced amplifier

A common application for 90° hybrids is the balanced amplifier circuit shown in Figure 15.8. As long as the two amplifiers are identical they can have arbitrary input and output impedances but the overall circuit will have input and output impedances of Z_0.

To see how this happens, suppose that the hybrids are 50-ohm devices but that the input impedance of the amplifiers is not 50 ohms. Imagine that the interconnections are made using 50-ohm transmission line. The input lines have equal lengths and the output lines have equal lengths. The amplifiers are identical so the two signals have equal phase changes upon reflection. An input signal is split by the input hybrid; half the power will be incident on the top amplifier and half on the bottom amplifier. But reflections from the amplifiers will be out of phase by 180° when they arrive back at the input of the hybrid because the signal on the upper path will have made a round trip through the 90° arm of the hybrid. The two reflections therefore cancel and there is no net

Figure 15.8. A balanced amplifier has constant input and output impedances.

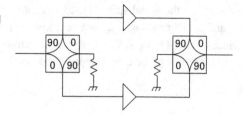

reflection. The input impedance of the overall amplifier will be the characteristic impedance of the hybrid (the value of the resistors, if transformer hybrids are used). The output side works the same way, and this combination of two arbitrary but identical amplifiers produces an amplifier with ideal constant input and output impedances.

15.4 How to analyze circuits containing hybrids

Let us work out an example problem to illustrate the way voltages are added in the manner of transmission line analysis. In Figure 15.9, a 50-ohm hybrid has arbitrary impedances, Z_2 and Z_3 terminating Ports 2 and 3. Port 4 is terminated in 50 ohms, so no power exiting Port 4 will be reflected back into the hybrid. We want to find the impedance seen looking into Port 1. Rather than work directly with impedances, we work with the equivalent reflection coefficients, $\Gamma = (Z - Z_0)/(Z + Z_0)$. Once we have found Γ_{in}, it can be converted to Z_{in}. The input signal is denoted by the forward arrow at Port 1. We will give it unity amplitude.

Figure 15.9. An example circuit problem: finding the impedance looking into Port 1.

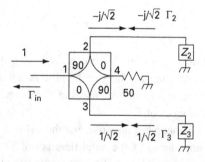

The signal incident on Z_2 is therefore $(1/\sqrt{2})e^{-j\pi/2} = -j/\sqrt{2}$ where $1/\sqrt{2}$ is the reduction in amplitude due to the equal power split and $-\pi/2$ is the phase shift of the 90° path between Ports 1 and 2. The signal reflected back into Port 2 is $[-j\sqrt{2}]\Gamma_2$, i.e., the signal incident on Z_2 multiplied by the reflection coefficient of Z_2. This reflected signal will split as it enters Port 2 and its contribution to the wave leaving Port 1 will be its amplitude multiplied by $(1/\sqrt{2})e^{-j\pi/2}$ as it is split and phase shifted by the 90° path or $[-j(\sqrt{2})]\Gamma_2 \times -j(\sqrt{2}) = -\Gamma_2/2$.

Figure 15.10. A balanced amplifier built with 180° hybrids.

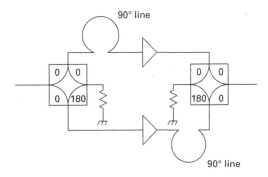

Figure 15.11. Conversions between 90° and 180° hybrids

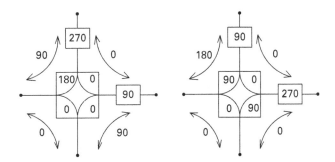

Similarly, the contribution to Γ from the reflection at Z_3 is $\Gamma_3/2$, and the overall reflection is given by $\Gamma = -\Gamma_2/2 + \Gamma_3/2$, which solves the problem. The reader can work out the general case where Port 4 is also terminated by an arbitrary impedance, Z_4. Hint: let V_2, V_3, and V_4 be the wave amplitudes flowing out of Ports 2, 3, and 4. The waves flowing into these ports will therefore have amplitudes $\Gamma_2 V_2$, $\Gamma_3 V_3$, and $\Gamma_4 V_4$. By examination of the circuit, write three equations for V_2, V_3, and V_4, the three unknowns.

A balanced amplifier can also be built with 180° hybrids if two 90° lines are added as shown in Figure 15.10.

This use of cables to allow substitution of 180° hybrids for 90° hybrids in this circuit can be taken much farther; any hybrid can be converted to any other hybrid by adding lengths of transmission line to the ports. Figure 15.11 shows how to convert a 180° hybrid into a 90° hybrid and vice versa.

15.5 Power combining and splitting

An obvious way to power N loads from a single 50-ohm source is to transform the impedance of each load to $50N + j0$ ohms and then connect the transformed

Figure 15.12. Hybrids used as power combiners.

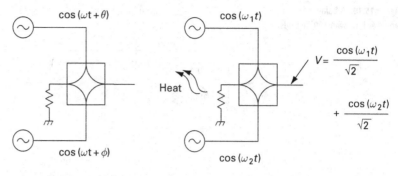

(a) Combining equal frequencies   (b) Combining different frequencies

loads in parallel across the generator. However, if any load changes, the power delivered to the other loads will change. A similar argument applies to combining the power from several sources: if any source changes amplitude or phase, the combined output will change. Hybrids provide a way to combine or split power without using impedance transformation and in a way that isolates multiple sources or multiple loads from each other.

To use a hybrid as a two-input power combiner, the unused port is terminated, either externally, as shown in Figure 15.12 or internally, as shown in Figure 15.6. The two signals to be combined should have equal amplitudes, as well as the correct phase relationship, to steer the total available power into the desired port. If the phase difference is changed by 180°, all the power will flow into the terminated port. Note that if one source fails, the other source will not know it; it still sees a matched load. This provides a fail-safe circuit, although the power output will drop by 75%.

When a hybrid is used to combine two signals of different frequencies, as in Figure 15.12(b), half the power of each signal will always be lost in the terminated port. Circuits to combine signals of different frequencies without loss are known as diplexers. (How would you make one?)

Power splitting is, of course, just time-reversed power combining. When hybrids are used as splitters and the source impedance is equal to the port impedance, the signal at any output port will remain constant when the loads on the other output ports are removed, shorted, or changed in any way.

15.5.1 Hybrid trees

Trees of hybrids are often used to make multi-input combiners or multiple-output splitters, as shown in Figure 15.13. High-power transmitters often use such tree structures to combine the power of many low-power solid-state amplifiers.

These trees have the same advantages as single hybrids, when used for power combining or splitting. A 2^N-to-1 combiner/splitter has $2^N - 1$ internal hybrids.

Figure 15.13. A tree of hybrids.

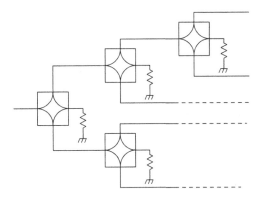

If you buy a three-way TV antenna splitter, you may find that each output provides only one quarter, rather than one third of the input power. What would have been a fourth output port is internally terminated. Or, if the splitter contains two hybrids, one output port can supply half the input power while the other two output ports can each supply one quarter of the input power.

15.6 Other hybrids

There are many ways to make hybrids without transformers. Most of them are circuits whose elements are transmission lines or capacitors and inductors. Unlike the ideal transformer hybrid, these hybrids are all frequency dependent – perfect hybrids only at their center frequency. However, most have a useful bandwidth of about an octave, which is usually sufficient for RF and microwave applications. You can use straightforward circuit analysis to analyze these hybrids.

15.6.1 Wilkinson power divider (or combiner)

This hybrid, shown in Figure 15.14 (in a 50-ohm version), uses two quarter-wave pieces of 70.7 $(50\sqrt{2})$ ohm transmission line. It has only three external ports; the fourth port is internally terminated, that is, connected to a load equal to its characteristic impedance.

It is easy to see that power applied to Port 1 will divide between Ports 2 and 3. By symmetry, the voltages at Ports 2 and 3 must be identical so no power is dissipated in the internal termination. Fifty-ohm loads at Ports 2 and 3 are transformed by the 90° cables to 100 ohms. The parallel connection of 100 ohms and 100 ohms at Port 1 produces the desired 50-ohm input impedance. The Wilkinson divider is usually classified as a 180° hybrid since its outputs have the same phase, even though this phase is 90° rather than 0°.

Figure 15.14. Wilkinson power divider.

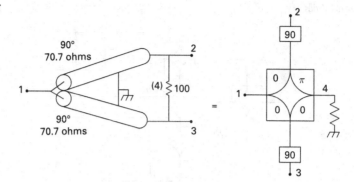

15.6.2 Ring hybrid

The ring hybrid, shown in Figure 15.15, uses four pieces of transmission line. To have 50-ohm ports, the hybrid must be built from 70.7-ohm transmission line.

Figure 15.15. Ring hybrid.

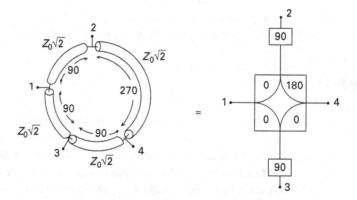

Branch-line hybrids

These are ladder networks made of quarter-wave lengths of transmission line. The simplest is shown in Figure 15.16. More complicated versions have more branches and provide more bandwidth.

Figure 15.16. Branch line hybrid.

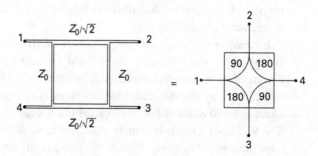

15.6.3 Lumped element hybrids

Two examples of 50-ohm lumped element hybrids are shown in Figure 15.17. These two circuits are obtained by replacing the Z_0 and $Z_0/\sqrt{2}$ arms of the simple branch line hybrid with the pi (or T) lumped circuit impedance inverters discussed in Chapter 13.

Figure 15.17. Two lumped-element hybrids.

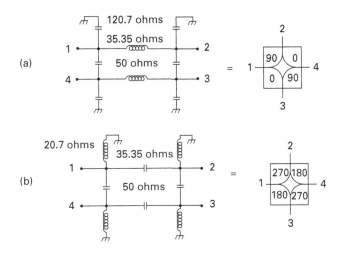

15.6.4 Backward coupler

The *backward coupler* is shown in Figure 15.18 in shielded pair and stripline versions. The coupled transmission lines have both electric and magnetic coupling. When power flows from left to right, between Ports 1 and 2, coupled power flows left (backwards) out of Port 3. This type of coupler can be designed so that the voltage coupling coefficient, c, between Ports 1 and 3 is anywhere between 0 and 1. If designed for $c = 1/\sqrt{2}$, the coupler is a hybrid (a 3-dB directional coupler). This coupler can be analyzed in terms of its common mode and differential characteristic impedances, Z_{CM} and Z_{DIFF}, which are defined as follows. If the two inner conductors are driven together as a single (though split) center conductor, the characteristic impedance of the line is Z_{CM}. When the two inner conductors are regarded as a balanced shielded transmission line, its characteristic impedance is Z_{DIFF}. Two equivalent parameters are defined: $Z_{EVEN} = 2 Z_{CM}$ and $Z_{ODD} = Z_{DIFF}/2$. Analysis using simultaneous forward and reflected common mode and differential mode signals (see Problem 15.8) produces the formulas:

$$Z_{ODD} = Z_0([1 - c]/[1 + c])^{1/2} \text{ and } Z_{EVEN} = Z_0([1 + c]/[1 - c])^{1/2} \quad (15.1)$$

Figure 15.18. Backward couplers (coupled transmission line hybrids).

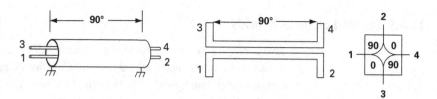

where Z_0 is the desired port impedance. However, these are only indirect design formulas, since electromagnetic analysis is needed to find physical dimensions to produce the required values for Z_{ODD} and Z_{EVEN}.

The *magic-T* of Figure 15.19 is a four-port waveguide junction that combines an *E*-plane tee junction with an *H*-plane tee junction. This waveguide hybrid, whose operation is surprisingly simple, is discussed in Chapter 16.

Figure 15.19. Waveguide magic-T hybrid.

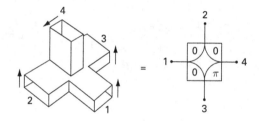

Problems

Problem 15.1. Verify for the transformer hybrid of Figure 15.2 that Port 2 is isolated from Port 3, i.e., show that with a generator connected to Port 3, the voltage at Port 2 will be zero provided the impedance terminating Port 1 is twice the impedance terminating Port 4.

Problem 15.2. Explain why the duplex telephone circuit and the two-way telephone repeater could be built with either 90° or 180° or any other hybrids.

Problem 15.3. (a) Calculate the overall gain of the balanced amplifier of Figure 15.8 if the (identical) individual amplifiers each have gain G_0. (Answer: G_0)

(b) Calculate the overall gain if one of the interior amplifiers is dead, i.e., has zero gain. (Answer: $G_0/4$.)

Problem 15.4. The 50-ohm hybrid shown below is properly terminated by 50-ohm resistors on two of its ports. The third port is terminated by a 25-ohm resistor. If a 50-ohm generator is connected to the remaining port, what fraction of the incident power will be reflected back into the generator? (Answer: 1/36.)

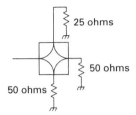

Problem 15.5. In the figure below, a 50-ohm hybrid is fed power from two amplifiers. The signals from these amplifiers have the same frequency and the same phase but the upper amplifier supplies 1 watt while the lower amplifier supplies 2 watts. In this seemingly unbalanced configuration, how much power reaches the load resistor? (Answer: 2.91 watts.)

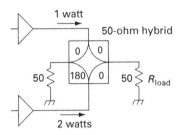

Problem 15.6. Four identical 1-watt amplifiers and six hybrids are used to make a 4-watt amplifier. If one of the interior amplifiers fails, how much power will be delivered to the load? (Answer: 2.25 watts.)

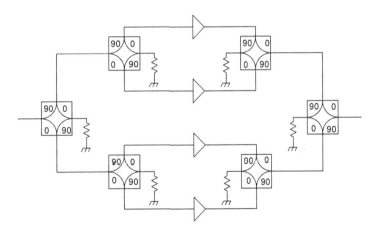

Problem 15.7. Design an op-amp circuit to replace the telephone hybrids in Figure 15.3(b). Put a low-value resistor in series with the line so that a differential

amplifier will produce a voltage proportional to the current $\times Z_0$. Use op-amps to produce signals that are the sum and difference of this voltage and the line voltage.

Problem 15.8.   Analyze the coupled-line hybrid to derive Equations (15.1) and (15.2). Let $V_a(z)$ and $V_b(z)$ denote respectively the voltages on the bottom conductor and top conductors. You can write these voltages as the superposition of a common-mode ("even") wave and its reflection plus a differential-mode ("odd") wave and its reflection. Putting $z = 0$ at the right-hand end, confirm that these voltages take the form

$$V_a(x) = V_e e^{-jkx} + V_e \Gamma_e e^{jkx} + V_0 e^{-jkx} + V_0 \Gamma_0 e^{jkx}$$

$$V_b(x) = V_e e^{-jkx} + V_e \Gamma_e e^{jkx} - V_0 e^{-jkx} - V_0 \Gamma_0 e^{jkx}$$

where, as usual, $k = 2\pi/\lambda$. The reflection coefficients, Γ_e and Γ_0, are calculated in terms of the even and odd characteristic impedances, $Z_e = 2 Z_{CM}$ and $Z_0 = Z_{DIFF}/2$, in the usual way, i.e., $\Gamma_{e,0} = (Z_0 - Z_{e,0})/(Z_0 + Z_{e,0})$ (see Chapter 10). Note that Z_0 is the *load* impedance for the even and odd waves.

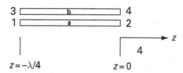

Assume a unity amplitude wave incident at Port 1, i.e., $V_a(-\lambda/4) = 1$. The voltage at the isolated port must be zero, i.e., $V_b(0) = 0$. Use these two equations to find V_e and V_0. Then use V_e and V_0 to show that the coupled voltage, $V_3 = V_b(-\lambda/4)$ is given by $c = (Z_e - Z_0)/(Z_e + Z_0)$. Finally, impose the (match) condition that $I_1 = 1/Z_0$, i.e., $I_a(-\lambda/4) = 1/Z_0$, to show that $Z_e Z_0 = Z_0^2$.

Reference

Montgomery, C. G., Dicke, R. H. and Purcell, E. M., *Principles of Microwave Circuits*, Volume 8 of the MIT Radiation Laboratory Series, New York: McGraw Hill, 1948. Reprinted London: Peter Peregrinus, 1987.

16 Waveguide circuits

In this chapter we examine rectangular metal waveguides and, in particular, their most common mode of operation, the fundamental "TE_{10}" mode. We will also see how the concepts developed for two-conductor transmission lines apply to waveguides and look at waveguide versions of some low-frequency components.

The ability of a hollow metal pipe to transmit electromagnetic waves can be demonstrated by holding it in front of your eye. You can see through it, so, at least, it passes electromagnetic waves of extremely short wavelengths. From a purely dimensional analysis, you would guess correctly that the longest wavelength a pipe could transmit must be of the order of the pipe's transverse dimensions. It turns out that, for propagation in a rectangular pipe, the free-space wavelength, c/f, must be less than twice the longer transverse dimension and, for a circular pipe, less than 1.706[1] times the diameter. Waveguides have less loss and more power handling capacity than coaxial lines of the same size and they need no center conductor nor insulating structures to support a center conductor. Metal waveguides are used most often in the range from 1000 MHz to 100 GHz, where they have practical dimensions. Waveguides for optical frequencies are coated glass fibers.

16.1 Simple picture of waveguide propagation

A common RF engineering argument for the plausibility of transmitting electromagnetic waves through a hollow metal pipe is shown in Figure 16.1, where a two-conductor transmission line evolves into a waveguide. Quarter wave shorted stubs are added to the line. Since a shorted quarter-wave line presents an open circuit, these stubs do not short the line. More stubs are added to both sides until a rectangular pipe is formed.

This plausibility argument, while not rigorous, does illustrate some important properties of waveguide propagation in the fundamental mode (the

---

[1] The factor 1.706 is $\pi/1.841$, where 1.841 is the smallest root of the equation $d/dx\, J_1(x) = 0$ and $J_1(x)$ is the first-order Bessel function of the first kind, a function whose shape resembles $\sin(x)/(x+1)^{1/2}$.

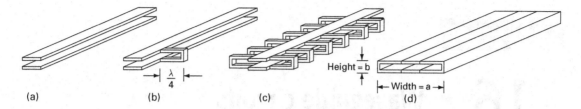

Figure 16.1. Transmission line-to-waveguide evolution. Shorted quarter-wave stubs do not short the transmission line.

simplest and lowest frequency mode): the electric field, which is essentially vertical between the conductors in Figure 16.1(a), becomes perfectly vertical in the waveguide, though its magnitude must fall to zero at the waveguide's sides, since the metallic walls short out any tangential electric field. And, just as the conductors of the transmission line of Figure 16.1(a) can have any separation and still support wave propagation, the waveguide of Figure 16.1(d) can have any height. The width, however, is critical. The total width of the guide must be at least $\lambda/2$ to accommodate a quarter-wave stub on each side and still have nonvanishing conductor strips, as shown in Figure 16.1(c). This means that there is a low-frequency cutoff; wave propagation is not possible if the wavelength is greater than $2a$ where a is the waveguide width (the longer dimension).

Standard waveguide designations indicate the shape and size of the guide. WR430, for example, denotes "Waveguide, Rectangular," 4.3 inches (10.9 cm) wide. The standard width-to-height ratio is two-to-one. (While the height of the guide can be made arbitrarily small, the waveguide will become increasingly lossy because, for a given power, the currents increase.) One of the largest standard waveguide sizes, WR2300, with a width of 23 inches (58.4 cm), has a low-frequency cutoff of 257 MHz. One of the smallest, WR3, with a width of 0.03 inches (0.076 cm), has a low-frequency cutoff of 197 GHz.

For a standard (width = 2 × height) waveguide, the fundamental mode, called the TE_{10} mode, is the only possible mode for frequencies above the low-frequency cutoff and below twice the low-frequency cutoff. Other modes exist above this one-octave range. At frequencies where higher modes are possible, these modes can be unintentionally excited at sharp bends, robbing power from the desired mode. This power does not couple properly to circuit elements designed for the fundamental mode and dissipates in the walls. Whenever possible, a microwave system designer therefore tries to use only the fundamental mode.

The essential details of this most important mode are derived and discussed below.

16.2 Exact solution: a plane wave interference pattern matches the waveguide boundary conditions

Exact solutions for the E and B fields within waveguides of arbitrary shape are normally deduced through a head-on assault using Maxwell's equations.

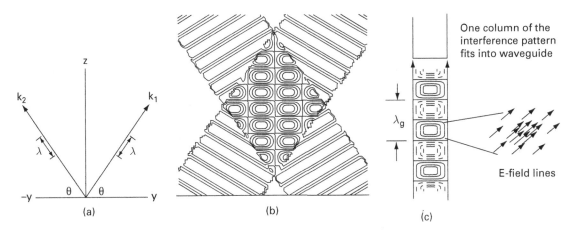

Figure 16.2. Superposition of two plane waves (a) produces an interference pattern (b) streaming in the z-direction. One (or more) columns of that pattern satisfies waveguide boundary conditions (c). E is in the x-direction (coming out of the page).

However, for rectangular waveguides, an exact solution can be obtained indirectly by setting up two plane waves in empty space.

A single plane wave satisfies Maxwell's equations inside a waveguide, but cannot satisfy the boundary conditions at the metal walls. However, the superposition of two properly chosen plane waves forms a traveling interference pattern which does satisfy the boundary conditions and is therefore a valid solution to the waveguide problem. This solution technique can be compared with the "image charge" method in electrostatics, often introduced as a technique to solve for the electric field when a point charge sits alone above an infinite metal sheet. An equal but opposite charge is placed at the mirror image point behind the sheet and the sheet is removed. The electric field lines connecting the charges pass perpendicularly through the x–y plane, satisfying the boundary condition that the E field must be perpendicular when intersecting a conducting surface. The superposition of the two fields, that of the actual charge and that of the image charge, is the solution to the original problem.

Here we construct a solution by superposing two plane waves, identical except for their propagation directions. Both plane waves will be polarized in the x-direction, i.e., their electric fields are in the x-direction. Since electromagnetic waves in free space are transverse waves, the propagation vectors, k_1 and k_2,[2] corresponding to these waves must both lie in the y–z plane, as shown in Figure 16.2(a).

The first wave, with propagation vector k_1, travels in the NNE direction, while the other, k_2, travels NNW. In (b), the plane waves are drawn as streams with finite width. Contour lines of the electric field are perpendicular to the directions of propagation. Figure 16.2(b) shows that, in the area where the streams overlap, the sum of the individual electric fields produces an *interference pattern* consisting of columns of cells which stream northward in the z-direction. If we could watch two waves come together in the ocean, we

---

[2] By definition, the propagation vector, k, is in the direction of travel, i.e., perpendicular to the wavefront, and has magnitude $|k| = 2\pi/\lambda$.

would see them produce this interference pattern. As the streams leave the overlap region, they recover their original plane wave form. In the interference pattern, the contour lines of constant E resemble squared-off ovals. (Remember that E is always perpendicular to the page; the oval-like figures are contours of field strength; they are not field lines.)

The pattern formed by a column of cells (Figure 16.2c) solves the waveguide problem if the width of the cells is equal to a, the width of the waveguide. Why is this a solution? First, the electric field at the side walls is zero at all times, satisfying the boundary condition that, at a conducting wall, there can no parallel electric field. Second, the electric field is always in the x-direction, so it is perpendicular where it intersects the top and bottom walls, satisfying the boundary on those walls as well. Third, each plane wave and therefore their sum, is a solution to Maxwell's equations in empty space, i.e., the interior of the waveguide. What about all the columns of cells outside the boundary of the waveguide? We can forget them, just as we ignore the electric field on the image charge side of the x–y plane in the electrostatic example.

Let us apply a little algebra to find the wave's propagation vector, cutoff frequency and phase velocity. Let k denote the magnitude of k_1 and k_2 so that $k_1 = k\cos(\theta)\,\hat{y} + k\sin(\theta)\,\hat{z}$ and $k_2 = -k\cos(\theta)\,\hat{y} + k\sin(\theta)\,\hat{z}$. The electric field is the sum of the fields of the two waves, i.e.,

$$E(r, t) = \frac{-E}{2j}\,e^{j(\omega t - k_1 \cdot r)} + \frac{E}{2j}\,e^{j(\omega t - k_2 \cdot r)}, \tag{16.1}$$

where the vector r denotes position in the y–z plane and E is a constant equal to the twice maximum electric field of each wave. The amplitudes, $-E/(2j)$ and $E/(2j)$, have been chosen so that $y = 0$ will be a column boundary and also to phase the E field to be maximum at $z = 0$ when $t = 0$. Substituting the expressions for k_1 and k_2, we have

$$\begin{aligned} E &= E\left(-e^{(j\omega t - k(z\cos\theta + y\sin\theta))} + e^{j(\omega t - k(z\cos\theta - y\sin\theta))}\right)/(2j) \\ &= E\sin(ky\sin\theta)e^{j(\omega t - kz\cos\theta)}. \end{aligned} \tag{16.2}$$

As always, it is the real part of E that is the actual electric field. Note that all the y dependence is contained in the expression $\sin(k\sin(\theta)\,y)$. This is independent of t, so, in the y-direction, the diffraction pattern is a standing wave. The z and t dependence, however, are contained in the wave factor $e^{j(\omega t - kz\cos\theta)}$, so the entire diffraction pattern propagates in the $(+z)$-direction with an effective wavevector $k_{\text{guide}} = k\cos\theta$. For a wave of a given frequency, we can find the value of θ that satisfies the side-wall boundary condition. Suppose that the bottom wall of the waveguide extends from $y = 0$ to $y = a$, i.e., the guide width is a. The boundary condition at $y = 0$ is satisfied for any value of θ, since $\sin(0) = 0$. But the boundary condition at $y = a$ demands that $\sin(k\sin(\theta)\,a) = 0$. This will be satisfied if $k\sin(\theta)\,a = n\pi$, where n is an integer. Here we will let $n = 1$, so that $\sin(\theta) = \pi/(ka)$ and $\cos(\theta) = (1 - [\pi/(ka)]^2)^{1/2}$. Thus the E field in the waveguide is

Figure 16.3. Electric field configuration in the TE$_{10}$ mode.

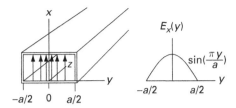

given by (the real part of) $E = E \sin(\pi y/a)e^{j(\omega t - k_g z)}$ where $= k\cos(\theta) = k(1-[\pi/(ka)]^2)^{1/2}$. Figure 16.3 shows how the magnitude of the E field is a maximum at the center line of the waveguide and falls to zero at the side walls.

16.2.1 Guide wavelength

From the expression for k_g, we see that the spatial period along the waveguide will be $2\pi/k_g$. Solving for this length, known as the *guide wavelength*, we find

$$\lambda_{\text{guide}} = \frac{\lambda}{\sqrt{1 - \left(\frac{\lambda}{2a}\right)^2}}. \tag{16.3}$$

16.2.2 Magnetic field

Just as the electric field in the guide is the superposition of the electric fields of two plane waves, the magnetic field is the superposition of their magnetic fields. For a plane wave, the magnetic field is perpendicular to both the electric field and the propagation vector,

$$\boldsymbol{B} = \hat{k} \times \mathbf{E}/c, \tag{16.4}$$

where c is the speed of light. The magnetic fields of our two plane waves have z-components as well as y-components, so the magnetic field in the waveguide is not purely transverse with respect to the direction of propagation. In this TE$_{10}$ mode and all other TE modes, only the electric field is purely transverse. There are also TM modes, in which only the magnetic field is purely transverse. Waveguides, unlike coaxial cable, have no TEM modes, in which both E and H fields are transverse.

We can use Equation (16.4) to find the magnetic of field each plane wave and then sum them to get the field in the waveguide. The result is

$$\frac{B_y}{E} = \frac{k_g}{\omega} \sin\left(\frac{\pi y}{a}\right) e^{j(\omega t - k_g z)} \tag{16.5}$$

and

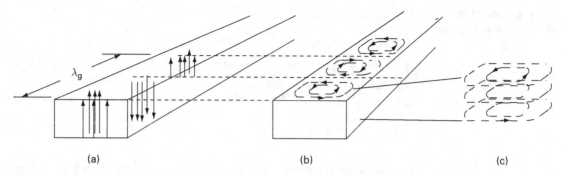

(a)   (b)   (c)

Figure 16.4. Electric field lines (a) and magnetic field Lines (b) The electric lines are bundles of vertical vectors while the magnetic lines are stacks of concentric loops (c.).

$$\frac{B_z}{E} = \frac{\pi}{j a \omega} \cos\left(\frac{\pi y}{a}\right) e^{j(\omega t - k - gz)}. \tag{16.6}$$

The form of this B field is shown in Figure 16.4(b).[3]

Note that the magnetic field lines are stacked concentric loops in the y-z plane with no component normal to the walls. You can use Equations (16.5) and (16.6) to find the exact shape of these loops (see Problem 16.4).

16.2.3 Wall currents

Wall currents, which flow on the inside surfaces of the waveguide, are determined by the tangential magnetic field. The currents are perpendicular to the B field and their magnitude (in amperes/meter) is given by B/μ_0 (the permeability of free space, μ_0, is equal to $4\pi \cdot 10^{-7}$). The wall currents are indicated in Figure 16.5. These currents converge or diverge from areas on the broad wall

Figure 16.5. Wall currents (solid lines) in relation to the magnetic field (dashed lines).

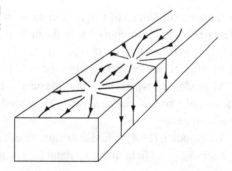

---

[3] The reader familiar with Maxwell's equations can quickly derive Equations (16.5) and (16.6) from Equation (16.2) by applying the curl E equation, which here becomes $j\omega B_y = -\partial E_x / \partial z$ and $j\omega B_x = \partial E_x / \partial y$.

where positive charge is arriving or leaving. The *E*-field lines start and end at these charge patches. Note that the currents on the narrow walls are perfectly vertical because the tangential magnetic field has no *x*-component.

The fields and currents shown in Figures 16.2–16.5 are, of course, snapshots at an instant in time. As the wave propagates, these patterns move uniformly along the *z*-axis with a (phase) velocity given by $v_{ph} = \omega/\beta$.

16.3 Waveguide vs. coax for low-loss power transmission

Consider a situation requiring a low-loss transmission line. Let us compare a standard 2:1 aspect ratio waveguide to a cylindrical coaxial line. To minimize the loss we will make both as large as possible, but here we will impose the restriction that they are also small enough so that modes higher than the fundamental mode cannot propagate. Appendix 16.1 shows that the diameter of this lowest-loss coaxial line and the height of the lowest-loss waveguide are very close to $\lambda/2$ and that the coaxial line will have 2.4 times the loss of the waveguide and will carry only 23% as much power before breakdown.

16.4 Waveguide impedance

There are several ways to define an impedance for a waveguide. One way is to define the voltage to be the potential difference between the top and bottom walls at the middle of the guide and the current to be the integrated current across the top wall. The ratio of voltage to current gives an impedance. Another definition uses voltage and power flow. Still another method uses the ratio of electric field to magnetic field at the center of the guide. The various definitions give $Z_0 = 377$ ohms (impedance of free space) within a factor of 2. But regardless of how impedance is defined, there is no ambiguity in the concept of reflection coefficient. Recall that a shunt capacitance on an ordinary (TEM) transmission line produces a reflection coefficient on the negative *j*-axis of the Smith chart. The same kind of reflection is produced in a waveguide by a short vertical post or a horizontal iris. These obstructions are therefore called "capacitive posts" or "capacitive irises." An iris across the narrow dimension of the guide causes a reflection on the positive *j*-axis so is called an "inductive iris." Figure 16.6 shows examples of inductive and capacitive irises. (The equivalent circuit for a thin iris is just a single shunt susceptance.)

Figure 16.6. Waveguide irises: (a) inductive iris; (b) capacitive iris.

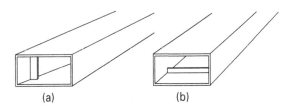

(a)   (b)

The combination of an inductive and a capacitive iris (a thin wall with a hole) is equivalent to a parallel resonant circuit. You can see how these resonant irises could be spaced at quarter-wave intervals in a waveguide to make a coupled-resonator filter.

16.5 Matching in waveguide circuits

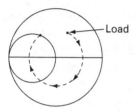

Figure 16.7. A load's reflection coefficient located on the Smith chart.

Impedance matching in waveguide circuits can be done with the same techniques used for ordinary transmission lines. Suppose we are using a waveguide to supply power to some device, maybe a horn antenna, and that we have an instrument – a reflectometer or network analyzer – that can measure the reflection coefficient looking into the waveguide. We can locate the reflection coefficient on the complex reflection plane (Smith chart) as shown in Figure 16.7.

As we move down the guide, away from the load, the reflection coefficient circles the center of the chart and eventually arrives at the unity conductance circle. We locate this position on the guide and install the appropriate inductive or capacitive iris. In practice the tuning process is sometimes very simple: we find the point at which we need to add shunt capacitance. If the reflected wave is small (not a severe mismatch) we do not have to add much capacitance so we get out the ball-peen hammer and dent the broad side of the guide. An expert learns just how hard to swing the hammer.

A note on matching: suppose we join two dissimilar waveguides (perhaps of different sizes) at a junction, which could be some kind of elbow, coupling, butt joint, etc. Assume that the system is nominally lossless, i.e., all metal. We want to match the junction so that a wave coming from either direction will suffer no reflection. We carry out the above procedure on one side of the junction. Do we have to then match the other side? No, the job has been done. Time-reversal produces an equally good solution to Maxwell's equations in which all the power flows in the opposite direction. Of course this applies just as well to ordinary (TEM) transmission lines as it does to waveguides. This simple argument fails for lossy junctions because the time-reversed solution requires the absorptive material to *produce* power, but a stronger argument, based on the reciprocity theorem leads to the same conclusion.

16.6 Three-port waveguide junctions

Two kinds of waveguide T-junctions (three-port junctions) are shown in Figure 16.8.

The *series-T* gets its name from the fact that the voltage of the input guide divides between the two output guides. This works out well because the half-height output guides have half the impedance of the input guide and the junction is inherently matched. (The half-height guides could be increased to full-width in a gradual taper that would not cause much reflection.) The *shunt-T* applies the full input voltage across each of the output arms – not such a natural as the series-T.

Figure 16.8. Waveguide T-junctions.

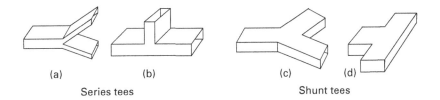

     (a)   (b)   (c)   (d)

           Series tees   Shunt tees

16.7 Four-port waveguide junctions

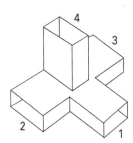

Figure 16.9. Waveguide Magic T.

The *Magic T* hybrid can be built using a procedure that itself seems like magic. We start with the bare waveguide junction (nothing hidden inside) as shown in Figure 16.9.

First we match Port 1, i.e., eliminate reflections from Port 2 when the other ports are feeding matched, i.e., reflectionless, loads. To do this we start by putting matched loads on Ports 2 and 3. (We do not have to put a load on Port 4 since, by symmetry, it is isolated from Port 1.[4]) With the loads in place we measure the reflection at Port 1 and install the necessary iris (or dent) somewhere down line 1. Then we do the same process on Port 4. That's it. The two matches and the isolation by virtue of symmetry are sufficient. We now have a perfectly matched Magic T hybrid.

Simple narrowband transitions from coax to waveguide have mostly been built with empirical methods. With the aid of three-dimensional finite element simulation programs, wideband transitions have been designed. In general, the designer first looks at the fields on both sides and finds a mechanical structure that causes the main features of the fields to line up. The remaining reflection should be small and can be tuned out with a small iris or other structure whose complexity depends on the desired bandwidth.

Rectangular waveguides, like TEM lines, can carry only one signal in each direction. But square or round guides can have two independent waves; they are both fundamental mode waves but they have different polarizations. To launch or recover these two waves independently requires an *orthomode coupler*, which has no TEM counterpart. The simplest orthomode couplers use coaxial or waveguide connections mounted at right angles on the sides of the square or round guide. Some couplers produce circular rather than rectangular polarizations. Wideband orthomode transitions are always needed for radio astronomy and their development is an active field.

---

[4] The *E* field is vertical as a wave enters Port 1. Would it point left or right as it emerged from Port 4? Since the geometry is symmetric, there is no reason to favor right or left. Hence, no wave emerges from Port 4.

Appendix 16.1 Lowest loss waveguide vs. lowest loss coaxial line

For lowest loss we will make the waveguide and the coaxial line as large as possible, but, as explained above, with the restriction that each be capable of supporting only its fundamental mode. Our lowest loss $TE_{1,0}$ waveguide will be made with its width equal to the wavelength. (If it is any wider, the second mode, $TE_{2,0}$, becomes possible.) We will make the height equal to half the width, i.e., the usual aspect ratio. For air-filled coaxial line at the frequency where non-TEM modes become possible, the inner and outer radii, r_i and r_o, satisfy the inequality $(r_o + r_i)\,\pi \geq 1.03\lambda$.[5] The equal sign applies when $r_i / r_o = 1/3.6$. This ratio also provides the lowest loss air-filled coaxial line for a given outer diameter (see Appendix 16.2). Note that the characteristic impedance, $Z_0 = 60 \ln (r_o / r_i)$, will be 77 ohms for this lowest-loss coaxial cable. Using this ratio of diameters, the maximum outer diameter is given by $r_o = 1.03\lambda\pi^{-1}/(1+1/3.6) = 0.26\lambda$. These relative waveguide and coax cross-sections are shown in Figure 16.10.

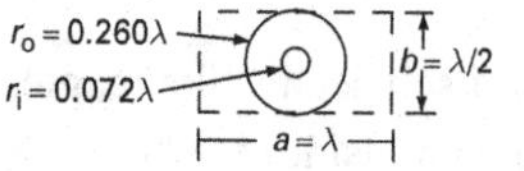

Let us compare the losses. The amplitude of a wave propagating in the $(+x)$-direction on any lossy line is proportional to $\exp(-\alpha x)$ where α, the loss factor, has units of inverse meters. The power is therefore proportional to $\exp(-2\alpha x)$ and the fractional power loss per meter is 2α. Note that $20 \log(e)\alpha = 8.686\alpha$ dB/meter. Because of the skin effect, the loss of a line is proportional to its surface resistance which is given by

$$R_S = \sqrt{\frac{\omega\mu}{2\sigma}} \quad \text{ohms per square} \tag{16.7}$$

where σ is the bulk dc conductivity and ω is the (angular) frequency. (For copper, $R_s = 2.61 \times 10^{-7}$ ohms$/\sqrt{\text{Hz}}$; for aluminum, $R_s = 3.26 \times 10^{-7}$ ohms$/\sqrt{\text{Hz}}$). The loss factor for air-filled coax line is given by

$$\alpha_{\text{coax}} = \frac{R_s}{2Z_0}\left(\frac{1}{2\pi r_o} + \frac{1}{2\pi r_i}\right). \tag{16.8}$$

Our lowest loss coax line has $Z_0 = 77$, $r_o = 0.26\lambda$ and $r_i = 0.072\lambda$ so its loss becomes

$$(\alpha_{\text{coax}})_{\text{min}} = 0.0183\frac{R_s}{\lambda}. \tag{16.9}$$

For air-filled rectangular waveguide in the fundamental mode the loss factor is given by

$$\alpha_{\text{WG}} = \frac{R_s}{377}\frac{\left(1 + \dfrac{2b}{a}\left(\dfrac{\lambda}{2a}\right)^2\right)}{b\sqrt{1 - \left(\dfrac{\lambda}{2a}\right)^2}} \tag{16.10}$$

$r_o = 0.260\lambda$
$r_i = 0.072\lambda$
$b = \lambda/2$
$a = \lambda$

Figure 16.10. Relative cross-sections of lowest-loss waveguide and coaxial cable.

[5] Reference [2], p. 42.

where 377 ohms is $\sqrt{\mu_0/\epsilon_0}$, the "impedance of free space." In our case $a = 2b = \lambda$ so

$$(\alpha_{\mathrm{WG}})_{\mathrm{min}} = \frac{R_s}{\lambda}\frac{5}{337\sqrt{3}} = 0.0076\frac{R_s}{\lambda}. \tag{16.11}$$

The loss of the coax is therefore higher than that of the waveguide by a factor of 0.0183/0.0076 or about 2.4. What about power handling capacity? The breakdown of either the waveguide or the coax depends on the maximum E field, E_{max}. (For air at sea-level pressure, E_{max} is about 30 000 volts/cm.) For rectangular waveguide in the fundamental mode the power is related to the maximum E field by

$$\frac{\mathrm{Pwr}}{E_{\mathrm{max}}{}^2} = \frac{1}{4 \cdot 377}ab\frac{\lambda_0}{\lambda_g} = 6.63 \times 10^{-4}ab\frac{\lambda_0}{\lambda_g} \tag{16.12}$$

where Pwr is in watts, E_{max} is in volts/cm, a and b are in cm, λ_0 is the free-space wavelength, and λ_g, the guide wavelength, is given by

$$\lambda_g = \frac{\lambda_0}{\sqrt{\left(1 - \left(\frac{\lambda_0}{\lambda_{\mathrm{cutoff}}}\right)^2\right)}}. \tag{16.13}$$

For our waveguide $\lambda_0/\lambda_{\mathrm{cutoff}} = 1/2$ so $\lambda_g = 2\lambda_0/\sqrt{3}$ and

$$\frac{\mathrm{Pwr}}{E_{\mathrm{max}}{}^2} = 5.74 \times 10^{-4}ab = 5.74 \times 10^{-4}\lambda\,\lambda/2 = 2.37 \times 10^{-4}\lambda^2. \tag{16.14}$$

Turning to the coax, the $\ln(r)$ dependence of voltage and the characteristic impedance, $Z_0 = (377/2\pi)\ln(r_o/r_i) = 60\ln(r_o/r_i)$ allow us to find the power in terms of the maximum E field:

$$\frac{\mathrm{Pwr}}{E_{\mathrm{max}}{}^2} = \frac{Z_0 r_i^2}{2 \cdot 60^2}. \tag{16.15}$$

In our case $Z_0 = 77$ and $r_i = 0.072\lambda$ so

$$\frac{\mathrm{Pwr}}{E_{\mathrm{max}}{}^2} = \frac{77}{2 \cdot 60^2}(0.072\lambda)^2 = 0.55 \times 10^{-4}\lambda^2. \tag{16.16}$$

We see that the waveguide can handle $2.37/0.55 = 4.3$ times the power of the minimum loss coaxial line. The waveguide is clearly better both for loss and power handling capacity. In high-power applications the waveguide has the additional advantage that there are no interior surfaces needing cooling and no mechanical spacers to center an inner conductor. (Insulating spacers in high-power coaxial lines must fit tightly; high voltage develops across any gap. This problem generally reduces the power-handling capacity of the coaxial line by something like an order of magnitude.)

Appendix 16.2 Coax dimensions for lowest loss, highest power, and highest voltage

Lowest loss

For a given outer diameter, the characteristic impedance of a coaxial line is increased by making the inner diameter smaller. For a given power, the current is decreased. But the smaller inner conductor has more resistance. The I^2R product, i.e., the dissipation, has a minimum when the ratio of diameters is 3.6. This follows from Equation (16.8) which can be rewritten as

$$\alpha_{\text{coax}} = \frac{R_s}{2 \cdot 60 \cdot 2\pi r_o} \frac{1}{\ln(x)} (1 + x) \tag{16.17}$$

where $x = r_o/r_i$. The minimum of $(1+x)/\ln(x)$ occurs at $x = 3.6$ so the characteristic impedance of lowest-loss air-filled coaxial line is $Z_0 = 60 \ln(3.6) = 77$ ohms.

Highest power

From Equation (16.15) we see that to maximize the power-handling capability of the coaxial line we must maximize the expression $Z_0 r_i^2$, i.e., we must maximize $\ln(x)/x^2$. The maximum occurs when $\ln(x) = 1/2$ so the characteristic impedance of the maximum power line is $Z_0 = 60/2 = 30$ ohms.

Maximum voltage

If the line is to withstand maximum voltage the optimum value of $\ln(x)$ is 1 and the characteristic impedance is 60 ohms. This also follows from Equation (16.15): if we express power as $V_{\max}^2/(2Z_0)$ then $V_{\max}$ is proportional to $Z_0 r_i$ or $\ln(x)/x$, which reaches a maximum at $x = e$.

Relative performance of 50-ohm coaxial line

The 50-ohm line commonly used in RF work ($x = r_o/r_i = 2.3$) strikes a compromise between lowest loss, highest power and highest voltage. For loss, we compare $(1+x)/\ln(x)$ for $x = 2.3$ and $x = 3.6$ to see that the 50-ohm line will have only 10% more loss than a 77-ohm line with the same outer diameter. For power handling, we compare $\ln(x)/x^2$ and find that the 50-ohm line can carry 62% as much power as a 30-ohm line with the same outer diameter. Finally, for voltage we compare $\ln(x)/x$ and find that the 50-ohm line can handle 98% as much voltage as a 60-ohm cable with the same outer diameter.

Problems

Problem 16.1. Suppose a car enters a long tunnel which is essentially a rectangular metal tube 10 meters wide by 5 meters high. The car radio becomes silent inside the tunnel. Was the radio more likely tuned to an AM station or an FM station?

Problem 16.2. Examine the waveguide current distribution shown in Figure 16.5 (for the fundamental mode) and draw a sketch showing the position(s) in which a narrow slot could be cut through the waveguide wall without affecting its operation.

Problem 16.3. Describe an experimental setup that could be used to demonstrate the waveguide E-field and B-field distributions shown in Figure 16.4.

Problem 16.4. In the discussion just above Equation (16.2), let $n = 2$ instead of 1. For this choice of n (the TE_{20} mode), find the cutoff wavelength and the guide wavelength. Sketch the electric and magnetic field lines.

Problem 16.5. Use Equations (16.5) and (16.6) to find the mathematical shape of the magnetic field loops. Hint: the slope of a field line, dz/dy, is given by B_z/B_y. Set up the equation $dz/B_z = dy/B_y$ and note that the left side contains only z while the right side contains only y. They can therefore be integrated separately. Remember to add a constant of integration.

References

[1] Collin, R. E., *Foundations for Microwave Engineering*, New York: McGraw Hill, 1992.
[2] Montgomery, C. G., Dicke, R. H. and Purcell, E. M., *Principles of Microwave Circuits*, London: Peter Peregrinus, 1987 (originally Volume 8 of the MIT Radiation Laboratory Series, New York: McGraw Hill, 1948).
[3] Moreno, T., *Microwave Transmission Design Data*, Sperry Gyroscope Corp, 1948, reprinted by Dover Publications, 1958.
[4] Ramo, S., Whinnery, S. R. and Van Duzer, T. *Fields and Waves in Communication Electronics*, 3rd edn., New York: John Wiley, 1994. (Original edition was Ramo, S. and Whinnery, S. R., *Fields and Waves in Modern Radio*, New York: John Wiley, 1944).

17 Small-signal RF amplifiers

In this chapter we discuss the amplifiers used commonly in the front-end and IF stages of receivers and in antenna-mounted preamplifiers. The maximum output power of these amplifiers is typically from 0.01 W to 0.1 W (10–20 dBm). The power amplifiers discussed in Chapter 3 use the full range of the transistor conductance to "push" or "pull" the output voltage to any value from zero to the dc supply voltage(s). Small-signal amplifiers, on the other hand, are class-A amplifiers in which the signal voltages are small, compared with the dc bias voltages. The small ac signals add to dc bias voltages, so the output signal, δV_{out}, produced by an input signal δV_{in} is given by $\delta V_{\text{out}} = [dV_{\text{out}}/dV_{\text{in}}]\, \delta V_{\text{in}} + 1/2\, [d^2 V_{\text{out}}/dV_{\text{in}}^2]\, (\delta V_{\text{in}})^2 + \cdots$ The ac voltage gain is, therefore, $dV_{\text{out}}/dV_{\text{in}}$, evaluated at the quiescent bias point. When operated over only a small range of δV_{in}, the higher derivatives of V_{out} versus V_{in} make only small contributions, and the amplifier is essentially linear. Key characteristics of these amplifiers are gain, bandwidth, input and output impedances, linearity (those higher derivatives), and internally generated noise.

17.1 Linear two-port networks

Small-signal amplifiers are linear amplifiers; the output signal should be a faithful reproduction of the input signal.[1] A general definition of small-signal amplifiers could be that they are amplifiers built entirely of nominally linear elements (which include resistors, capacitors, inductors, transmission lines, and transistors operated over a small differential range), from which it follows that the overall circuit will also be nominally linear. An amplifier, being an example of a two-port network (or simply a "two-port"), has an input terminal, an output

---

[1] While small-signal amplifiers are linear almost by definition, an important exception is the limiting amplifier or *limiter*. In these amplifiers, the gain decreases for increasing signal levels. A cascade of limiters can have an output level almost independent of input level. A limiter is used ahead of an FM detector if the particular FM detector is sensitive to amplitude variations as well as frequency variations.

terminal, and a common terminal (ground). The operation of any linear two-port can be described by four variables: the input and output voltages and currents. Any two of these variables can be considered independent variables ("input" or "cause"). The remaining two variables are then dependent variables ("output" or "effect"). If, for example, V_1 and V_2 are chosen as the dependent variables, the two-port is described by the equations:

$$V_1 = Z_{11}I_1 + Z_{12}I_2 \qquad (17.1)$$

$$V_2 = Z_{21}I_1 + Z_{22}I_2. \qquad (17.2)$$

For this choice of dependent variables, the four coefficients are known as the *Z parameters*. We are implicitly dealing with ac circuit analysis, so these four parameters generally are complex and are functions of frequency. The important point to note is that, for a given frequency, *any* two-port network (amplifier, filter, transmission line, etc.) can be completely described by just four complex numbers. By convention, the current at either terminal is positive when it flows *into* the terminal. Note that the output variables are linear functions of the input variables, since the input variables appear raised only to the first power.[2] By inspection of Equation (17.1) we see that Z_{11} is the network's input impedance when the output current is zero, i.e., when the output is open circuited. The *forward transfer impedance*, Z_{21}, is the open-circuit output voltage divided by the input current – a "transimpedance." If we are given the load impedance, we can use Equations (17.1) and (17.2) to calculate the power gain of an amplifier (Problem 17.1). The *reverse transfer impedance*, Z_{12}, is a measure of reverse feedthrough. If the RF amplifier preceding an unbalanced mixer in a super-heterodyne receiver has reverse feedthrough, some power from the local oscillator will get to the antenna and be radiated. But what is more important, reverse feedthrough can cause an amplifier to oscillate for certain combinations of input and output terminations. This two-port formalism provides more than just a top-level description of an amplifier. It is the basis for amplifier circuit analysis and design, since the active devices (transistors) inside the amplifier can themselves be represented as two-port networks, whose parameters are furnished by the manufacturer on data sheets. Another equivalent set of parameters is the *Y*-parameters, defined by

$$I_1 = Y_{11}V_1 + Y_{12}V_2 \qquad (17.3)$$

$$I_2 = Y_{21}V_1 + Y_{22}V_2. \qquad (17.4)$$

Conversion formulas between parameter sets are easily derived. For example,

---

[2] The dependence of the output variables on only the first power of the input variables follows from a general definition of linearity: If an input a causes an output A and an input b causes an output B, then an input C_1a+C_2b, where C_1+C_2 are constants, will result in an output C_1A+C_2B.

$$Y_{11} = \frac{I_1}{V_1}\bigg|_{V_2=0.} = \left[Z_{11} - \frac{Z_{21}Z_{12}}{Z_{22}} \right]^{-1}. \tag{17.5}$$

The widely used *S-parameters*, which are the subject of Chapter 28, form another equivalent four-parameter set, for which the variables are linear combinations of the voltages and currents, and correspond to input and output waves at each port. Two of the parameters, S_{11} and S_{22}, are reflection coefficients, while the other two are transmission coefficients. A characteristic impedance, Z_0, usually 50 ohms, is implicit. Again, conversion formulas are readily derived, for example,

$$S_{11} = \frac{Y_{22} + Z_0^{-1} - Z_0(Y_{11}Y_{22} + Y_{11}Z_0^{-1} - Y_{12}Y_{21})}{Y_{22} + Z_0^{-1} + Z_0(Y_{11}Y_{22} + Y_{11}Z_0^{-1} - Y_{12}Y_{21})}. \tag{17.6}$$

In this introductory chapter, we discuss amplifiers in terms of voltages and currents, in the interest of presenting the basic concepts in terms totally familiar to the reader.

17.2 Amplifier specifications – gain, bandwidth, and impedances

The small-signal gain (forward and reverse), bandwidth, input impedance and output impedances could be called "linear specifications" because they can all be calculated from the amplifier's Z-parameters or from any of the other equivalent sets of parameters. The gain and bandwidth of an amplifier are ultimately limited by the characteristics of the transistor(s). Transistors have unavoidable built-in reactances: there are at least two capacitors in even the simplest transistor circuit models (approximate equivalent circuits). Elaborate models for microwave transistors can contain a dozen capacitors and inductors. Amplifiers designed for narrowband use (fractional bandwidths of 20% or less) use input and output matching networks to absorb or "cancel" these reactances. At higher frequencies, the shunt capacitive reactances become lower. The matching networks must then necessarily have higher loaded Qs which means that bandwidth decreases. This limitation is fundamental; no matter how complicated the matching network, gain must be traded for bandwidth. Negative feedback around a transistor will lessen the effect of its reactances. But feedback decreases the gain so again there is a tradeoff between gain and bandwidth. In some applications the input and output impedances of an amplifier are critical. For example, if a narrowband filter is placed between two amplifiers, the amplifiers must present the proper impedances to the filter if the intended passband shape is to be realized. The frequency dependences of the input and output impedances of an amplifier are, of course, related to the bandwidth, since the frequency response is normally determined by mismatch (i.e., reflection).

17.2.1 Amplifier stability

An amplifier is required to be stable (not oscillate) in its working environment. A 100-MHz amplifier, for example, will not be satisfactory if it oscillates, even at a very different frequency, say 1 GHz. Oscillation invariably takes the circuit into large-amplitude excursions and the combination of amplification and oscillation is highly nonlinear. An amplifier that remains stable when presented with *any* combination of (passive) source and load impedances (but no external feedback paths) is said to be *unconditionally stable*. Unconditional stability is not always necessary. An IF amplifier in a receiver needs only to be stable in its never-changing working environment. The input RF amplifier in a short-wave radio, however, might be connected to any random arbitrary antenna so it should be unconditionally stable, at least with respect to input impedance. General-purpose commercial modular amplifiers are usually designed to be unconditionally stable. Using these, a system designer can realize a needed transfer function by cascading amplifiers, filters, etc., and know that the combination will be stable. Stability, like gain and input and output impedances, is predictable from the two-port parameters. To find whether an amplifier will be unconditionally stable, it is necessary and sufficient to show that the real parts of the input and output impedances for any frequency are positive for any passive load and source impedances. Suppose that an analysis shows that for some combinations of load impedance and frequency, the real part of the input impedance is negative, but never more negative than -5 ohms. Then adding a series resistor of more than 5 ohms to the input of the amplifier would make it unconditionally stable. Such resistive remedies, however, always decrease gain and increase internally generated noise. The reverse transfer parameter (Z_{12}, Y_{12}, S_{12}, ...) plays a key role in stability. For example, a sufficient (but not necessary) condition for unconditional stability is that its reverse transfer parameter be equal to zero. It is important to note, however, "unconditional stability" simply means that the two-port cannot be provoked into oscillation by varying its termination impedances. A multistage amplifier circuit could contain an oscillating internal stage and still have input and output impedances with positive real parts for all frequencies and arbitrary terminations. (Of course a multistage amplifier will be unconditionally stable if every stage is unconditionally stable.)

17.2.2 Overload characteristics

Any amplifier will become nonlinear at high enough signal levels, if only because the output runs up against the "rail" of the dc power supply. But before this occurs, transistor nonlinearity comes into play. A straightforward specification of an amplifier's upper power limit is the *1-dB compression point*. This is the value of the output power at which the gain has dropped by 1 dB, i.e., the point at which the output power is 79.4% of what would be predicted on the basis of low-power gain measurements.

Intermodulation

Small departures from linearity, even when the amplifier is far below compression, become a concern in a receiver when the passband of an RF
amplifier contains two or more signals at frequencies f_a, f_b, f_c, ... that are
much stronger than the desired signal (the signal that will be isolated downstream by a narrow bandpass filter). Nonlinearity can produce mixing prod-
ucts at frequencies of $n_a f_a + n_b f_b + n_c f_c + \cdots$ where $n_i = 0, \pm 1, \pm 2, \ldots$ In
receivers, the most troublesome of these products are the third-order products
$2f_a - f_b$ and $2f_b - f_a$. (Third-order products will be inevitably produced if the
output voltage of an amplifier contains even a small term proportional to the
cube of the input voltage.) The special problem with these particular products
is that they can fall within the IF passband. To see this, suppose f_2 and f_3 are
the frequencies of signals close enough to a desired frequency, f_1, that they
will pass through the broadband front-end of a receiver. The local oscillator is
tuned to convert f_1 to the center of the IF passband. But the third-order
products $2f_3 - f_2$ and/or $2f_2 - f_3$, being very close to f_1, can also fall within the
narrow IF passband and interfere with the desired signal. Second-order
intermodulation is not so troublesome since the products have frequencies
far outside the IF passband.

A standard measurement of intermodulation is the *two-tone test*, which uses
two closely spaced signals of equal amplitude, A. On a log-log plot of output
power versus input power each of these fundamental signals will fall on a 45°
line, with slope = 1. The third-order products, however, will fall on a line with
slope = 3 because the power in the third-order products is proportional to the
cube of the power in each of the input signals. The *third-order intercept* is
the point at which the third-order product would have as much power as each of
the fundamental signals. Usually the number given for the intercept point is the
output signal strength. The second-order intercept is defined the same way.
Figure 17.1 shows a third-order intercept point of about +37 dBm. Generally an

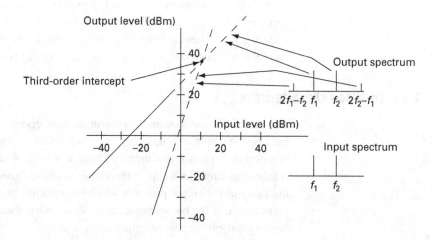

Figure 17.1. Two-tone test to
specify amplifier linearity.

amplifier cannot be driven all the way to the intercept points; they are extrapolations from measurements made at much lower input levels. (The output strengths of the fundamental and the second- or third-order product need only be measured at one input level. Lines with slopes of one, two, or three are then drawn through them to locate the intercepts.)

Dynamic range

Every amplifier adds some noise to the signal. (Later we will discuss amplifier noise in some detail.) Very weak signals will be buried in this noise and lost. The *dynamic range* of an amplifier is therefore determined at the low end by the added noise and at the high end by nonlinearity. In order to handle strong signals, a receiver should keep mixing products small by having as little amplification as possible prior to the narrowest bandpass filter. We will see, on the other hand, that if a receiver begins with a mixer or with a narrowband filter, the loss in these elements adds noise and will render the receiver less sensitive than if the first element after the antenna had been a low-noise amplifier. A trade-off must often be made between sensitivity and dynamic range. Power dissipation is obviously important for battery-operated equipment where milliwatts may count. But to achieve high dynamic range, a small-signal amplifier may have a fairly high-power quiescent point and have to dissipate as much as several watts of power.

17.3 Narrowband amplifier circuits

Amplifiers for frequencies below about 30 MHz look very much like resistance-coupled audio amplifiers. The load resistors are replaced by shunt inductors which cancel the transistor capacitances, which would otherwise tend to be short circuits at RF. These resonant circuits make a narrowband amplifier. Often an even narrower bandpass is desired; the inductors are given smaller values and are shunted with external capacitors (effectively increasing the transistor capacitances). Focusing on one stage of an amplifier (or an amplifier of one stage), the fundamental design decisions are transistor selection and circuit configuration, i.e., common-emitter, common-base, or common-collector. (Here, and usually elsewhere, *emitter, base,* and *collector* can also mean *source, gate, and drain.*) The choice of a transistor will be based on the ability to provide gain at the desired frequency, noise, and perhaps linearity. The orientation of the transistor might be common-emitter when maximum gain is required, common-base when the device is being pushed near its upper frequency limit or when the isolation between input and output is critical, or common-collector when very low output impedance is needed. As far as noise goes, it turns out that the three orientations are equivalent when used in a high-gain multistage amplifier.

17.4 Wideband amplifier circuits

Most wideband amplifiers use feedback. An unbypassed emitter impedance provides series feedback. An impedance between collector and base provides shunt feedback. Commercial modular general-purpose amplifiers use resistive series and shunt feedback. These amplifiers are quite flat up to one or two GHz and have input and output impedances close to 50 ohms over the whole range. Resistive feedback is simple but degrades the noise performance of an amplifier. Wideband low-noise amplifiers often use feedback networks made only of lossless elements, i.e., reactors. The Miller[3] effect multiplies the effective input capacitance in a common-emitter amplifier. This capacitance can be neutralized, at least in a narrowband amplifier. Wideband amplifiers often use the *cascode* circuit in which a common-emitter input stage drives a (low impedance) common base stage. Another good high-frequency circuit, the differential pair, uses an emitter follower stage (high input impedance and low output impedance) to drive a common-base stage.

17.5 Transistor equivalent circuits

An amplifier designer needs a precise electrical description of the transistor(s). For analysis, it is sufficient to have tables of the small-signal parameters of the transistor(s). These tables are usually given in data sheets from the manufacturer; they can be produced using a vector network analyzer. A table of numbers, however, is an awkward representation for design (synthesis) and a common tactic is to represent the transistor by an (approximately) equivalent model circuit of resistors, capacitors, inductors, voltage-controlled voltage generators, voltage-controlled current generators, etc. An exact equivalent circuit for a single frequency can be constructed directly from the small-signal parameters corresponding to that frequency (Problem 17.4). This might be an adequate model for the design of a narrowband amplifier but remember that even an amplifier intended for only a narrow frequency range must be stable at *all* frequencies. For this reason, and also to aid in the design of wideband amplifiers, models are constructed to represent the transistor over a wide frequency range. Normally the topology of an equivalent circuit is based on the construction and physics of the transistor. The element values are determined by least-square fitting programs that make the small-signal parameters of the model agree as closely as possible with the measured small-signal

[3] The collector signal in a common-emitter amplifier has a larger magnitude (due to amplification) than the base signal. It also has the opposite sign. The voltage across the transistor's inherent base-to-collector capacitance is therefore larger than the base voltage. As a result, more current flows in this capacitance than if the collector were grounded. The value of the capacitance is, in effect (Miller effect), multiplied.

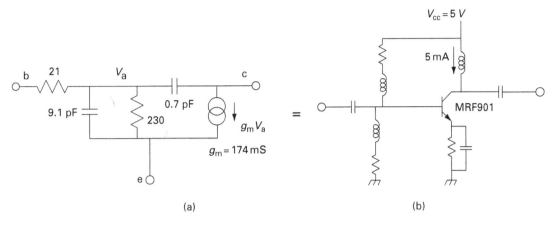

Figure 17.2. Simple equivalent circuit model for the MRF901 bipolar transistor (100 MHz-2 GHz).

parameters of the actual transistor over the desired frequency range. Agreement can always be improved by adding more elements to the model, but an overly-complicated model will block the intuition of the designer. Equivalent circuits have from one to perhaps twenty parameters. Figure 17.2(b) shows a simple "hybrid-π" equivalent circuit model for a common high-frequency transistor. The component values in the model circuit were determined by least-square fitting to the data sheet values over the range from 100 to 2000 MHz. Since the equivalent circuit models a *biased* transistor, it is actually equivalent to the circuit like that of Figure 17.2(a), which includes a power supply and dc biasing components. The biasing components include three resistors to set the dc collector current. A bypass capacitor grounds the emitter at RF frequencies. Blocking capacitors keep the dc bias voltages from interacting with circuitry outside the transistor and vice versa. An RF choke (inductor) provides enough reactance that practically no signal current can flow from the collector to V_{cc} (RF ground). Two more two chokes prevent RF currents from flowing in the base bias resistors. (Usually these resistors have high enough values that these chokes are unnecessary.)

17.6 Amplifier design examples

Designing a small-signal amplifier for microwave frequencies can be difficult, especially when the design must meet specifications for frequency response, gain, stability, input and output impedances, and noise. Computer-aided design is often used. But, to provide an example, let us design a simple amplifier using the transistor model of Figure 17.2. The amplifier is to be driven from a 50-ohm source and is to drive a 50-ohm load. The only other specifications imposed are that the power gain be at least 10 dB at 430 MHz, and that the amplifier not oscillate while connected to the specified source and load. Let us try a very simple common-emitter circuit consisting of the transistor with a matching inductor in series with the collector (rather than going with a two-element

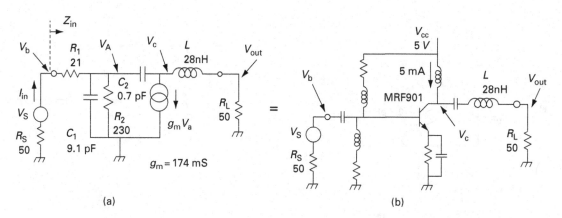

Figure 17.3. Common-emitter amplifier design example: (a) small signal equivalent circuit; (b) full circuit.

matching circuit on each side). The small-signal equivalent circuit of the amplifier is shown in Figure 17.3(a), while the complete amplifier, including bias circuitry, is shown in Figure 17.3(b).

The circuit of Figure 17.3(a) can be analyzed to find V_c in terms of V_b. The output voltage, V_{out}, is then simply $V_c R_L/(R_L + j\omega L)$. Note first that the currents to ground must sum to zero or

$$\frac{V_A - V_b}{R_1} + \frac{V_A}{R_2} + V_A j\omega C_1 + V_A g_m + \frac{V_c}{j\omega L + R_L} = 0. \tag{17.7}$$

Next, consider the current flowing to the left through C_2. This gives us the equation

$$(V_c - V_A)j\omega C_2 = \frac{V_A - V_b}{R_1} + \frac{V_A}{R_2} + V_A j\omega C_1. \tag{17.8}$$

Solving these two equations for V_c as a function of V_b, we find

$$V_c = \frac{-V_b(g_m - j\omega C_2)/R_1}{j\omega C_2(1/R_1 + 1/R_2 + j\omega C_1 + g_m) + (1/R_1 + 1/R_2 + j\omega(C_1 + C_2))/(R_L + j\omega L)}. \tag{17.9}$$

From inspection of Figure 17.3, we see that $V_{out} = V_c R_L/(R_L + j\omega L)$ and that the input current, I_b, is given by $I_b = (V_b - V_A)/R_1$. Since we have already found V_c as a function of V_b, Equation (17.7) (or 17.8) gives us V_A, which lets us calculate I_b and also the input impedance, $Z_{in} = V_b/I_b$. The input power (drive power delivered to the amplifier) is given by $|I_b|^2 \, \text{Re}(Z_{in})$. The power gain of the amplifier is then calculated from

$$\text{Power Gain} = \frac{\text{power out}}{\text{power in}} = \frac{|V_{out}|^2/R_L}{|I_b|^2 \text{Re}(Z_{in})}. \tag{17.10}$$

These expressions are easy to evaluate using a program such as Mathcad or MATLAB. Using Equation (17.9), we find that the power gain at 430 MHz reaches a maximum of 17.25 dB for $L = 28\,\text{nH}$. This is within 0.25 dB of the

Figure 17.4. Power gain vs. frequency for common-emitter amplifier design example.

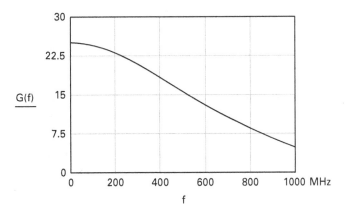

maximum power gain available from this transistor (which could be obtained by using a two-element output matching network rather than just a series inductor). Figure 17.4 shows the power gain versus frequency. The input impedance is 13.3–17.0j, from which you can calculate that the output power can be increased by 2.18 dB by the addition of an input matching network to transform the 50-ohm source impedance to 13.3+17.0j.

Analysis of even this simple amplifier requires a fair amount of algebraic effort. Problem 17.5 shows you how to add the hybrid-π transistor model to the circuit analysis program of Problem 1.3. This addition will let you plot the frequency response and find input and output impedances of this amplifier or any arbitrary cascade of transistors and other two-port devices.

We can check the stability of this amplifier when connected to the specified 50-ohm source and load impedances by verifying that any arbitrary set of initial conditions (two capacitor voltages and one inductor current, at $t=0$) results in transient currents that decay exponentially rather than grow exponentially. Because the circuit is linear, we know that there will be transient solutions in which the time dependence goes as $e^{j\omega t}$, where ω is a *characteristic frequency*. If ω has a positive imaginary part, the solution decays exponentially. To find the characteristic frequencies, we set the source voltage, V_s, equal to zero, and assume that V_A and V_B in Equations (17.7) and (17.8) are proportional to $e^{j\omega t}$. Solving these equations results in a third-order polynomial in ω which must be equal to zero. The three roots of the polynomial turn out to be $\omega_1 = j\,2.64E9$, $\omega_2 = j2.16E9$, and $\omega_3 = j18.1E9$. This shows that all three of these transient solutions will decay exponentially, since their imaginary parts of the three frequencies are all positive.[4] For example, $e^{j\omega_1 t} = e^{-2.64E9\,t}$, which is an exponential decay with a time constant of 0.38 nsec. A superposition of these three particular solutions that satisfies an arbitrary initial set of initial conditions will,

---

[4] Here, the three roots are purely imaginary. In general, they also have real parts, corresponding to decaying *oscillations*. For example, if the load resistance is changed from 50 to 500 ohms, the three roots become j2.44e8 and ±6.89e9 + j1.94e10.

therefore, produce a decaying transient and this circuit will be stable with the 50-ohm source and load impedances.

Let us check stability if we add an input matching network consisting of a series inductor with 17 ohms of reactance to cancel the $-j17.0$ input reactance and a transformer to step the 50-ohm source down to 13.3-ohm input resistance. With the addition of a fourth reactive element, the circuit will now have four characteristic frequencies. Straightforward algebraic manipulations produce a fourth-order polynomial in ω whose four roots are $\pm4.18E9 +j7.83E8$, j2.28E10, and j2.48E10. Since the imaginary parts are positive, the amplifier remains stable with this input matching network and the amplifier's *available gain* (power available from the amplifier divided by power available from the source) becomes equal to the power gain. Chapter 28 discusses how stability is usually evaluated in the context of S parameters.

Figure 17.5(a) shows an amplifier using the same transistor model arranged in a common-base configuration; the base is grounded and the input signal is applied to the emitter. The equivalent small-signal circuit is shown in (b).

You can analyze this circuit with the methods used above for the common-emitter configuration. Write one equation that sets to zero the sum of all the currents away from node "A." Do the same for node "c" (the collector terminal) and solve the two equations simultaneously for V_c as a function of V_e (the emitter voltage).

Figure 17.6 shows a model for a microwave GaAS FET, which can be used in place of the BJT model for the same kind of amplifier analysis. Manufacturers'

Figure 17.5. A common-base RF amplifier.

Figure 17.6. High-frequency GaAs FET model.

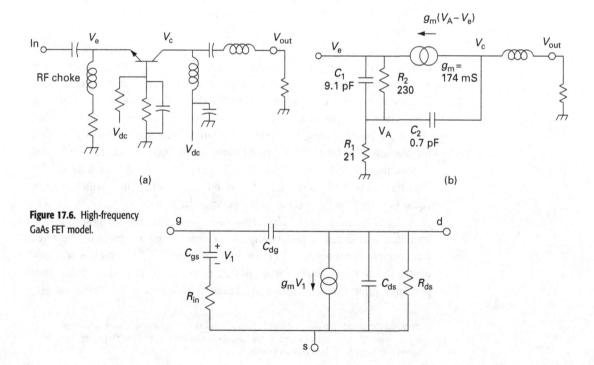

(a)

(b)

data sheets often include model circuits along with two-port data. There are many excellent textbooks devoted entirely to the amplifier design.

17.7 Amplifier noise

The output signal from any amplifier will always include some random noise generated within the amplifier itself. Most of the hiss from a radio receiver is due to noise generated by atmospheric electricity. But if the antenna is disconnected, the noise does not entirely disappear. The remainder is being generated within the receiver. Physical mechanisms that cause this noise include thermal noise (discussed below) and shot noise, which is noise due to the randomness in the flow of discrete charges – electrons and holes in transistors. The first stage in most receivers is an RF amplifier, and its noise usually dominates any other receiver-generated noise. This is easy to see; since this stage usually has considerable gain, its output power will be much greater than the noise power contributed by the second stage, so the second stage will hardly change the signal-to-noise ratio. In the same way, noise contributed by the third stage is even less important, and so forth.

Thermal noise

Thermal noise is such a universal phenomenon that it provides the very vocabulary for the definition of terms such as *receiver noise temperature* and *antenna noise temperature*. Let us examine the fundamentals of thermal noise. Any object, being hotter than absolute zero, converts thermal energy (molecular vibrational energy) into electromagnetic radiation: at infrared radiation for ordinary temperatures, but also visible radiation, if the object is red hot. Likewise, a resistor can convert thermal energy into electrical power. If two resistors are connected in parallel, each one delivers a tiny amount of electrical power (in the form of a random voltage waveform) to the other. If they are at equal temperatures, the power flow is equal in both directions.

How much power can a resistor generate by virtue of being hot? Answer: Any resistor can deliver kT watts per hertz, that is, kT is the spectral density of the power that a resistor of R ohms will deliver to a matched load (a load of impedance $R+j0$). Here k is Boltzmann's constant (1.38×10^{-23} joule/kelvin – 1.38×10^{-23} watt seconds/kelvin) and T is the absolute temperature. It is useful to remember that for $T_0 = 290$ K, which is universally taken as a standard reference temperature, the power available from a resistor, referred to 1 mW, is $= -114$ dBm/MHz, since $10 \log (1.38 \times 10^{-23} \times 10^3 \times 290 \times 10^6) = -114$.

To demonstrate that a resistor should produce this power, kT watts/Hz, an argument appropriate to radio engineering considers an antenna surrounded by a blackbody, i.e., an antenna within a cavity whose walls are at temperature T_1. We will be concerned with the spectral density at a particular spot frequency so

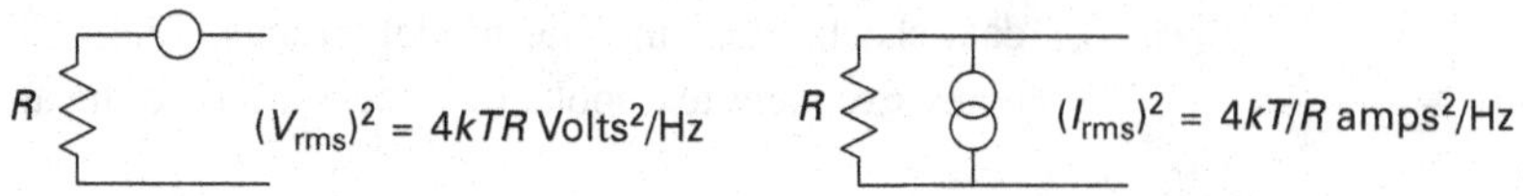

Figure 17.7. Equivalent circuits for a resistor as a noise source.

R $(V_{rms})^2 = 4kTR$ Volts^2/Hz R $(I_{rms})^2 = 4kT/R$ amps^2/Hz

we can specify that the antenna be resonant, i.e., that it have a purely resistive impedance, $R + j\,0$, at that frequency. Let a transmission line connect the antenna to a resistor R which is outside the blackbody but also at temperature T_1. We know the antenna will intercept blackbody radiation and that power will be transmitted through the line to the external resistor. We also know from thermodynamics that, in this isolated system, it is impossible for the resistor to get hotter than T_1; heat cannot flow from a colder to a hotter object. The only way to resolve this is for the resistor to produce an equal amount of power, which travels back to the antenna and is radiated back into the cavity. We can use some antenna theory to calculate the power. All antennas are directive; when used to receive, they have more effective area to intercept power from some directions than from other directions. But for any antenna, the average area turns out to be $\lambda^2/4\pi$ where λ is the wavelength. Blackbody radiation flux at long wavelengths is given by the Rayleigh–Jeans law, brightness $= 2kT/\lambda^2$ watts/m^2/Hz/steradian. This includes power in two polarizations. Since any antenna can respond to only one polarization, we use half the Rayleigh–Jeans brightness to calculate the power the antenna puts on the transmission line:

$$P = \int \frac{B(\theta,\phi)}{2} A(\theta,\phi)\,\mathrm{d}\Omega = \frac{kT}{\lambda^2}\frac{\lambda^2}{4\pi}4\pi = kT. \tag{17.11}$$

This value, kT, is then also the power a resistor of R ohms will deliver to another resistor of R ohms. It follows that the open-circuit noise voltage from a resistor is therefore $\langle V_n^2\rangle = 4kTR$ volt^2/Hz. Figure 17.7 shows the Thévenin and Norton equivalent circuits of a resistor as a noise generator.

17.8 Noise figure

At any given frequency, a figure of merit for a receiver, an amplifier, a mixer, etc., is its *noise figure*, whose definition is as follows:

Noise figure is the ratio of the total output noise power density to the portion of that power density engendered by the resistive part of the source impedance, with the condition that the temperature of the input termination be 290 K.

Noise figure is a function of frequency and of source impedance but (as we will see later) is independent of output termination. Consider Figure 17.8.

The voltage source V_n represents the thermal noise voltage from R_s, the resistive part of the source impedance. The source, V_s, is the actual signal voltage, if any. The internal noise of the amplifier can be considered to result

Figure 17.8. Equivalent circuit
of an amplifier and signal source.

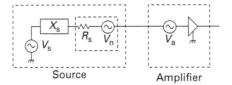

from another equivalent input noise source, V_a, at the input of the amplifier.
With this model, the noise figure, as defined above, can be written in terms of V_n
and V_a: $F = (V_n^2 + V_a^2)/V_n^2$. Note that, because the amplifier noise is represented
by an equivalent generator at the input side, this expression does not contain G,
the gain. Since V_n^2 is proportional to the source temperature, T_0, it is natural
to assign the amplifier an equivalent noise temperature by writing the noise
figure as $F = (T_0 + T_a)/T_0$. This amplifier noise temperature is just $T_a = (F-1)$
$T_0 = (F-1) \times 290$ K. Conversely, the noise figure is given by $F = (290 + T_a)/290$.
An amplifier can have a noise temperature less than its physical temperature.
The dish-mounted amplifiers used for home satellite reception have typical
noise temperatures of 30 K. (Refrigeration, however, can help; FET amplifiers
for radio astronomy are often physically cooled to about 10 K and produce noise
temperatures of only a few kelvins.)

So far we have not mentioned the signal voltage, V_s. Equation (17.12) shows
that the noise figure also specifies the ratio of the input signal-to-noise ratio to
the output signal-to-noise ratio:

$$\frac{\text{Input SNR}}{\text{Output SNR}} = \frac{V_s^2/V_n^2}{V_s^2/(V_n^2 + V_a^2)} = (V_n^2 + V_a^2)/V_n^2 = F. \qquad (17.12)$$

Cascaded amplifiers

The noise from an amplifier of only modest gain will not totally dominate the
noise added downstream, so it is useful to know how noise figures add. Suppose
amplifier 1, with noise figure F_1 and gain G_1, is followed by amplifier 2 with F_2
and gain G_2. Suppose further that they are matched at their interface so that
$G = G_1 G_2$ and that the output impedance of amplifier 1 is equal to the source
impedance corresponding to the specified F_2. Figure 17.9 shows how to
compute the overall noise figure.

The noise figure of the cascade is $F_{12} = F_1 + (F_2 - 1)/G_1$. It is interesting to
calculate the noise figure of an infinite cascade of identical amplifiers as it is a
lower limit to the noise we would get from any shorter cascade. You can verify
that T_{infinity}, the equivalent noise temperature of the infinite chain, is given by
$T_a/(1 - 1/G)$ where T_a and G refer to the individual identical amplifiers.

Finally, let us look at the overall noise figure of an amplifier preceded by an
attenuator, as shown in Figure 17.10. Suppose the gain of the attenuator is G_{attn}.
(The gain of an attenuator is less than unity; the gain of a 6-db attenuator, for
example, is 1/4.) Referred to the amplifier input, the noise power engendered by

Figure 17.9. Overall noise figure of cascaded amplifiers.

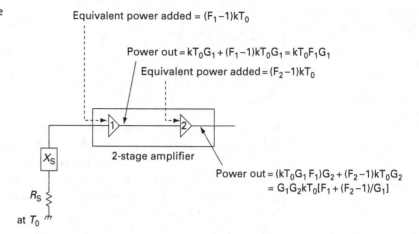

Figure 17.10. An amplifier preceded by an attenuator.

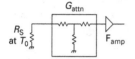

the source resistance is $T_0 G_{\text{attn}}$. Since the amplifier still sees its standard source impedance, its total noise, referred to the input, is still $(F_{\text{amp}})T_0$.

The overall noise figure is therefore

$$F_{\text{tot}} = \frac{F_{\text{amp}} T_0}{G_{\text{attn}} T_0} = \frac{F_{\text{amp}}}{G_{\text{attn}}} \tag{17.13}$$

so, if an amplifier is preceded by an M-dB attenuator ($G_{\text{attn}} = 10^{-M/10}$), the noise figure of the combination is M dB higher than the noise figure of the amplifier alone. We could just as well have derived this result by using the relation for cascaded devices. The noise figure of the attenuator, from the definition of noise figure, is $F_{\text{attn}} = kT_0/(G_{\text{attn}}kT_0) = 1/G_{\text{attn}}$. The noise figure of the cascade becomes $F_{\text{tot}} = 1/G_{\text{attn}} + (F-1)/G_{\text{attn}} = F/G_{\text{attn}}$ as before.

17.9 Other noise parameters

In what we have considered so far, the noise produced by a device, a transistor, amplifier, etc. is specified by a single parameter, its noise figure. But the noise figure depends on the source impedance from which the device is fed, which makes this parameter something less than a complete noise description of the device. We will see in Chapter 24 that a total of *four* noise parameters are sufficient to describe a device. The noise figure for any given source impedance can then be calculated from these four parameters which are R_{opt}, X_{opt}, F_{min}, and R_n. The (complex) impedance $Z_{\text{opt}} = R_{\text{opt}} + X_{\text{opt}}$ is the source impedance that yields the minimum noise figure, F_{min}. The "noise resistance," R_n, is a parameter that determines how fast the noise figure increases as the source impedance departs from Z_{opt}. We will see in Chapter 24 that the noise figure for an arbitrary source impedance is given by

$$F = F_{\text{min}} + (R_n/G_s)|Y_{\text{source}} - Y_{\text{opt}}|^2. \tag{17.14}$$

(Here Y_{opt} is just $1/Z_{opt}$ and G_s is the real part of the source admittance.) We will also see that noise figure is somewhat deficient as a figure of merit. A piece of wire has $F = 1$ but is not a valuable amplifier since it has no gain. With a given transistor, circuit A might produce a lower noise figure than circuit B, but circuit A may have less gain. We will see that $T_{infinity}$, defined above, is the proper figure of merit.

17.10 Noise figure measurement

A straightforward determination of an amplifier's noise figure is possible if one knows its gain and has a spectrum analyzer suitable for measuring noise power density. Consider the common situation where we have an amplifier to be used in a 50-ohm environment. We connect a 50-ohm load to its input and use the spectrum analyzer to measure the output power density, S_{out}(watts/ Hz), at the frequency of interest. We know that the portion of this power density engendered by the input load is kTG, where G is the amplifier gain. The noise figure is therefore given by $F = S_{out}/(kTG)$. This assumes we have done the measurement at $T = 290$ K. If T was not 290, you can verify that $F = [S_{out} - Gk(\text{T} - 290)]/(290Gk)$. For low-noise amplifiers, a comparison method is used. This method requires a cold load and a hot load, i.e., two input loads at different temperatures, T_{hot} and T_{cold}. The amplifier is connected to a bandpass filter (whose shape is not critical) and then to a power meter (which needs to have only relative, not absolute accuracy). The ratio of power meter readings, hot to cold, is called the Y-factor. The noise temperature of the amplifier is then given by:

$$T_a = (T_{hot} - YT_{cold})/(Y - 1). \qquad (17.15)$$

Problems

Problem 17.1.   Derive an expression for the power gain (output power/input power) for the two-port network described by Equations (17.1) and (17.2) when the load impedance is Z_L.

Problem 17.2.   For a general two-port network, derive expressions for the Y parameters in terms of the Z parameters.

Problem 17.3.   (a) A certain amplifier with 20 dB of gain has a third-order intercept of 30 dBm (one watt at the output). If the input consists of 0 dBm (0.001 watt) signal at 100 MHz and another 0 dBm signal at 101 MHz, what will be the output power of the third-order products at 102 MHz and 99 MHz?

(b) Same as (a) except that the input signal at 100 MHz increases in power to 10 dBm (0.1 watt) while the input signal at 101 MHz remains at 0 dBm.

Problem 17.4.   The Z-parameter description of a two-port corresponds in a one-to-one fashion to the equivalent circuit shown below in (a). Another circuit is shown in (b). Find expressions for Z_A, Z_B, Z_C and V in terms of the Z parameters to make the two circuits equivalent.

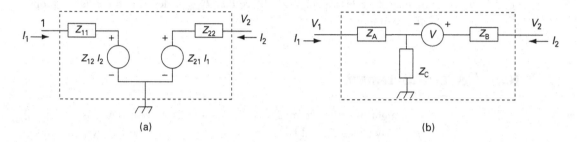

(a)   (b)

Problem 17.5.   The figure below at (a) shows a small-signal hybrid-π model for a common-emitter BJT transistor. (This model contains one more component than the model of Figure 17.2.)

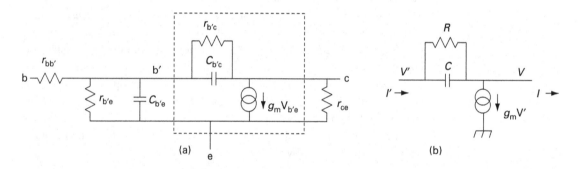

(a)   (b)

The components form a simple ladder network, except for the portion inside the dashed line box. Show, for this box, the circuit in (b), that the relations between the input voltage and current and the output voltage and current are given by

$$V' = \frac{V}{1 - g_m Z} + \frac{IZ}{1 - g_m Z}$$

and

$$I' = \frac{V g_m}{1 - g_m Z} + \frac{I}{1 - g_m Z}.$$

Use these equations to make this box a new circuit element in the analysis program of Problem 1.3. The program will then be able to analyze BJT common-emitter amplifiers. (Note that the other two resistors and the other capacitor in this hybrid-π model can be included in a circuit as if they were external components.)

Example answer: For the MATLAB example given in Problem 1.3, add the element, "HY_PI" by inserting the following sequence of statements in the "elseif chain":

```
elseif strcmp(component,HY_PI)==1
ckt_index=ckt_index+1; gm=ckt{ckt_index}; %gm
ckt_index=ckt_index+1; R=ckt{ckt_index};%R
ckt_index=ckt_index+1; C=ckt{ckt_index}; %C
Z=R/(1+1j*w*R*C);
Iold=I; I=(I+gm*V)/(1-gm*Z);V=(V+Iold*Z)/(1-gm*Z);
```

Problem 17.6. Show that the noise figure of an infinite cascade of identical amplifiers is given by $F_\infty = (F - 1/G)/(1 - 1/G)$. Assume that the amplifiers have a standard input and output impedance such as 50 ohms, that G is the gain corresponding to this impedance, that F is the noise figure corresponding to a source of this impedance, and that F_∞ is also to be with respect to this standard impedance. Hint: use the formula for a cascade of two amplifiers and the standard "infinite-chain-of-anything" technique – adding another link does not change the answer.

(This problem is not just academic. With only a few amplifiers the gain will be high enough to make the noise figure very close to F_∞ which is the best possible combination of the given amplifiers.)

Problem 17.7. Consider the balanced amplifier circuit shown below. The 3-dB, 90° hybrids are ideal. The amplifiers are identical and all impedances are matched. The individual amplifiers have power gain G and noise figure F_0. The hybrids are perfect, i.e., they have no internal loss and are perfectly matched.

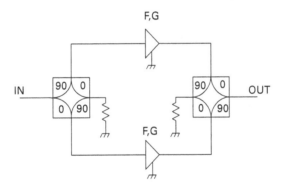

a. Show that the overall noise figure of this circuit is equal to the noise figure of the individual amplifiers.
b. If one amplifier dies, i.e., provides zero output, what is the overall noise figure?

Problem 17.8. We derived the overall noise figure, $F_{tot} = F (G_{attn})^{-1}$, for an amplifier preceded by an attenuator when the physical temperature of the attenuator is T_0, the standard 290 K reference temperature. Assuming now that the attenuator is at some different physical temperature, T_1, show that the overall noise figure is given by $(G_{attn})^{-1}$ $[G_{attn} + (T_1/T_0)(1 - G_{attn}) + F - 1]$.

References

[1] Carson, R. S., *High Frequency Amplifiers*, New York: John Wiley, 1975.

[2] Gonzalez, G., *Microwave Transistor Amplifier Analysis and Design*, Englewood Cliffs, N.J.: Prentice Hall, 1984.

[3] Krauss, H. L., Bostian, C. W. and Raab, F. H., *Solid State Radio Engineering*, New York: John Wiley, 1980.

[4] Vendelin, G. D., Pavio, A. M. and Rohde, U. L., *Microwave Circuit Design Using Linear and Nonlinear Techniques*, New York: John Wiley, 1990.

18 Demodulators and detectors

In communications equipment, "detection" is synonymous with demodulation, the process of recovering information from the received signal. The term detector is also used for circuits designed to measure power, such as square-law microwave power detectors. In this chapter we discuss various AM, FM, and power detector circuits. The demodulator is the last module in the cascade of circuits that form a receiver and at this stage the frequencies (IF or baseband) are relatively low. For this reason the detector (and, sometimes, the final IF bandpass filter) was the first receiver section to evolve from analog to digital processing, notably in broadcast television receivers. Receivers for ordinary AM and FM have, for the most part, continued to use traditional analog detectors, but receivers for the newer digital radio formats can also use their digital signal processors to demodulate traditional AM and FM broadcasts. Demodulators for OFDM and CDMA digital modulation formats are discussed in Chapter 22.

18.1 AM detectors

There are two basic types of AM detector. An envelope detector uses rectification to produce a voltage proportional to the amplitude of the IF voltage. A "product" detector multiplies the IF signal by a reconstituted version of the carrier. This detector is a mixer (see Chapter 5), producing sum and difference frequencies. The sum component is filtered away. The difference frequency component, at $f = 0$ (baseband), is proportional to the amplitude of the IF voltage.

18.1.1 AM diode detector

The classic diode envelope detector circuit for AM is shown in Figure 18.1. The input signal voltage, $V_{sig}\cos(\omega t)$, is usually provided by a tuned transformer at the output of the IF amplifier. This tuned circuit forms part of the IF bandpass filter.

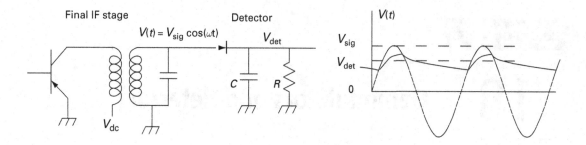

Figure 18.1. Diode envelope detector.

The diode and the parallel RC form a fading-memory peak detector. Except for the resistor, R, the output capacitor would remain charged to the maximum peak voltage of the input sine wave. The resistor provides a discharge path so that the detector output, V_{det}, can follow a changing peak voltage (AM). Since the input sine wave, an RF signal, has a much higher frequency than the amplitude modulation frequencies, the RC time constant can be made large enough so that the droop between charge pulses is much less than indicated in Figure 18.1. (Of course the time constant must also be small enough that output voltage can accurately follow a rapidly changing modulation envelope.) This detector, or any other envelope detector, is known as a *linear detector* since its output voltage is linearly proportional to the amplitude of the input sine wave.

Analysis assuming an ideal rectifier

Note that the detector of Figure 18.1 is identical to a simple half-wave rectifier capacitor-input power supply. As with the power supply, this circuit has poor regulation with respect to a changing load. But here we have a constant load resistance R (which we will assume includes the parallel input resistance of the subsequent audio amplifier). The equivalent circuit is shown in Figure 18.2. If the diode is modeled as a perfect rectifier (zero forward resistance and infinite reverse resistance) the analysis of this circuit is straightforward. The value of the

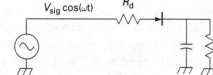

Figure 18.2. Diode envelope detector equivalent circuit.

diode's forward resistance, R_{d}, can be increased to account for source resistance. But here the high-Q resonant circuit at the detector input forces the waveform to remain sinusoidal. If we assume that C is large enough to make the output ripple negligible compared to the output voltage, the output voltage, V_{det}, can be calculated by noting that V_{det}/R must equal the average current through the diode, which is the average of $(V-V_{\text{det}})/R_{\text{d}}$ during the part of the input cycle when this expression is positive. The result is that V_{det} is proportional to the source voltage. (Curves showing output voltage vs. ωCR for various values of R_d/R are found in the power supply chapters of many handbooks.) The ratio of V_{det} to the peak source voltage is known as the detector efficiency. For a typical AM detector, $R \geq 10R_d$ and $\omega CR \geq 100$. This gives a detector efficiency greater than 65% and an rms ripple less than about 1% of the dc output. With the assumed ideal rectifier, however, R could be any value. The analysis in the following section shows the limitations imposed on R by a real diode.

Analysis with a real diode

Here we will discard the perfect rectifier in favor of the standard diode for which $I_{diode} = I_s \exp(V_{diode}/V_T - 1)$, where I_s is the reverse saturation current and V_T is the so-called thermal voltage, 0.026 volts.[1] In the equivalent circuit of Figure 18.2, we will now let R_d be zero, i.e., we will assume that any voltage drop across the diode's bulk resistance is negligible compared to the drop across the junction. As before, the analysis to find V_{det} consists in equating V_{det}/R to the average current through the diode:

$$\frac{V_{det}}{R} = \langle I_{diode} \rangle = \frac{1}{2\pi} \int_0^{2\pi} I(\theta) d\theta \qquad (18.1)$$

where

$$I(\theta) = I_s \{ \exp[(V \cos \theta - V_{dc})/V_T] - 1 \}. \qquad (18.2)$$

This pair of equations is equivalent to

$$\frac{V_{det}}{0.026} = \ln \left(\frac{V_s I(V)}{2\pi(V_{det} + V_s)} \right), \qquad (18.3)$$

where V_s, a "saturation voltage," is defined by $V_s = I_s R$ and

$$I(V) = \int_0^{2\pi} \exp(V \cos \theta / 0.026) d\theta. \qquad (18.4)$$

Using a desk-top computer math utility to solve Equation (18.3) results in a set of curves (Figure 18.3) showing V_{det} vs. V for various values of V_s. Note that the detector output is very nonlinear for low-amplitude input signals when V_s is less than about 0.01 V. Suppose, then, that we pick $V_{sat} = 0.01$ V. A germanium diode or zero-bias Schottky diode might have a saturation current of 10^{-6} A, which would

---

[1] The thermal voltage is given by kT/e, where e is the charge of an electron, k is Boltzmann's constant, and T is an assumed temperature, 300 K.

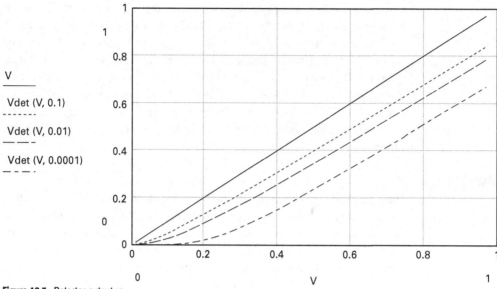

$\dfrac{V}{\overline{}}$

Vdet (V, 0.1)

Vdet (V, 0.01)

— — —

Vdet (V, 0.0001)

— — ·

Figure 18.3. Detector output vs. input voltage for several values of $V_s = I_s R$.

then require that $R = 0.01/10^{-6} = 10\,\mathrm{k}$ ohms, a convenient value. The low I_s of an ordinary silicon diode might require that R be more than 10^7 ohms, which would require the audio amplifier to have an inconveniently high input impedance. The power dissipated in the detector is V_{det}^2/R. This detector would typically produce, say, $2\,\mathrm{V}$ (to operate up in the linear range), which corresponds to a power dissipation of $2^2/10^4 = 0.4\,\mathrm{mW}$. For a given signal strength at the receiver input, the RF stages must have enough gain to produce $0.4\,\mathrm{mW}$.

AC-coupled diode detector

Sometimes circuit considerations require that the detector input be ac coupled. In this case a dc return must be furnished for the detector diode. Such a circuit is shown in Figure 18.4 where an RF choke (a large value inductor) provides this

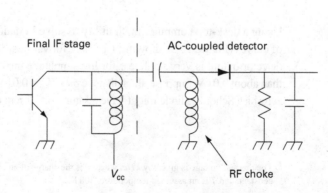

Final IF stage   AC-coupled detector

Figure 18.4. AC-coupled diode detector.

V_{cc} RF choke

dc return path, forcing the average voltage at the left side of the diode to be zero, as in the circuits of Figures 18.1 and 18.2.

18.1.2 Product detectors

Demodulation of SSB and Morse code (cw) signals is usually done by a *product detector* which mixes (multiplies) the IF signal with a locally generated carrier. The resulting difference frequency components become the demodulated signal, while the sum frequency components are discarded. This is shown in Figure 18.5. The free-running oscillator is historically known as a *beat frequency oscillator* (BFO).

Figure 18.5. Product detector for SSB and CW.

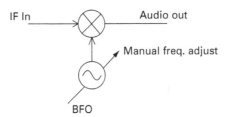

For SSB reception of voice signals, the frequency of the BFO is manually adjusted until the audio sounds approximately natural. For Morse code reception, the BFO is deliberately offset to produce an audible tone, the "beat note." This was first done by the radio pioneer, Reginald Fessenden. In his heterodyne detector, the predecessor to the Edwin Armstrong's superheterodyne, the incoming signal voltage was combined with the voltage from a local oscillator,[2] which was a small arc source, an early negative resistance oscillator.

A product detector can also be used for AM demodulation. Again the signal is *multiplied* by a locally generated carrier. Here the local carrier must have the frequency and phase of the received carrier. Any error in frequency creates a strong audio beat note with the carrier of the received signal and an error in phase reduces the amplitude. Nevertheless, the product detector overcomes the limitations of diode envelope detectors; the input signal levels do not have to be as high and there is no low-signal threshold below which the detector is useless. When the AM signal is consistently strong (usually the case for most broadcast listeners) the local carrier can be a hard-limited version of the input signal. This works because, in double sideband AM, the modulated signal has the same zero crossings as the unmodulated carrier. The product detector shown in Figure 18.6 uses this method. This detector is commonly used for the video detector in analog television receivers.

---

[2] In his heterodyne patent of 1902, Fessenden proposed that the transmitting station send *two* signals, closely spaced in frequency. At the receiver, the nonlinear detector would produce an audible beat note. One of these signals could be continuous (not keyed). Instead, it became practical to produce the continuous signal at the receiver site using a *local* oscillator.

Figure 18.6. Product AM detector.

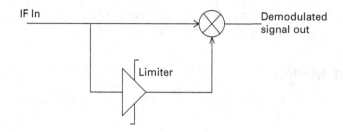

Figure 18.7. Synchronous AM detector.

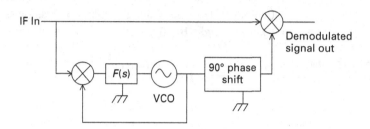

The *synchronous detector*, shown in Figure 18.7, is an improved product detector circuit in which a phase lock loop is used to generate the local carrier. The carrier of the input signal provides the reference signal for the loop. In a practical circuit, a limiting amplifier can be used at the reference input to make the loop dynamics independent of the signal level.

The PLL gives the synchronous detector a flywheel effect: the narrowband loop maintains the regenerated carrier during abrupt selective fades (common in short-wave listening). This prevents distortion, common in short-wave receivers, caused by momentary dropouts of the carrier. The PLL provides, in effect, a bandpass filter so narrow that its output cannot change quickly. Note the 90° phase shift network; if the phase detector is the standard multiplier (mixer), the VCO output phase differs from the reference phase by 90°. Without the network to bring the phase back to 0°, the output of the detector would be zero.

18.1.3 Digital demodulation of AM

The analog AM demodulators discussed above can be implemented as well in digital circuitry. Envelope detection via full-wave rectification of the IF signal can be done by simply taking the absolute value of the numbers produced by the A-to-D converter. For digital processing, the IF signal is often converted to two baseband signals, I and Q, which result from mixing the IF signal with $\cos(\omega_{IF}t)$ and $\sin(\omega_{IF}t)$. In this case, the amplitude of the signal is given by $(I^2+Q^2)^{1/2}$, which can also be computed digitally, though not as easily as $|V|$. Synchronous detection can be done.

18.2 FM demodulators

A variety of circuits, often called *discriminators*, have been used to demodulate FM. Most of these circuits are sensitive to amplitude variations as well as frequency variations, so the signal is usually amplitude limited before it arrives at the FM detector. This reduces the noise output, since amplitude noise is eliminated (leaving only phase noise). In addition, it ensures that the audio volume is independent of signal strength.

18.2.1 Phase lock loop FM demodulator

We have already pointed out that a phase lock loop may be used as an FM discriminator. As the loop operates, the instantaneous voltage it applies to the voltage-controlled-oscillator (VCO) is determined by the reference frequency, which here is the signal frequency. The linearity of the VCO determines the linearity of this detector, shown in Figure 18.8.

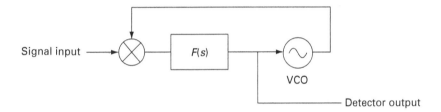

18.8. Phase lock loop FM detector.

18.2.2 Tachometer FM detector

A tachometer FM detector or "pulse counting detector", shown in Figure 18.9, is just a one-shot multivibrator that fires on the zero crossings of the signal. Each positive zero crossing produces a constant-width output pulse. The duty cycle of the one-shot output therefore varies linearly with input frequency so, by integrating the output of the one-shot, we get an output voltage that varies linearly with frequency.

Figure 18.9. Tachometer FM detector.

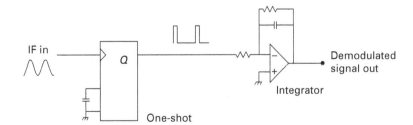

18.2.3 Delay line FM detector

The *delay line discriminator* is often used in C-band satellite television receivers to demodulate the FM-modulated video and sound. The IF frequency in these receivers is typically 70 MHz. Figures 18.10 shows a quarter-wave delay line (which could be a piece of ordinary transmission line) which delays the signal at one input of the multiplier. If the input signal is $\cos((\omega_0+\delta\omega)t)$, then the signal at the output of the delay line is $\cos((\omega_0+\delta\omega)(t-\tau)) = \sin((\omega_0+\delta\omega)t - \tau\delta\omega)$ and the baseband component at the output of the multiplier is $-\sin(\tau\delta\omega)$. For small $\tau\delta\omega$ this is just $-\tau\delta\omega$. The output voltage is thus proportional to the frequency offset, $\delta\omega$. If the delay line is lengthened by an integral number of half-wave lengths, the sensitivity of the detector is increased, i.e., a given shift from center frequency produces a greater output voltage.

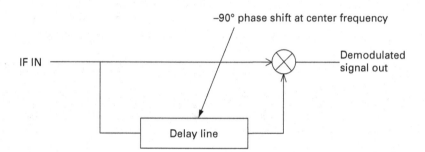

Figure 18.10. Delay line FM detector.

FM quadrature demodulator

Figure 18.11. Quadrature FM detectors. (a) An LC network is used as the delay element. The multiplier provides the necessary resistive termination, (b) A voltage divider provides the necessary phase shift (see Problem 18.3).

The *quadrature FM demodulator*, shown in Figure 18.11, is the same as the delay line discriminator except that an LC network is used to provide the delay, i.e., a phase shift that varies linearly with frequency. These circuits are commonly used in integrated circuits for FM radios and television sound; the LC networks or an equivalent resonator is normally an off-chip component.

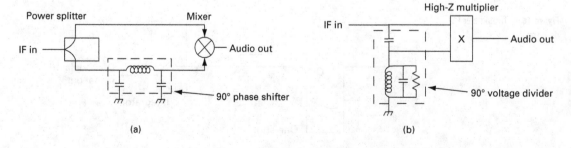

18.2.4 Slope FM detector

Slope detection, in which FM modulation is converted to AM modulation, is the original method to demodulate FM. The amplitude response of the IF bandpass filter is made to have a constant slope at the nominal signal frequency. An input signal of constant amplitude will then produce an output signal whose amplitude depends linearly on frequency. A simple envelope detector can detect this amplitude variation. An AM receiver can slope-detect an FM signal if detuned slightly to put the FM signal on the upper or lower sloping skirt of the IF passband filter. A refined balanced slope detector (Figure 18.12) uses two filters with equal but opposite slopes. The filter outputs are individually envelope-detected and the detector voltages are subtracted. This makes the output voltage zero when the input signal is on center frequency, f_0, and also linearizes the detector by cancelling even-order curvature, such as an $(f-f_0)^2$ term, in the filter shape.

The Foster–Seeley "discriminator," a classic FM detector, is an example of a balanced slope detector, though this is hardly obvious from the circuit, shown in Figure 18.13.

Note that the transformer has a capacitor across both its primary and its secondary. If the transformer had unity coupling, a single capacitor on one side or the other would suffice. The fact that there are two capacitors tells us the

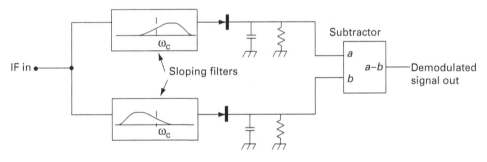

Figure 18.12. Balanced slope FM detector.

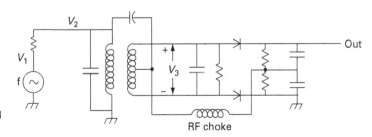

Figure 18.13. Foster–Seeley FM detector.

coupling is not unity; leakage inductance is an element in this circuit. In fact, the leakage inductance, and the magnetizing inductance, are used in a phase shift network, shown in Figure 18.14(a), used to produce the FM-to-AM conversion.

In this circuit, L_1 and L_2 are, respectively, the leakage and magnetizing inductances of the transformer. At the center frequency, f_c, V_2 lags V_3 by 90°. As the frequency increases from f_c, this lag increases. Mostly because of the change in relative phase over the operating region, the magnitude of the vector sum $V_2 + 0.5V_3$ decreases with frequency, while the magnitude of the vector sum $V_2 - 0.5V_3$ increases with frequency. These two combinations correspond to the outputs of the sloping filters in Figure 18.12.

Figure 18.15 is an equivalent circuit of the Foster–Seeley detector using an equivalent circuit for a transformer made up of the leakage inductance, the magnetizing inductance, and an ideal transformer. We will assume for convenience that the primary and secondary inductances and the coupling coefficient have values that give the ideal transformer a 1:1 ratio (see Problem 14.9).

Comparing Figures 18.15 and 18.13, you can see how the top and bottom halves of the transformer secondary are used to add $0.5V_3$ and $-0.5V_3$ to V_2. The

Figure 18.14. (a) Phase shift network: (b) magnitude vs. frequency of $V_2+0.5V_3$ and $V_2-0.5\,V_3$.

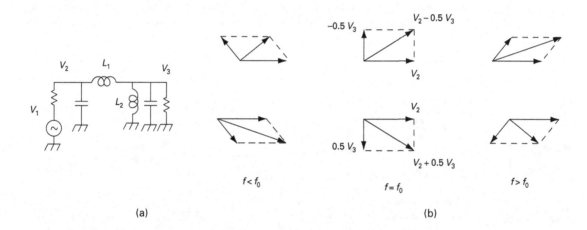

(a)   (b)

Figure 18.15. Foster–Seeley detector equivalent circuit.

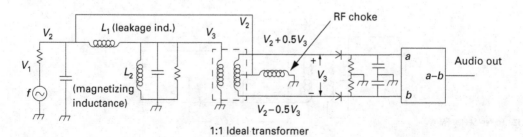

diode detectors produce voltages equal to the magnitudes of $V_2 + 0.5V_3$ and $V_2 - 0.5V_3$ and these magnitude voltages are subtracted, as in Figure 18.12, to produce the output. The RF choke provides a dc return for the diodes. Note that the subtractor can be eliminated by moving the grounding point in the secondary circuit to the bottom of the transformer's secondary winding. The dc blocking capacitor that bridges the transformer is only needed if there is a dc voltage on the primary winding, as when the signal source is the collector of a transistor biased through the primary winding. Finally, note that the capacitor and resistor in parallel with the magnetizing inductance L_2 in the equivalent circuit are actually located on the secondary side of the transformer which is equivalent to being on the left-hand side of the 1:1 ideal transformer.

18.2.5 FM stereo demodulator

Stereo FM, a compatible add-on dating to the 1960s, transmits an L+R (left plus right) audio signal in the normal fashion, and this signal is used by monaural receivers (or stereo receivers switched to "mono"). At the transmitter, The L−R audio is multiplied by a 38 kHz sine wave to produce a DSBSC (double sideband suppressed carrier) signal. This signal, well above the audio range, is added to the L+R audio signal, together with a weak 19 kHz sine wave "pilot" signal. The sum of these three signals, an example of *frequency division multiplexing* (FDM), drives the VCO (or equivalent) to produce the FM signal. At the receiver, the sum signal is demodulated by any ordinary FM demodulator. After demodulation, the L−R signal is brought back down to baseband by a product detector, i.e., a multiplier with a 38 kHz L.O. This L.O., which must have the correct phase (and therefore also the correct frequency), is derived by putting the pilot signal through a frequency doubler. This is shown in the block diagram of Figure 18.16.

Figure 18.16. FM stereo broadcast receiver broadcasting block diagram.

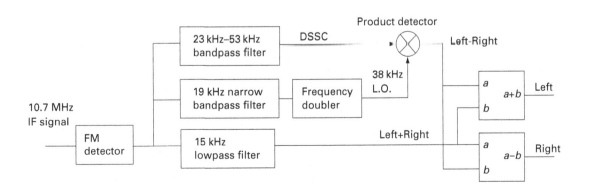

Note that the L–R signal has been doubly demodulated: first by the FM detector and then by the AM product detector. In this stereo system, the L–R signal is more susceptible to noise than the L+R signal. For "full quieting" stereo reception, about 20 dB more signal strength is required than for equivalent monaural reception. For this reason, FM stereo receivers are provided with manual or automatic switches to select monaural (L+R only) operation.

18.2.6 Digital demodulation of FM

When the IF signal has been converted into baseband I and Q signals, the phase of the IF sample is given by $\tan(\theta(t)) = Q(t)/I(t)$. Since frequency is the time derivative of phase, we take the differential of this expression: $(1+\tan^2\theta)\delta\theta = \delta Q/I - Q\delta I/I^2$. Substituting Q/I for $\tan\theta$ yields

$$\delta\theta = \frac{I\delta Q - Q\delta I}{I^2 + Q^2}.$$ (18.5)

For each sampled I, Q pair, $\delta\theta$ is calculated from this pair and the previous pair. The resulting values of $\delta\theta$ are proportional to $d\theta/dt = \omega$. (Note that this could also be done with analog circuitry. Only two analog multipliers would be needed if the IF signal has passed through a limiter, making $(I^2 + Q^2)$ a constant.)

18.3 Power detectors

Square-law detectors are not used as demodulators but are used in laboratory instruments that measure power (wattmeters and rms voltmeters). If we are measuring a sine wave we know that the rms voltage is equal to the peak voltage divided by $\sqrt{2}$. When we know the shape of the waveform, a true square-law meter is not necessary. Even noise power can sometimes be estimated with other than a square-law device. The power of a Gaussian random noise source can be measured by averaging the output of a V^{2n} law device or a $|V|^n$ law device where $n = 1, 2, \ldots$ The square-law device, however, is always the optimum detector in that it provides the most accurate power estimate for a fixed averaging time. When we need the optimum power measuring strategy or if we need to measure the rms voltage of an unknown waveform, we must average the output from a true square-law device. "True rms voltmeters" built for this purpose use a variety of techniques to form the square of the input voltage. Some instruments use a network of diodes and resistors to form a piecewise approximation of a square-law transfer function. Other instruments use a thermal method where the unknown voltage heats a resistor. The temperature of the resistor is monitored by a thermistor while a servo circuit removes or adds dc (or sine wave ac) current to the resistor to keep it at a constant temperature. The diode network

Figure 18.17. (a) Simple diode
power detector; (b) preferred
detector.

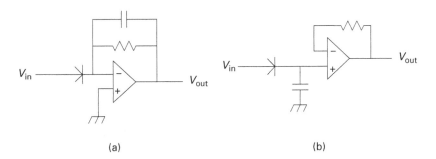

Figure 18.17. (a) Simple diode power detector; (b) preferred detector.

requires large signals and the thermal method has a very slow response. Gilbert cell analog multipliers can be used to square the input voltage, but they are limited to relatively low-frequency signals. Generally, when the power is very low and/or the frequency is very high, and/or a very wide dynamic range is needed, the square-law detector uses a semiconductor diode. A simple (but not particularly recommended) circuit is shown in Figure 18.17. In this circuit the average current through the diode is converted to a voltage by the op-amp. The virtual ground at the op-amp input ensures that V_{in} is applied in full to the diode.

It is important to remember that the I vs. V curve for a diode does not follow a square law, even in a limited region. Rather, the law of the diode junction is exponential: $I = I_s[\exp(V/V_T) - 1]$ where V_T, the thermal voltage, is 26 mV. The diode is used at voltages much smaller than V_T so it is permissible to write $I = I_s[\, V/V_T + 1/2(V/V_T)^2 + 1/6(V/V_T)^3 + 1/24(V/V_T)^4]$. Obviously we can restrict the input voltage enough to neglect the last term. But what about the first term, the linear term? This dominant term will provide a current component that has the same frequency spectrum as the input signal; if that spectrum extends down to zero the output of the detector will be corrupted with extra noise. The third-order term will also have a baseband component. But if the input signal is a bandpass signal, the first-order and third-order terms are high-frequency signals that will be eliminated by whatever lowpass filter is applied to do the averaging (the capacitor in the above circuit). The simple square-law diode detector, then, is appropriate for measuring signals whose frequency components do not extend down into the baseband output spectrum of the detector circuit. (You can think of various two-diode balanced circuits to cancel the linear component but remember that the two diodes must be matched very closely to start with and then maintained at the same temperature.) The diode and op-amp circuit shown above serves to explain why diode detectors are used for bandpass signals but not for baseband signals. A better circuit – the preferred circuit – is shown in Figure 18.17(b). Here the sensitivity is not dependent on I_s and the circuit has a much better temperature coefficient. In this circuit no dc current can flow in the diode. The capacitor, however, ensures that the full ac signal voltage is applied to the diode. Expanding the exponential relation for the diode current and taking only the dc components we have

$$0 = \frac{-V_{dc}}{V_T} + \frac{\frac{1}{2}\langle V_{in}^2 \rangle}{V_T^2} \tag{18.6}$$

and

$$V_{dc} = \frac{\frac{1}{2}\langle V_{in}^2 \rangle}{V_T} \tag{18.7}$$

which shows that the dc output voltage is indeed proportional to the average square of the input voltage.

Problems

Problem 18.1. Assume that in the envelope detector circuit shown below, the diode is a perfect rectifier (zero forward resistance and infinite reverse resistance) and that the op-amps are ideal.

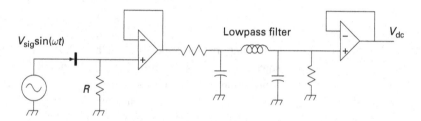

(a) Calculate the efficiency, V_{dc}/V.
(b) Calculate the effective load presented to the generator (the value of a resistor that would draw the same average power from the source).

Problem 18.2. Draw a vector diagram to show why an envelope detector will produce an audio tone when it is fed with the sum of an IF signal and a (much stronger) BFO signal.

Problem 18.3. Show that the voltage divider network in Figure 18.11(b) can produce an output voltage shifted 90° from the input voltage when the input voltage is at a specified center frequency, ω_0. Determine the position of the output phase when the input signal is slightly higher or slightly lower than ω_0.

Problem 18.4. When an interfering AM station is close in frequency to a desired AM station, an audio tone "beat note" is produced, no matter whether the receiver uses an envelope detector or a product detector. (In the case of an envelope detector, the beat note is produced because the amplitude of the vector sum of the two carriers is effectively modulated by an audio envelope. In the case of the product detector, the carrier of the undesired station acts as a modulation sideband and beats with the BFO.) Will the same thing happen with FM? Suppose two carriers (i.e., cw signals), separated by say, 1 kHz, appear in the IF passband of an FM receiver. Let their amplitudes be in the ratio of, say,

1:10. Draw a phasor diagram of the sum of these two signals. Does the vector sum have phase modulation? Will the receiver produce an audio tone? What happens in the case when the amplitudes of the two signals are equal?

Problem 18.5. Try the following experiment with two FM receivers. Tune one receiver to a moderately strong station near the low-frequency end of the band. Use the other receiver (the local oscillator) as a signal generator. (This receiver must have continuous rather than digital tuning.) Turn its volume down and hold it close to the first receiver so there will be local oscillator pickup. Carefully tune the second receiver 10.7 MHz higher in frequency until an effect is produced in the sound from the first receiver. What is the effect? Can you use this experiment to confirm your answer to Problem 18.4?

References

[1] Gosling, W., *Radio Receivers*, London: Peter Peregrinus, 1986.
[2] Landee, R. W., Davis, D. C. and Albrecht, A. P., *Electronics Designers' Handbook*, New York: McGraw-Hill, 1957.
[3] Rohde, U. and Whitaker, J., *Communication Receivers – DSP, Software Radios, and Design*, 3rd edn, New York: McGraw Hill, 2000.

19 Television systems

Television system dissect the image and transmit the pixel information serially. The image is divided into a stack of horizontal stripes ("lines") which are scanned left to right, producing a sequence of pixel (picture element) brightness values. The lines are scanned in order, one after the other, from top to bottom. Brightness values for each pixel are transmitted to the receiver(s). The image is reconstructed by a display device, whose pixels are illuminated according to the received brightness values. This chapter presents television technology in historical order: (1) the electromechanical system that Nipkov patented in 1884 but which was not demonstrated until 1923; (2) all-electronic television, made possible by the development of cathode ray picture tubes and camera tubes; and (3) digital television, which uses data storage and processing in the receiver, allowing the station to update the changing parts of the image, rather than retransmit the entire image for every frame. With the lowered data rate, the bandwidth needed previously to transmit one analog television program can now hold multiple programs.

19.1 The Nipkov system

Electronic image dissection and reconstruction were first proposed in the Nipkov disk system, patented in 1884, which used a pair of rotating disks, as shown in Figure 19.1. The camera disk dissected the image while the receiver disk reconstructed it. The receiver screen, a rectangular aperture mask, was covered by an opaque curtain containing a pin hole, illuminated from behind by an intensity-modulated gas discharge lamp. The position of the pin hole was analogous to the position of the illuminated spot on a CRT. This scanning pinhole was actually a set of N pinholes, arranged in a spiral on an opaque disk that rotated behind the aperture mask. Only one hole at a time was uncovered by the aperture. As the active hole rotated off the right-hand side of the aperture, the next hole arrived at the left-hand side, displaced downward by one scanning line. Identical holes in the transmitter disk

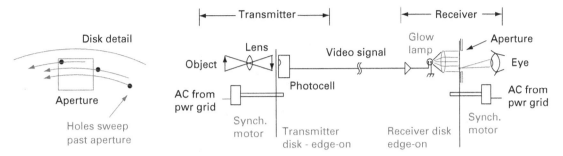

Figure 19.1. Nipkov rotating disk television system.

allowed light from the original image to hit a photocell. Of course the rotating disks had to be synchronized, but when the disks were driven by synchronous motors on the same ac power grid, it was only necessary to find the correct phase.

This primitive low-resolution system was finally demonstrated in 1923 after the invention of the photoelectric cell, vacuum tube amplifier, and neon glow lamp. While the eye views only a single illuminated spot, the persistence of human vision retains an image on the retina of the eye long enough to make the image appear complete if the entire screen is scanned at a rate more than about 20 times per second.[1] Some early experimental broadcasts were made with this very low resolution system in the U.S. on 100-kHz wide channels in the 2–3 MHz range.

19.2 The NTSC system

All-electronic television broadcasts were first made in Germany in 1935 and in England in 1936. The system used in the United States was proposed by the National Television Standards Committee (NTSC) of the Radio Manufacturers Association (RMA). Commercial broadcasting using the NTSC began on July 1, 1941. NBC and CBS both started television service that day in New York City. These stations and four others (Philadelphia, Schenectady, Los Angeles, Chicago) maintained broadcasts throughout World War II to some 10–20 thousand installed receivers. Compatible color broadcasting was added to the NTSC standard in the early 1950s. An engineering tour de force, this system effectively transmits simultaneous red, green, and blue images through the original 6-MHz channels in such a way that monochrome receivers are unaffected.

The NTSC standard specifies a horizontal-to-vertical aspect ratio of four-to-three for the raster (German for screen) with 525 horizontal lines, each scanned in 62.5 microseconds. About 40 of these lines occur during the vertical retrace interval, so there are some 525–40 or 485 lines in the picture. If the horizontal

---

[1] The florescent "phosphorous" material on a television CRT faceplate provides addition persistence.

resolution were equal to the vertical resolution, the number of horizontal picture elements would be $4/3 \times 485 = 646$. The NTSC standard specifies somewhat less horizontal resolution: 440 picture elements. The horizontal retrace of a CRT requires about ten microseconds, so the active portion of each line is $62.5 - 10 = 52.5$ microseconds. The maximum video frequency is therefore given by $\frac{1}{2} \times 440/52.5 = 4.2 \, MHz$. (A video sine wave at 4.2 MHz would produce 220 white stripes and 220 black stripes.)

19.2.1 Interlace

The NTSC frame rate is 30 Hz, i.e., the entire image is scanned 30 times each second. The line rate is therefore $525 \times 30 = 15 \, 750 \, Hz$. Interlaced scanning is specified. This is shown in Figure 19.2. Lines 1–262 and the first half of line 263 make up the first field. The second half of line 263 plus lines 264–525 make up the second field. The lines in the second field fit between the lines of the first field. Each field takes 1/60 sec, fast enough that the viewer perceives no flicker. If all 525 lines were scanned in 1/60 sec, the signal would require twice the bandwidth and the CRT beam deflection circuitry would require more power. Interlacing provides full resolution for fixed scenes but creates artifacts on moving objects (see Problem 19.5).

Figure 19.2. Interlaced scanning.

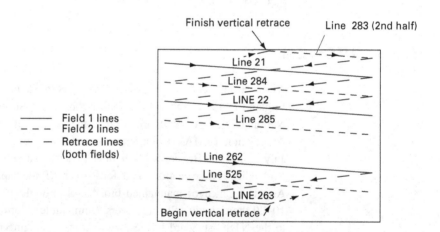

19.2.2 The video signal

The video signal amplitude-modulates the video transmitter. Synchronization pulses are inserted between every line of picture information. The NTSC system uses negative video modulation so that less amplitude denotes more brightness. The principal reason for using this polarity is that impulse interference creates black dots rather than more visible white dots on the screen of the receiver.

19.2.3 Synchronization

A horizontal sync pulse is inserted in the retrace interval between each scanning line and is distinguished by having a higher amplitude than the highest amplitude picture information, i.e., the sync pulse is "blacker" than the black level already blanking the beam during the retrace. The composite video, picture information plus synchronizing pulses, is shown in Figure 19.3. This waveform, which would be observed at the output of the video detector after lowpass filtering to remove the 4.5 MHz sound, shows three successive scan lines.

Television receivers have a threshold detector in the synchronization circuitry in order to look only at the tips of the sync pulses, i.e., the portion that is above the black level, and therefore totally independent of the video information. The burst of eight sinusoidal cycles at 3.579 545 MHz on the "back porch" of each synch pulse provides a reference for the color demodulator, described later. A vertical sync reference is provided by a series of wider sync pulses that occur near the beginning of the vertical blanking period, i.e., every 1/60 second at the end of every field.

Figure 19.3. NTSC video waveforms.

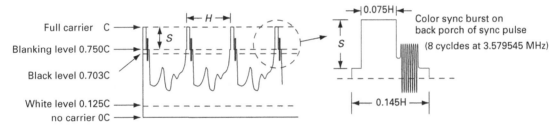

19.2.4 Modulation

Radio transmission of video information (television) requires that we modulate a carrier wave with the composite video signal. The NTSC system uses full-carrier AM modulation for the video. Since the NTSC video signal extends to 4.2 MHz, ordinary double-sideband AM would require a bandwidth of 8.4 MHz. To save bandwidth, the lower part of the lower sideband is removed at the transmitter by filtering, allowing a 6-MHz channel spacing. The resulting *vestigial sideband* signal consists of the entire upper sideband, the carrier, and a vestige of the lower sideband, as shown in Figure 19.4.

At the receiver, a low-frequency video component, because it is present in both sidebands, would produce twice the voltage that would be produced by a high-frequency video signal, present only in the upper sideband. This problem is corrected by using an IF bandpass shape that slopes off at the lower end such as shown in Figure 19.5. At the video carrier frequency the amplitude response is ½. (This response is as good as and simpler to obtain than the dotted curve where all of the double sideband region is reduced to ½.)

Figure 19.4. NTSC 6-MHz channel allocation.

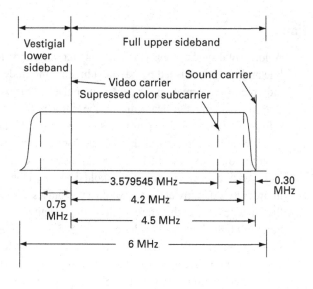

Figure 19.5. IF response to equalize the vestigial sideband.

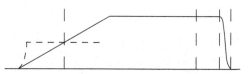

For this IF equalization to work correctly, this IF filter should also have linear phase response. When surface acoustic wave (SAW) filters became available to determine the IF bandpass shape, this requirement was easily satisfied.

19.2.5 Sound

The audio or "aural" signal is transmitted on a separate carrier, 4.5 MHz higher in frequency than the video carrier. The NTSC system uses FM modulation for the audio component. The maximum deviation is 25 kHz, i.e., the maximum audio amplitude shifts the audio carrier 25 kHz. Normally the audio transmitter is separate from the visual transmitter and their signals are combined with a diplexer to feed a common antenna.

19.2.6 NTSC color

The NTSC standard for compatible color television was adopted in 1953. Like color photography, color television is based on a tricolor system. When two or more colors land at the same spot on the retina, their relative intensities determine the perceived hue. When three complete images in three suitably chosen primary colors are superimposed, the eye perceives a full-color image. The particular red, green, and blue standards specified in the NTSC standard were based on practical phosphors used for the dots on the faceplate of the CRT. Since color television requires the simultaneous transmission of three images, it is interesting to see how the color system can shoehorn three images into the 6-MHz channel originally allocated for a single monochrome image and do it in a way that made color broadcasts compatible with existing monochrome receivers. The solution to the bandwidth problem takes advantage of the fact that the monochrome video signal leaves empty gaps across the 6-MHz band. There is considerable redundancy in a typical picture. In particular, any given line is usually very much like the line preceding it (producing strong correlation at 62.5 microseconds) so the video signal is similar to a repetitive waveform with a frequency of 15 750 Hz, the horizontal scan frequency. If the lines were truly identical, the spectrum would be a comb of delta functions at 15 750 Hz and its harmonics. Since one line differs somewhat from the next, these delta functions are broadened, but the spectral energy is still clumped around the harmonics of the horizontal scanning frequency, leaving relatively empty windows which can be used to transmit color information. (Note that if the entire picture is stationary then the spectrum is a comb of delta functions spaced every 30 Hz and essentially all the bandwidth is unused.)

Instead of transmitting the red, blue, and green (RGB) signals on an equal basis, three linear combinations are used. One of these, the *luminance* signal, Y, is chosen to be the brightness signal that would have been produced by a monochrome camera: $Y = 0.299R + 0.587G + 0.114B$. The other two linear combinations are $I = 0.74(R-Y) - 0.27(B-Y)$ and $Q = 0.48(R-Y) + 0.41(B-Y)$.

The luminance signal directly modulates the carrier, just as in monochrome television, and monochrome receivers respond to it in the standard fashion. Each of the other two signals, I and Q, modulates a subcarrier (just as the Right-minus-Left audio signal modulates a 38-kHz subcarrier in FM stereo). The color subcarriers have the same frequency, 3.579 545 MHz, but they differ in phase by 90°. Figure 19.6 shows how two independent signals are transmitted and recovered using this quadrature AM modulation (QAM).

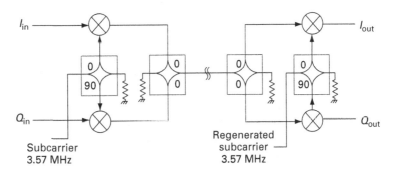

Figure 19.6. Quadrature AM modulator/demodulator lets the two color signals occupy one band.

The reference signal necessary to regenerate these local carriers is sent as a burst of about eight sinusoidal cycles at 3.579 545 MHz on the back porch of each horizontal sync pulse as shown in a previous figure. This "color burst" is used in the receiver as the reference for a phase lock loop. The color information, like the luminance information, is similar from line to line so its power spectrum is also a comb whose components have a spacing equal to the horizontal scanning frequency. The subcarrier frequency is chosen at the middle of one of the spectral slots left by the luminance signal so the comb of color sidebands interleaves with the comb of luminance sidebands – a frequency multiplexing technique known as spectral interlacing.

Compatibility is achieved because the spectral interlacing greatly reduces visible cross-talk between chrominance and luminance information. To see this, consider a very simple signal, a uniform color field such as an all-yellow screen. Since this field has a color, i.e., it is not black, white, or gray, there will be nonzero I and Q signals. In this example, since the color information is constant, the I and Q signals together are just a sine wave at the color subcarrier frequency. Their relative amplitudes determine the hue while their absolute amplitudes determine the saturation. One would expect this 3.58-MHz video component to produce vertical stripes. And the beam, as it sweeps across the screen, does indeed get brighter and dimmer at a 3.58-MHz rate, trying to make some 186 stripes in the 52.5 microsecond scan. But, on the next line, these stripes are displaced by exactly one half-cycle. The result is that the entire screen, rather than having 186 vertical stripes, has a fine-gridded checkerboard or "low visibility" pattern. Colored objects viewed on a monochrome receiver can be seen to have this low visibility checkerboard pattern. Note that there are some unusual situations where spectral interleaving does not work. If the image itself is like a checkerboard with just the right grid spacing the luminance signal will fall into the spectral slots allocated for the chrominance signal and vice versa. A herringbone suit for example, will often have a gaudy sparkling appearance when viewed on a color receiver. NTSC receivers eventually were equipped with comb filters to separate the chrominance and luminance signals, but at the expense of some vertical resolution.

The low-visibility principle is applied not only to avoid luminance–chrominance cross-talk but also to reduce the effect of the beat between the 4.5-MHz sound carrier and the color subcarrier. To take advantage of the low-visibility principle, television standards were modified slightly when color television was introduced. The sound carrier remained the same at 4.5 MHz above the video carrier. The relation between the horizontal scanning frequency, f_h, and the color subcarrier frequency, f_{SC}, was picked to be $f_{SC} = 227.5 f_h$. Then the sound subcarrier-minus-color subcarrier beat was likewise made an odd number of half-multiples of the horizontal scanning frequency: $4.5\,\text{MHz} - f_{SC} = 58.5 f_h$. Putting these two relations together determines the horizontal frequency, 15 734.264 Hz, and the color subcarrier frequency, 3.579 545 MHz. The number of scanning lines remained at 525 so the vertical frequency changed from 60 Hz to 262.5f_h = 59.940 Hz. With these choices, the sound carrier is at 286 times the horizontal frequency. This would produce a high visibility pattern but the sound carrier is above the nominal video band and can be filtered out easily.

Color television receiver

The block diagram of Figures 19.7 shows the overall organization of an NTSC color television receiver. The first block is simply an AM radio receiver for VHF, UHF, and cable frequencies (about 52–400 MHz).

The carrier of the selected channel is translated to an IF frequency of 45.75 MHz and the IF bandwidth is about 6 MHz. The shape of the IF passband must be quite accurately set to equalize the vestigial sideband, to provide full video bandwidth, and to reject adjacent channels. This originally required careful factory adjustment of many LC tuned circuits but was later determined by the geometry in a single SAW filter. The IF signal contains the composite video (luminance and sync) plus the sound and color information. The sound and color signals, which are essentially narrowband signals around 4.5 MHz and 3.57 MHz, ride on top of the luminance signal. A 4.5-MHz bandpass filter isolates the sound signal in the block labeled "4.5-MHz FM Receiver." FM stereo sound uses the demodulator described in Chapter 18. The sound sub-carrier is at 31.5 kHz so that the oscillator in the demodulator can be synchronized to the horizontal sweep trigger at 15.75 kHz. A 4-MHz lowpass filter eliminates the sound signal from the video. The resulting video signal, the "Y" signal, contains the brightness information and would produce the correct picture if sent to a monochrome picture tube.

Chrominance processor

The 3.57-MHz color burst on the back porch of the horizontal sync pulses provides a reference for the phase locked L.O. in this QAM demodulator. Not shown on the diagram is an electronic switch, the burst gate, which is controlled by the synchronization circuitry to apply the reference signal only during the burst period in order to improve the signal-to-noise ratio of the loop. The local carrier is fed to a product detector (multiplier) to demodulate the I signal. A 90° phase shift network proves a second local carrier, shifted in phase for a second product detector to demodulate the Q signal.

Sync processor

A comparator, with its threshold set at the black level, strips away the video signal, producing a clean train of sync pulses. As explained above, a simple RC differentiator then provides horizontal reference pulses and an RC integrator provides vertical reference pulses. A VCO, phase-locked to the horizontal reference pulses, provides a flywheel stabilized horizontal time base. The VCO operates at twice the horizontal frequency and is divided by 2 to provide 15 734 Hz for the PLL phase detector and to drive the horizontal deflection

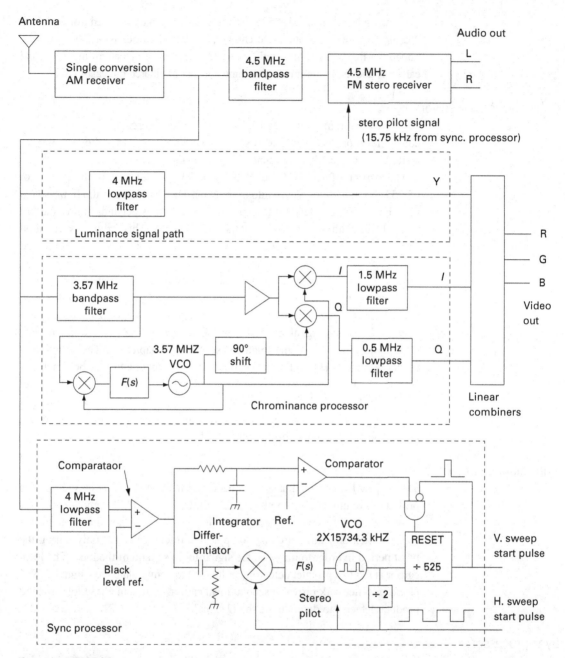

Figure 19.7. NTSC color television receiver block diagram.

circuitry. The VCO is also divided by 525 to provide an equally stable vertical time base. This divider must operate with the right phase for the picture to have the correct vertical alignment so the divide-by-525 counter has a reset input which will be triggered when the counter output has failed to coincide with several consecutive vertical reference pulses.

19.3 Digital television

The first digital television broadcasting in the U.S. began in 1998, following the 1996 adoption by the Federal Communications Commission (FCC) of the system developed by the Advanced Television Standards Committee (ATSC). Television stations were assigned new channels for digital broadcasting but have also continued their NTSC analog broadcasts during the digital phase-in period which, at this writing, is scheduled to end in 2009.

The ATSC standard incorporates MPEG-2 program compression (itself a standard, ISO/IEC 13818), which is based on the temporal and spatial redundancy of the video program material.[2] MPEG-2 is also used for digital television in Europe (DVB) and Japan (ISDB), for direct television broadcasts from satellites, and for DVDs. Digital processing and storage make it possible to exploit the redundancies of the program material to compress up to six standard-definition (480-line) programs or at least two high-definition (720 or 1080-line) programs into the same 6-MHz wide channel needed for a single NTSC program. (The same redundancies are exploited to a much smaller degree in the NTSC system with its temporally-interlaced fields and its frequency-interleaved luminance and chrominance signals.) The net data rate for the ATSC system is 19.3 Mbits/sec through a 6-MHz channel, after overhead for error correction. Let us estimate how much compression is involved when four standard-definition programs with typical frame-to-frame motion are compressed down to a total bit rate of 19.3 Mbits/sec. We will assume these programs are in the "480i" format, where 480×704 pixels are transmitted at a 30-Hz rate. Let us further assume there are three 8-bit numbers, a luminance value and two chrominance values, for each pixel. The overall data rate would be four channels $\times 480 \times 704 \times 8$ bits $\times 3$(colors) $\times 30$ Hz $= 973$ Mbits/sec. Dividing by 19.3 Mbits/sec, the compression factor is 50.4. When audio is included in the calculation, the factor increases slightly.

Digital television uses a three-step process. First, the video and audio program material, in raw digital bit streams, is compressed to packet streams having much lower bit rates. The video and audio packets, together with packets of ancillary data and null packets (padding) are merged together to form a stream of *transport packets*, each one containing 187 data bytes. The packets have headers to identify the program (since a stream may contain several independent programs) and the type of packet: video, audio, or, ancillary information, or null. Second, the stream of packets is run through a two-stage forward error correction (FEC) encoder which adds redundant bits to each packet so that the receiver can detect and correct transmission errors caused by noise and

---

[2] The ATSC system does not use the audio part of MPEG-2 for audio, but instead uses AC-3 (Dolby digital) compression, which provides each program with up to five full audio channels plus a low-frequency subwoofer channel for "surround sound."

interference. Third, the bits in the stream of expanded packets produce the RF signal for transmission. This signal is single-sideband suppressed-carrier AM (SSBSC) except that a vestige of the lower sideband remains (as in the NTSC standard) and the carrier is not completely suppressed; a pilot carrier is inserted as a reference for synchronous detection in the receiver. At the receiver, the three steps occur in the opposite order. The bit stream is demodulated from the RF signal. The FEC-encoded packets are then decoded, correcting transmission errors and producing the original packets of compressed data. These packets are then separated, according to program, and each program is decompressed into the original audio and video bit streams.

19.3.1 MPEG encoding

The picture is encoded (compressed) at the lowest level in blocks of eight pixels by eight pixels. (Note that "compression" and "decompression" of the program data are usually called coding and encoding, the same terms used for the processes used to anticipate and correct transmission errors.) Video usually has a high degree of spatial correlation, i.e., adjacent pixels tend to have similar brightness and color. When an $(n \times n)$-pixel picture is encoded using a two-dimensional discrete Fourier transform (DCT), the resulting coefficient matrix can be inverse-tranformed to exactly reconstruct the picture. But if the coefficients are coarsely quantized, i.e., rounded to use fewer bits before the inverse transformation, the reconstructed picture is found to retain most of its original quality.[3] Moreover, because of spatial correlation, these coefficients tend to concentrate in one corner of the coefficient matrix. Away from this corner, the coefficients have low values. Thus, a great many of the coefficients can be represented as zeros and the rest by numbers of only a few bits, so DCT encoding providing a significant amount of compression.

Except for scene changes, the differences from one frame to the next are mostly due to motion of elements within the picture (subject motion) or motion of the picture as a whole (camera motion). From frame to frame, a given block will therefore mostly just shift its position somewhat. In the ATSC system, for each 16 × 16 block (*macroblock*) in a new frame, the MPEG encoder determines the displaced macroblock in the previous frame that provides the best match by minimizing the sum of the absolute values of the differences of the pixel brightness values. The position of the displaced macroblock is specified by a *motion vector*, e.g., 1,1 would indicate that the best match macroblock is shifted one pixel up and one pixel to the right. These motion vectors are part of the video update information. The rest of the information consists of DCT-tranformed *differences* between the pixel values in these displaced "prediction" macroblocks and the new pixel values in the block being updated. At the

[3] The same principle is used for audio compression. Blocks of audio voltage samples are transfomed and the resulting coefficients are compressed and transmitted.

encoder, the macroblock is subdivided into four 8×8 blocks, and the 64 pixel differences for each block which, again, have considerable spatial correlation, are transformed using an 8×8 DCT.[4] (It is computationally more efficient to transform small blocks, and the 8×8 size was found adequate to provide the desired compression.) When most of the coefficients in a transformed block are negligible, the resulting stream of digital numbers consists mostly of zeros, and run-length encoding (specification of the number of consecutive zeros) is used to increase the degree of compression. The number of consecutive zeros is increased by using a certain zigzag readout ordering of the coefficient differences. In addition, variable length coding is used for the nonzero transform differences; the viewer can tolerate coarser quantization for the high spatial frequencies.

Note that not all the frames can be predicted from previous frames; channel surfers need the picture to change promptly and, of course, a change of scene requires all-new data. Occasional *refresh frames* (tagged as such) are therefore inserted into the stream of prediction/correction frames. In a refresh frame, every block is transmitted. Between refresh frames, only changing macroblocks need to be transmitted. As in the NTSC system, the two color signals can have lower spatial resolution than the luminance signal, just as a black and white photograph can be "colorized" using a relatively broad brush. In the ATSC system, each of the two color elements has half the resolution (1/4 the number of pixels) as the luminance element, further compressing the video signal. The output from the MPEG encoder, together with the audio material, is a string of 188-byte packets: 187 data bytes plus one sync byte. Note that the packets are asynchronous; their lengths and mix is program dependent. Video and audio are both time-stamped so that they can be aligned by the decoder. After leaving the MPEG encoder, the bits in the 187 data bytes are "pseudo-randomized" by multiplication with the bits from a pseudo-random shift register sequence. This step, which is undone in the receiver, just ahead of its MPEG decoder, ensures that the spectrum is flat, even if the output of the encoder is interrupted, and makes it easier for the receiver to find the transmission packet sync byte which is added during the forward error correction encoding.

19.3.2 Forward error correction

The ATSC system applies two levels of forward error correction. "Forward" means that bits added in the encoding process make it possible to correct errors rather than simply detect them. Of course, this works only up to a limit; when the signal-to-noise ratio drops below a threshold, uncorrectable errors begin to produce visible artifacts and beyond this, the signal soon suffers total

---

[4] The DCT can be done by multiplying the (8×8) pixel or pixel difference matrix by an (8×8) transform matrix. The inverse of this matrix provides the inverse transformation. However, there are faster methods, analogous to the FFT algorithm for calculating the DFT.

Figure 19.8. ATSC transmission packet.

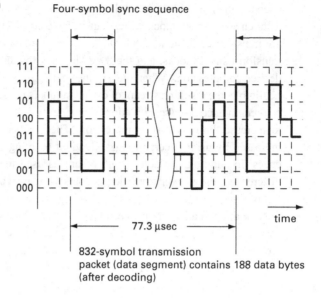

Four-symbol sync sequence

832-symbol transmission packet (data segment) contains 188 data bytes (after decoding)

77.3 μsec

time

degradation – the well known "threshold effect" in digital communication links. The bit-randomized MPEG packets are first processed by a Reed–Solomon encoder which adds 20 parity bytes to the 187 data bytes increasing each packet to 207 bytes. This Reed–Solomon encoding makes it possible to correct up to 10 defective bytes per packet, no matter how many bit errors these bytes contain. The next stage is an interleaving (shuffling) of bytes within blocks of 52 of the 207-byte packets. This interleaving makes it less likely that a wide noise burst will leave more than 10 bad bytes in any packet after de-interleaving has been done at the receiver. The second level of FEC encoding is a convolutional or trellis encoding, which is effective against white noise. The trellis encoder produces three output bits for every two input bits, bringing each packet up to $207 \times 3/2 \times 8 = 2484$ bits, or $2484/3 = 828$ three-bit (eight-level 8-VSB) symbols. The trellis encoding is a flow process; each bit triplet leaving the trellis encoder is calculated from only the last four pairs of input bits. Finally, a four-symbol sync sequence is added for a grand total of 832 8-VSB symbols per transmitted packet, as shown in Figure 19.8.

With the concatenated Reed–Solomon and Trellis error correction coding, the threshold of visibility for the ATSC system occurs at a S/N ratio of 14.9 dB. At this threshold, the data segment (832-symbol transmission packet) error rate is about 1.93×10^{-4} or 2.5 segment errors per second.[5] The FCC requires that over the area of service, a broadcaster's ATSC signal must exceed the threshold by 7 dB. Compared to NTSC, ATSC broadcasting requires less than 1/10 as much

---

[5] These numbers are taken from ATSC Document A/54A, "Recommended Practice Guide to the Use of the ATSC Digital Television Standard."

transmitter power. (Nearly all the ATSC transmitted power bears information, while much of the NTSC power goes into the carrier and horizontal sync pulses.)

19.3.3 RF modulation

The three-bit symbols are converted by an A-to-D converter into an eight-level analog signal which amplitude modulates the carrier wave. The carrier is suppressed except for a low-amplitude pilot. Figure 19.9 shows a modulator circuit whose input is the stream of transmission packets, each containing 832 three-bit symbols, including the four-symbol segment sync. A filter eliminates all but a vestige of the lower sideband.

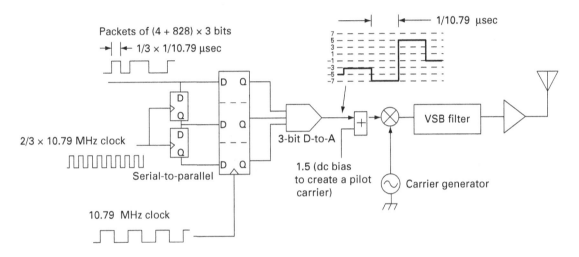

Figure 19.9. ATSC 8-VSB
modulator.

Figure 19.10 compares the spectrum of an ATSC signal with that of an NTSC spectrum.

Figure 19.10. Comparison of
the ATSC spectrum (a) and the
NTSC spectrum (b).

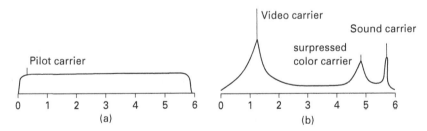

19.3.4 HDTV receiver

The ATSC standard provides 18 different video formats, ranging from high-definition 1920×1080 pixels, interlaced with 60 fields/second or noninterlaced with 30 or 24 frames/second, down to 640×480 pixels, interlaced with 60 fields/second or noninterlaced with 60, 30, or 24 frames/second. The receiver identifies the video format being used and decodes the video. It then scales and interpolates the decoded image to produce a picture that fits the native resolution of the display screen, unless the program happens to match the native resolution. When the screen does not have the same aspect ratio (16:9 or 4:3) as the program, the receiver will produce a "letter box" display, with the picture between black horizontal strips, or a "keyhole" display, with the picture between black vertical strips, or the view may be able to opt for a distorted picture that fills the screen. Figure 19.11 is a simplified block diagram for an ATSC television receiver.

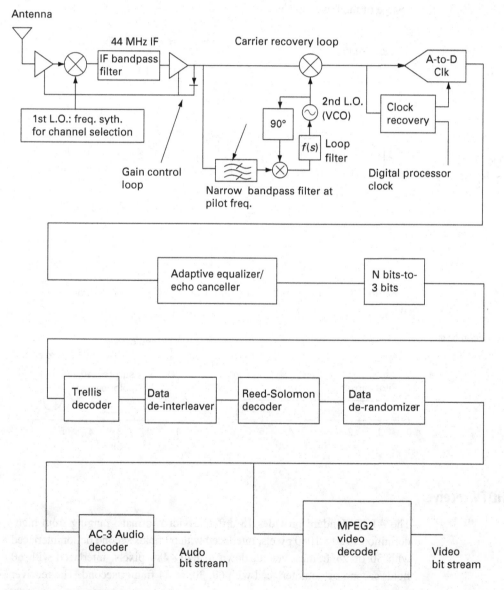

Figure 19.11. An ATSC television receiver block diagram.

The conventional RF sections are all in the first row of blocks. The signal is first mixed down to a first-IF frequency, often around 44MHz. Just as in an NTSC receiver, the L.O. for the first mixer selects the channel. The amplified IF signal is synchronously detected by multiplying it by a sine wave at the carrier frequency. This sine wave is produced by a VCO referenced to the pilot carrier in a phase lock loop. Without the 90° phase shifter in the loop, the VCO output would be 90° out of phase with the pilot carrier and there would be no output from the synchronous detector. (If the consortium that developed the ATSC standard had opted for quadrature AM (QAM), there would have been two final mixers producing I and Q signals, rather than just a single mixer producing the I signal.) The detected signal, which is a zero-frequency IF "baseband" signal, is fed to an A-to-D converter. While the amplitude of the signal has nominally only eight levels (three bits), at this point it is still dithered by multipath echoes and noise, so the A-to-D converter may have eight or ten bits. The clock rate of this converter may be a multiple of the symbol rate in order to recover the symbol clock with high precision. The clock recovery block finds the four-symbol segment sync sequence and phase locks the recovered symbol clock to it.

The equalizer is an adaptive filter which serves two purposes. It corrects the overall frequency response (transfer function) to minimize intersymbol interference (see Chapter 22) and it cancels echos – multipath interference. Both of these functions are essential in order to reproduce the transmitted signal well enough to distinguish one of eight possible amplitude values for each symbol. The usual form of the equalizer is a transversal FIR filter with programmable tap weights (amplitude and phase). The weighted outputs from the taps are summed to form the filter output. Every 313th transmission packet contains a "training" sequence, which is a reference waveform used by the receiver to determine the proper weights for the equalizer filter taps. Once the equalizer and the overall gain are correctly set, the signal can be sampled at the symbol rate × 1 to reproduce the data stream shown in Figure 19.8. Subsequent processing steps then undo the various stages of encoding. A trellis decoder (Viterbi decoder) is followed by a block that de-interleaves the bytes. Then the Reed–Solomon encoding is decoded. Finally, the pseudorandom coding is undone and the data can proceed to the MPEG decoder, which outputs synchronized video and audio.

Problems

Problem 19.1. Suppose the NTSC signal from a television receiving antenna consists of the direct signal (via the line-of-sight path to the transmitting station) and a weak secondary signal (via reflection from a metal tower off the line-of-sight). If the path taken by the secondary signal is 1 km longer than the path of the direct signal, what will be the position of the "ghost" image on the screen of the receiver?

Problem 19.2. Motion pictures from film shot at 24 frames/second are transmitted as 60 field/second television images using a technique called *3:2 pull down*. A film frame is held in place while two television fields are transmitted. The subsequent film frame is then pulled down and held in place while three television fields are transmitted. Show that this scheme results in an average film rate of 24 frames/second.

Problem 19.3. Consider a high-definition television monochrome signal with 1920×1080 pixels and progressive scan (no interlace) at 60 frames/sec. If the brightness of each pixel is specified by an 8-bit number, and if no compression is used, what is the data rate? Answer: 995 Mbps. What compression ratio is needed if this signal is to be transmitted over a standard width television channel at a rate of 19.3 Mbits/second?

Problem 19.4. Suppose an NTSC test pattern consists of five vertical bars of equal width but different colors. Let the bars all have the same luminance (brightness) and be only lightly colored (unsaturated). Sketch the waveform for one line of video.

Problem 19.5. Interlaced scanning provides more resolution for fixed scenes but can produce artifacts with moving objects. Think of a situation where interlaced scanning could make a moving object vanish.

Problem 19.6. Shannon's channel capacity theorem states that data, if suitably encoded for error correction, can be sent with an arbitrarily low error rate if the net data rate (bits/sec.) does not exceed the channel capacity, given by $C = B\log_2(1+S/N)$. Compare the ATSC bit rate, 19.3 Mbits/second with the theoretical capacity of the 6-MHz channel at the threshold S/N ratio of 14.9 dB.

References

[1] A/53: ATSC Digital Television Standard, Parts 1–6, 2007 Advanced Television Systems Committee, 1750 K Street, N.W., Suite 1200 Washington, D.C. 20006 (available at www.atsc.org).
[2] Benson, K. B., Editor, *Television Engineering Handbook*, New York: McGraw-Hill, 1986. (Updated version of Fink, D. G., Editor, "*Television Engineering Handbook*," New York: McGraw-Hill, 1957.)
[3] Jackson, K. G. and Townsend, G. B. Editor, *TV & Video Engineer's Reference Book*, London: Butterworth Heineman, 1991. Modern handbook from Britain.
[4] *NAB Engineering Handbook*, 10th edn, 2007, National Association of Broadcasters. National Association of Broadcasters, *Engineering Handbook*, 8th edn., 1993.
[5] Recommended Practice: Guide to the Use of the ATSC Digital Television Standard, including Corrigendum No. 1, Advanced Television Systems Committee, 1750 K Street, N.W., Suite 1200 Washington, D.C. 20006 (available at www.atsc.org).

20 Antennas and radio wave propagation

While discussing transmitter and receiver circuitry we did not have to know much about antennas or propagation. It sufficed to know only that a voltage applied to the terminals of a transmitting antenna causes a proportional voltage to appear very shortly thereafter at the terminals of a receiving antenna. To be more exact, it was sufficient to know that everything between the terminals of the two antennas is equivalent to a linear two-port network. Here we will consider the transmission through this propagation link.

When an ac source (transmitter) is connected to an antenna (practically any metal structure) the resulting current has a component that is in phase with the applied voltage. The impedance of the antenna therefore has a real part, a resistance, and draws power from the source. If the antenna is efficient, most of the power flows away from the antenna in the form of (energy-bearing) electromagnetic waves and only a small fraction of the power will be dissipated by ohmic heating of the antenna itself. The impedance will also generally have a nonzero imaginary part, a reactance. If the reactance is zero at the operating frequency the antenna is said to be resonant, just as an *RLC* circuit is purely resistive at its resonant frequency. An external tuning network (an *antenna tuner*) can be used to cancel the reactance and also transform the resistance to a value that matches a receiver's input impedance or to a value that draws a desired amount of power from a transmitter.

20.1 Electromagnetic waves

As an electromagnetic wave propagates away from the transmitting antenna, it takes on a spherical wavefront. By the time it reaches a distant receiving antenna, the wavefront has a very large radius of curvature and is essentially a plane wave. The *E* and *H* vectors (electric and magnetic fields, measured respectively in volts/ m and amperes/m) both lie in the plane of the wavefront, i.e., they are transverse to the direction of propagation as shown in Figure 20.1. The fields are in phase with each other; they rise and fall together in space and time.

Figure 20.1. An electromagnetic wave – the **E** and **H** fields are transverse to the direction of propagation.

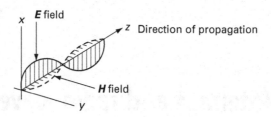

The ratio of the electric field strength to the magnetic field strength in free space is given by $E/H = (\mu_0/\varepsilon_0)^{1/2} = 120\pi = 377$ ohms. This ratio is known as the "impedance of free space." These fields are perpendicular to each other and their vector cross-product, $\boldsymbol{E} \times \boldsymbol{H}$, points in the direction of propagation. A continuous monochromatic wave has sinusoidal spatial components, $E_0 e^{-jkz}$ and $H_0 e^{-jkz}$, as shown in the figure. The wavenumber, k, is defined as $2\pi/\lambda$ radians per meter. The phase velocity, ω/k, is given by $c = (\mu_0\varepsilon_0)^{-1/2} = 3 \times 10^8$ m/s (the speed of light).

20.1.1 Propagation in a vacuum

Static electric and magnetic fields are always associated with sources, i.e., electric field lines terminate on electric charges and magnetic field lines encircle current filaments. But when an antenna launches an electromagnetic wave, the field lines break away from the sources, reconnecting into closed loops, and the wave becomes autonomous. For a vacuum, Maxwell's curl equations are

$$\nabla \times \boldsymbol{E} = -\mu_0 j\omega \boldsymbol{H}, \tag{20.1a}$$

$$\nabla \times \boldsymbol{H} = \varepsilon_0 j\omega E. \tag{20.1b}$$

The right-hand sides of these equations are like sources; the E-field is produced by the changing H-field and the H-field is produced by the changing E-field. You can verify the statements made above about plane waves by writing the E-field shown in Figure 20.1 as $\boldsymbol{E} = \hat{x}E_0 e^{j(\omega t - kz)}$ and substituting it into Equation (20.1a) to get $\boldsymbol{H}$. You can then verify that $\boldsymbol{H}$ satisfies Equation (20.1b).

Electromagnetic waves are produced by electric charges undergoing acceleration. Time-varying currents contain accelerating charges. A sinusoidally time-varying current distribution on an antenna launches sinusoidal electromagnetic waves. The time-averaged power density or "energy flux" of the waves is given by $S = \frac{1}{2}(E \times H) = \frac{1}{2}E^2/377$ watts/m^2, where E and H are the (peak) field amplitudes and the factor $\frac{1}{2}$ is the time average of $\cos^2(\omega t)$. At a receiving antenna, the fields from an incident wave produce currents that result in a voltage at the antenna terminals.

20.2 Radiation from a current element

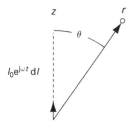

$I_0 e^{j\omega t}\, dl$

Figure 20.2. An elemental sinusoidal ac current element.

Just as Maxwell's equations can be manipulated to yield back Coulomb's law (which gives the E-field from an elementary point charge) and Ampere's law (which gives the H-field from a constant current element), they can also be stirred together to yield the E and H fields produced by a *time-varying* current element. The E and H fields produced by an antenna are superpositions of the E and H fields from every elemental ac current element on the antenna. Consider an ac current element in the z-direction, as shown in Figure 20.2.

This element has length δl and carries a current $I_0 \cos(\omega t)$ which, as usual, we express as $I_0 e^{j\omega t}$. At the observation point, r, the magnetic field is given by

$$\delta H = \frac{I_0 \delta l}{4\pi}\left(\frac{jk\sin(\theta)e^{j(\omega t - kr)}}{r} + \frac{\sin(\theta)e^{j(\omega t - kr)}}{r^2}\right), \qquad (20.2)$$

where $k = \omega/c$ and θ is the angle between the current element δl and the vector r. Equation (20.2) is Ampere's law, generalized for an elemental current with a sinusoidal time variation. The direction of δH is given by $\delta l \times r$ (into the page). The first term in the bracket falls off slowly as $1/r$ and is the radiation term. The second term falls off quickly as $1/r^2$ and corresponds to near-field stored energy; j is just a phase factor since $j = e^{j\pi/2}$. The normally implicit $e^{j\omega t}$ term is included here to highlight the wave term, $e^{j\omega t - kr}$. If we let the frequency approach zero, the radiation term vanishes (since k goes to zero), the second term becomes $\sin(\theta)/r^2$ and Equation (20.2) reduces to Ampere's law (also known as the Biot – Savart law) for calculating the magnetic field produced by a dc current element.

Figure 20.3 shows the radiation pattern (*antenna pattern*) obtained from the radiation term in Equation (20.2) (the first term in the bracket). This is a surface plot where the distance from the origin to any point on the surface is proportional to the power radiated in that direction, i.e., proportional to $|H|^2$ at a large constant value of r. We do not have to work out the E-field separately, since we already know that, in the far field, E is perpendicular to both H and k and that its magnitude (in volts/m) is 377 times the magnitude of H (in amps/m).[1]

In this chapter we are interested in radiation far from the antenna, so we can ignore the near field (second) term in Equation (20.2). At any observation point, the H-field is the integral of the contributions from every current element in any given antenna. For a wire antenna, we would evaluate a line integral. For a reflector antenna, we would have a surface integral, summing up the contributions from the current flowing on each element of the surface.

---

[1] If we do need to know E in the $1/r^2$ (near field) region, we can calculate the curl of H from Equation (20.3) and plug that into Equation (20.2) to get E.

Figure 20.3. Elementary dipole radiation pattern (surface of constant field strength).

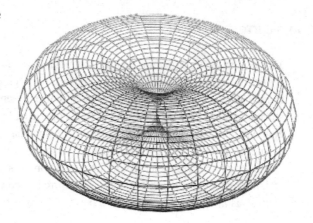

20.3 Dipole antenna

A short dipole antenna (shorter than, say, $\lambda/20$) will have the same radiation pattern as the elemental dipole since we can consider it to be a string of elemental dipoles, short enough that the phase paths from every one to the observation point are essentially the same. Thus, the string is equivalent to a single elemental dipole of strength $I\delta\, l = \sum I_i \delta l_i$. We can calculate the power radiated by a single elemental dipole by integrating the average energy flux density, $\frac{1}{2}EH$, over the surface of a bounding sphere to include the power radiated in every direction. Using the radiation term in Equation (20.2) for H, we have

$$\delta H = \frac{I_0 \delta l}{4\pi}\left(\frac{jk\sin(\theta)\mathrm{e}^{j(\omega t - kr)}}{r}\right),\tag{20.3}$$

$$\text{Total Pwr} = \int (1/2)EH\mathrm{d}S = \int (1/2)(\mu_0/\epsilon_0)^{1/2}H^2 r^2 \mathrm{d}\Omega$$

$$= (1/2)(\mu_0/\epsilon_0)^{1/2}(I_0 k\delta l/(4\pi))^2 \int \frac{\sin^2(\theta)}{r^2}r^2 \mathrm{d}\Omega$$

$$= (1/2)(\mu_0/\epsilon_0)^{1/2}(I_0 k\delta l/(4\pi))^2 \int_0^\pi 2\pi \sin^3(\theta)\mathrm{d}\theta \tag{20.4}$$

$$= (1/2)120\pi(I_0(2\pi/\lambda)\delta l/(4\pi))^2 8\pi/3$$

$$= 40\pi^2(I_0\delta l/\lambda)^2.$$

It turns out that the current distribution for a short dipole is a triangle function (actually the almost linear ramp portion of a sine function near zero), as shown in Figure 20.4.

Figure 20.4. (a) A short dipole has a triangular current distribution. (b) A half-wave dipole has a sinusoidal current distribution.

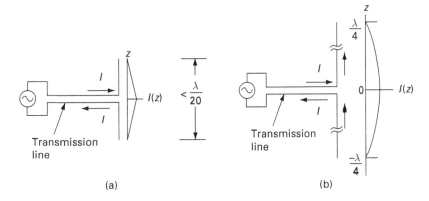

(a)   (b)

The average current for a short dipole is therefore half the input current, so the power radiated is

$$P_{\text{short dipole}} = 40\pi^2\left(\frac{I_{\text{in}}}{2}l/\lambda\right)^2, \tag{20.5}$$

where I_{in} is the peak input current and l is the length of the dipole. The input power must be equal to $\frac{1}{2}I_{\text{in}}^2R$, where R is the real part of the antenna's input impedance, known as the *radiation resistance*. For the short dipole, we can therefore write

$$R_{\text{short dipole}} = 2P_{\text{short dipole}}/I_{\text{in}}^2 = 20\pi^2(l/\lambda)^2. \tag{20.6}$$

For a short dipole with $l = 0.05\lambda$, $R = 0.49$ ohms. The imaginary part of the input impedance is negative and large, corresponding to a series capacitance with a low value. An antenna tuning network that increases the low value of R and tunes out the high value of capacitive reactance will need at least one high-value inductor. However, the limited Q of practical inductors will introduce losses that result in very low efficiency. A counterpart to the elemental dipole is the elemental current loop or *elemental magnetic dipole* (as opposed to *elemental electric dipole*). The loop is assumed to have constant current and its radiation pattern is the same as the pattern of the elemental dipole. A very small loop antenna has an extremely low radiation resistance and a high positive (inductive) reactance. It is an inefficient antenna in that its ohmic resistance is usually higher than its radiation resistance.

In practice, most dipoles have an overall length of $\lambda/2$. At this resonant length, the impedance of a thin wire dipole seen at the center gap is $73.1 + \text{j}0$ ohms, an entirely practical value. Resonant loops, with a circumference of the order of a wavelength, are also practical antennas. While transmission line arguments can

Figure 20.5. Fat monopole geometry.

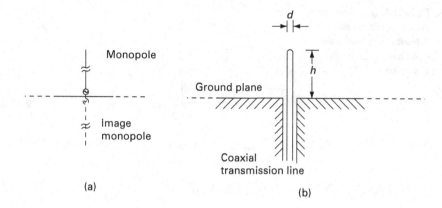

(a)   (b)

be used to show that the current distribution on a wire dipole is nearly sinusoidal, going to zero at the open ends, rigorous calculations of current distributions on arbitrary antennas require elaborate self-consistent treatments of the currents and fields. A benchmark problem is the fat dipole antenna shown in Figure 20.5(b). This "monopole" is really a dipole, since the ground plane induces an image monopole as shown in Figure 20.5(a).

This geometry is appealingly simple; the radiating element, as it enters the hole in the ground plane, becomes the center conductor of a coaxial line. You can find theoretical and experimental plots of input impedance of this antenna in most standard antenna textbooks. You can also model this dipole if you have access to a program for three-dimensional electromagnetic field simulation, such as Ansoft's HFSS.

20.4 Antenna directivity and gain

The radiation from any antenna is always stronger in some directions than in others; no antenna can be an isotropic radiator.[2] Let the energy flux (W/m^2) produced in a given direction be denoted as $S(\theta, \phi, r)$, where we are using standard polar coordinates. As before, the integral of S over an enclosing sphere will be the total radiated power transmitted, P_{RAD}:

$$\int S(\theta, \varphi, r) r^2 d\Omega = P_{RAD}. \tag{20.7}$$

---

[2] It is topologically impossible to have a constant tangential vector field on the surface of a sphere, a theorem sometimes expressed as "You cannot comb the hair smooth on a billiard ball." Thus, in the far field of an antenna (at a distance greater than the square of the largest physical dimension of the antenna divided by the wavelength), where the E and B fields are transverse (tangent to the surface of a sphere surrounding the antenna), the fields cannot be everywhere constant. The radiation pattern of an antenna, therefore, cannot be isotropic.

For any given direction, the ratio of the flux to the average flux is defined as the *directivity* of the antenna

$$D(\theta, \Phi, r) = \frac{S(\theta, \Phi, r)}{P_{\text{RAD}}/(4\pi r^2)}. \tag{20.8}$$

An equivalent statement is that the directivity is the factor by which the flux in the strongest direction exceeds the flux from a hypothetical isotropic antenna radiating the same total power. We know that the radiation flux eventually falls off as r^{-2}. For large r, then, the far-field directivity, D, is a function only of the polar coordinates θ and ϕ. Combining Equations (20.7) and (20.8), we see that the average directivity of an antenna is unity:

$$\frac{1}{4\pi} \int D(\theta, \Phi) d\Omega = 1. \tag{20.9}$$

Directive gain, $G(\theta, \phi, r)$, has the same definition as the directivity except that the radiated power is replaced by P_{inc}, the power incident on the antenna terminals:

$$G(\theta, \Phi, r) = \frac{S(\theta, \Phi, r)}{P_{\text{inc}}/(4\pi r^2)}. \tag{20.10}$$

If an antenna has no ohmic losses and its feedpoint impedance matches the transmission line impedance (so that no power is reflected), all the incident power will be radiated and the directive gain will be equal to the directivity. In most antennas used for transmitting, the losses are no more than a few percent and the distinction between directivity and gain is unimportant.[3] The maximum value of an antenna's directive gain is simply called the gain. A transmitter connected to an antenna having a gain of 20 dB will produce a directed signal 100 times more powerful than if it were connected to a hypothetical lossless isotropic radiator.

Since we have already calculated the total power radiated by an elemental dipole (Equation 20.4), we can easily calculate its directivity. The maximum flux density, calculated from Equation (20.3), is $S_{\text{max}} = (60/\pi)[\, I_0 k\, \delta\, l\, /(4\pi r)]^2$, where we have substituted 120π for $(\mu_0/\varepsilon_0)^{1/2}$. Dividing S_{max} by the average flux density, $S_{\text{avg}} = $ total power/$(4\pi r^2)$, yields a directivity of 1.5. As explained above, this will also be the directivity of a short dipole. The directivity of a half-wave dipole is 1.64 and its radiation pattern differs little from that of the elemental dipole.

---

[3] Small inefficient antennas are adequate for low-frequency receivers. Even though antenna losses add noise at the receiver input, signal strengths in the AM broadcast band and short-wave bands must already be considerably higher than this added noise in order to exceed the noise power from static (atmospheric electricity). Even with an inefficient antenna, the total power delivered to the receiver is much greater than the thermal noise added by the antenna and by the receiver itself.

20.5 Effective capture area of an antenna

The distance between transmitting and receiving antennas is generally so large that a plane wave can be assumed incident on the receiving antenna. The energy extracted from the incident wave is, of course, proportional to the average energy flux density, $\frac{1}{2}E \times H$ W/m^2. The proportionality constant is called the *effective area* (capture area) of the receiving antenna. It turns out, as shown below, that the effective area is proportional to gain:

$$A_{eff.} = G\lambda^2/4\pi. \tag{20.11}$$

Since gain and effective area are proportional, there is really no distinction between transmitting antennas and receiving antennas; the best transmitting antenna (most gain) is also the best receiving antenna (most capture area). A standard derivation of Equation (20.11) begins by applying the reciprocity theorem[4] to a system of two arbitrary antennas. The two antennas, as shown in Figure 20.6, need not be identical. We can suppose they are both matched to the same impedance value and that we have both a generator and a receiver that match this impedance. First we connect the generator to Antenna 1 and measure the power from Antenna 2.

If we now interchange the generator and load, the reciprocity theorem states that the power delivered to the load will be unchanged. Expressing this in terms of gain and effective area, we have $G_1 A_{eff\,2} = G_2 A_{eff\,1}$ or $G_1/A_{eff\,1} = G_2/A_{eff\,2}$, from which we see that the ratio of gain to effective area has the same constant value for any and all antennas. We can pick any conveniently simple antenna

Figure 20.6. Reciprocity: received power is unchanged when source and load are interchanged.

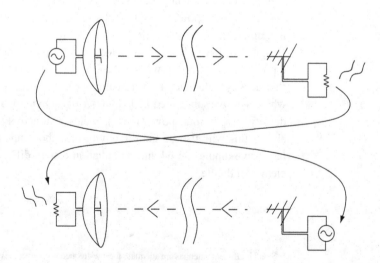

---

[4] The remarkable reciprocity theorem finds application in mechanics, optics, and acoustics as well as electrical engineering.

and use electromagnetic theory to calculate its gain and effective area. When this is done, it is found that the ratio of gain to effective area is $4\pi/\lambda^2$. A half-wave dipole antenna has a maximum gain of 1.64. Its effective area is therefore $1.64\,\lambda^2/4\pi$. If the dipole is made of thin wire, it has no real physical area, only a length, yet it has a nonzero effective area and can extract energy from an incident electromagnetic wave. Note that even a very short dipole has nearly the same effective area as a half-wave dipole. The effective area of a microwave dish antenna can approach its physical area, as we will see below, but normally it is from 50 to 80 percent of the physical area. This fraction, known as *aperture efficiency*, is usually set not by ohmic losses but rather by the illumination pattern of the primary feed antenna. A perfectly uniform illumination pattern (and no ohmic loss) produces an aperture efficiency of unity. The Arecibo radio telescope dish uses an aperture 700 ft in diameter (area = 35 800 m^2). Its aperture efficiency is about 70% at $\lambda = 12$ cm (the wavelength used there for radar astronomy) so its gain, using Equation (20.11), is $G = 4\pi(0.70 \times 35\,800)/(0.12)^2 = 22 \times 10^6$ or 73 dB.

20.6 Reflector and horn antennas

Radar antennas and satellite TV dishes are familiar examples of reflector antennas. From a transmitting standpoint, a primary or "feed" antenna illuminates the reflector. Currents induced in the metallic surface of the reflector become secondary radiators and their radiation forms the beam of the antenna. The larger the dish, the more directive the antenna, as we will see. These antennas are examples of "aperture antennas;" one can readily identify the aperture (usually circular) from which the radiation emanates, as if from a searchlight.

Let us use Equation (20.3) to calculate the radiation from a large flat rectangular metal plate on which there is a surface current density, J_S, which has the same amplitude and phase at every point.[5] For surface current, $J_S\mathrm{d}x\mathrm{d}y$ replaces $I_0\mathrm{d}l$ in Equation (20.3). To find the antenna pattern we integrate over the plate, summing up the contributions from each element of area. These are phasor contributions and the far field is actually an interference pattern. Once we have calculated the antenna pattern, we will integrate the power over a bounding sphere, as we did for the elemental dipole, to determine the total power radiated. With this, we can then calculate the gain.

Figure 20.7 shows the geometry for this antenna. The size of the plate is $2a$ by $2b$. Radiation will be strongest in the z-direction, perpendicular to the plate. We

---

[5] Such a current distribution could be established by illuminating the plate with radiation from a dipole far out in front of the plate, though this would not be an efficient feed antenna, since the plate would intercept only a small fraction of the dipole's radiation.

Figure 20.7. Rectangular aperture antenna geometry.

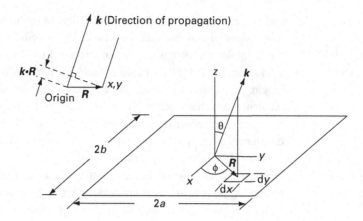

will assume that both a and b are much greater than λ, anticipating that this will form a very concentrated beam.[6]

The vector from the center of the plate to a distant observation point is denoted by r, and the vector from the center of the aperture to an arbitrary point x,y on the aperture is denoted by R. Again using the first term (the radiation term) of Equation (20.3), we see that the contribution to the distant H field from an element of area at the center of the plate is given by $\delta H = J_S\, dxdy\; jk(4\pi r)^{-1}\cos(\theta)e^{j(\omega t - kr)}$ where, as always, $k = 2\pi/\lambda$ and k is directed from the plate to the point r. Note that the $\sin(\theta)$ factor in Equation (20.3) is here $\cos(\theta)$ because the current is perpendicular, not parallel, to the z-axis. We will replace this $\cos(\theta)$ by unity, anticipating that, when both a and b are much greater than λ, the pattern will be highly concentrated around the $\theta = 0$ direction. Away from the origin, radiation to the point r from any other point on the plate will be shifted in phase by $k\cdot R$ radians, as shown by the inset in Figure 20.7. Using standard polar coordinates, this phase shift is given by

$$k\cdot R = xk_2 + yk_y = k\,\sin(\theta)[x\cos(\phi) + y\sin(\phi)]. \tag{20.12}$$

If we assume a uniform field over the aperture, the field at r, θ, ϕ can now be written as

$$H(r,\theta,\phi) = \frac{jkJ_S e^{j\omega(t-kr)}}{4\pi r}\int\limits_{-b}^{b}\int\limits_{-a}^{a} e^{jk\cdot R}dxdy$$

$$= \frac{jkJ_S e^{j\omega(t-kr)}}{4\pi r}\int\limits_{-b}^{b}\int\limits_{-a}^{a} e^{jk\theta(x\cos\phi + y\sin\phi)}dxdy, \tag{20.13}$$

[6] When used at a wavelength of 12 cm, the 210-m diameter illumination on the dish at Arecibo forms a beam whose width between half-power points is only 12/21 000 radians or 0.03 degree, a "pencil beam."

Figure 20.8. Radiation pattern (power in dB) from a rectangular aperture.

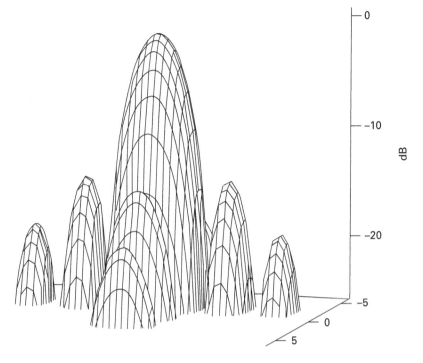

where we have replaced $\sin(\theta)$ by θ, again anticipating that the power will be negligible except for the region close to $\theta = 0$. Evaluating the integral we find

$$H(r,\theta,\phi) = \frac{jkJ_Sabe^{j\omega(t-kr)}}{4\pi r}\left(\frac{\sin(ka\theta\cos(\phi))}{ka\theta\cos(\phi)}\frac{\sin(kb\theta\sin(\phi))}{kb\theta\sin(\phi)}\right). \quad (20.14)$$

Since the average power density (watts/m^2) is given by $EH = \frac{1}{2}(\mu_0/\varepsilon_0)^{1/2}H^2$, the antenna pattern is proportional to the square of the term in brackets. Figure 20.8 shows the pattern, which has a strong central *main lobe* surrounded by *side-lobes*, which for this $(\sin(x)/x)^2$ power pattern are 13.3 dB below the peak of the main beam.

This plot uses rectangular "sky coordinates," u and v, measured in radians, where $u = \theta\sin(\phi)$ and $v = \theta\sin(\phi)$. The term in brackets is, therefore, $\sin(kau)/(kau) \sin(kbv)/(kbv)$. This plot was made for a square aperture; a rectangular aperture produces an elongated beam. For $a < b$, the beam is broader in the u-direction than in the v-direction. We can integrate the power density of the pattern to get the total average power:

$$\text{Pwr} = 1/2\sqrt{\frac{\mu_0}{\varepsilon_0}}\left|\frac{jJ_Skabe^{j\omega(t-kr)}}{4\pi r}\right|^2\iint r^2\left(\frac{\sin(kau)}{kau}\frac{\sin(kbv)}{kbv}\right)^2 du\,dv$$

$$= 1/2\sqrt{\frac{\mu_0}{\varepsilon_0}}\left(\frac{J_Skab}{4\pi}\right)^2\frac{\pi^2}{k^2ab}. \quad (20.15)$$

Figure 20.9. Horn and dish antenna aperture planes.

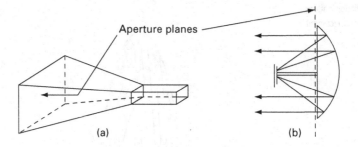

This let us calculate the gain:

$$G = 1/2\sqrt{\frac{\mu_0}{\epsilon_0}}|H(r,0,0)|^2 \div \frac{\text{Pwr}}{4\pi r^2} = \frac{4\pi(4ab)}{\lambda^2} = \frac{4\pi A}{\lambda^2}. \qquad (20.16)$$

This example antenna serves to establish the relationship between gain and effective area, $G = 4\pi\, A_{\text{eff}}\, /\lambda^2$, as long as we can argue that the effective area (capture area) is equal to the actual physical area. To argue this case, consider that a wave transmitted by this antenna is essentially a plane wave as it leaves the large aperture (as contrasted with the spherical wave emitted by a small antenna). Therefore, if we invoke time reversal, the wave entering the antenna would be a plane wave, just as if it had come from a distant transmitting antenna. Since the time-reversed wave is completely "absorbed" by the antenna, the effective area is equal to the physical area, $4ab$.

The integral in Equation (20.13) is just a two-dimensional Fourier transform of the aperture illumination. In this example, we used uniform illumination, so we have transformed the shape of the aperture with uniform weighting. The aperture can have any shape. For example, a circular plate of radius $a \gg \lambda$ produces a field pattern proportional to $J_1(ka\theta)/(ka\theta)$, where J_1 is the first-order Bessel function of the first kind. (The function $J_1(x)$ resembles $\sin(x)[x+1]^{-1/2}$). For any given aperture, uniform illumination produces the highest gain. But "tapered" illumination, where the current density is less near the edges of the aperture, is often used because it reduces the amplitude of the sidelobes.

A simple *microwave horn antenna* is essentially a waveguide funnel, fed at the small end, as shown in Figure 20.9. The other end fans out, usually in both transverse directions, to form a large aperture. The beam radiated from the aperture is comparable to the beam radiated from a dish antenna having equal area. The field distribution in the aperture plane is the same as the field at any cross-section in the waveguide, expanded to fill the aperture.

The horn antenna has no secondary radiators in its aperture plane. Of course there are currents on the inside walls of the waveguide and, in principle, Equation (20.3) can be used to calculate the radiation pattern far out in front of the horn. A much simpler way to find the radiation pattern is to use Huygen's principle, where the aperture *field* is regarded as a *source*, a two-dimensional array of wavelets, each of which re-radiates as if it were a current element.[7] If you know the E-field (or H-field) in the aperture plane, the squared magnitude of its two-dimensional Fourier transform yields the radiation pattern, just as the squared magnitude of the transform of the currents on a metal plate yielded the radiation pattern in the above example. This analysis technique is also applied to dish antennas. Ray-tracing methods (geometric optics) are used to find the phase and amplitude of the field at an aperture plane – usually just in front of the dish. This field is then regarded as a Huygen's source; its Fourier transform gives the far-field pattern without requiring an intermediate calculation of the currents on the reflector surface. This Fourier transform method is also applied to systems having multiple antennas, such as arrays of dish antennas, used together as an *interferometer* to form an extremely narrow beam.

20.7 Polarization

The elementary dipole of Figure 20.2 produces radiation with *linear polarization*; at any observation point the electric field vector has a fixed direction and an amplitude that oscillates sinusoidally. A dipole receiving antenna would have to be placed parallel to the incoming polarization. If placed perpendicular to the incoming E vector, it would pick up no signal. In all the preceding discussions, we have implicitly assumed that the polarization of a receiving antenna was matched to that of a transmitting antenna. *Circular polarization* can be produced with an antenna made of crossed dipoles. The dipoles are fed by signals that are equal in amplitude but 90° out of phase. This results in a transmitted E vector that rotates in the plane perpendicular to the direction of propagation. At any observation point, the magnitude of the E vector remains constant, but the vector rotates one turn for every period of the wave. A 90° hybrid can be used as a power splitter to feed the two dipoles. An identical antenna (a pair of crossed dipoles with a 90° hybrid as a combiner) makes a polarization-matched receiving antenna. Note that a single dipole can serve as a receiving antenna for circular polarization, but it collects only half the available power. If the incoming E field, viewed from the receiving position, rotates in the counterclockwise sense, the radiation has, by definition, *right-hand circular polarization (RCP)*. If the cables from the receiving antennas were to be interchanged at the hybrid, the resulting *LCP* antenna would have the wrong handedness and could receive no power from an RCP transmitting antenna. The most general polarization is elliptical, which is an admixture of linear and circular. Linear and circular polarizations are just special cases of elliptical polarization.

---

[7] See Reference [1] for a theoretical justification of Huygen's principle.

20.8 A spacecraft radio link

Consider the following example of a spacecraft telemetry link for which we wish to find the maximum range. Suppose we have a 1-W telemetry transmitter aboard a spacecraft and that the data rate requires a channel capacity corresponding to a signal-to-noise ratio of at least unity in a 1-Hz bandwidth. This link uses a frequency of 3 GHz (10 cm wavelength). The transmitting antenna on the spacecraft is a 2-m diameter dish. The ground station antenna is a 10-m diameter dish, as shown in Figure 20.10. Assume that both these dish antennas have an effective area equal to 60% of their physical apertures. Assume also that there is no pointing error, i.e., the antennas always point directly at each other and that the system temperature of the ground station receiver is 25 K. (The system temperature is the sum of the equivalent receiver noise temperature, the antenna noise temperature, and the sky noise temperature.)

1. What is the equivalent input noise power of the receiver? Boltzmann's constant, k, is 1.38×10^{-23} W/Hz/K, so the equivalent input noise power, kTB, is $1.38 \times 10^{-23} \times 25 \times 1 = 3.45 \times 10^{-22}$ W.
2. What is the effective area of the receiving antenna? The physical aperture is πR^2 so the effective area is $0.60 \ \pi R^2 = 0.60\pi 5^2 = 47.1 \ m^2$.
3. What is the gain of the transmitting antenna? The physical aperture is πR^2 and the aperture efficiency is 0.60 so the effective aperture is $0.60 \ \pi(1^2) = 1.88 \ m^2$. The gain is $4\pi A_{eff}/\lambda^2 = 4\pi(1.88)/(0.1^2) = 2369$.
4. What is the maximum range, R, in kilometers for the spacecraft to maintain the required signal-to-noise ratio? Here we simply set $P_{noise} = P_{received} = A_{rcvr} \ P_{trans}.G_{trans}/(4\pi R^2)$ from which we have $R^2 = (4\pi)^{-1} P_{trans}.G_{trans}/P_{noise}$. Using the parameters calculated above, we have

$$R = \sqrt{(4\pi)^{-1} \times 1 \times 2369/(3.45 \times 10^{-22})} = 7.4 \times 10^{11} m = 740 \times 10^6 km.$$

This is roughly the mean distance to Jupiter.

Figure 20.10. A spacecraft telemetry link.

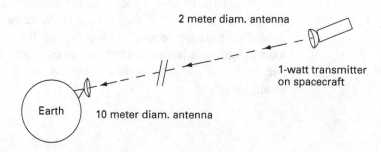

2 meter diam. antenna

1-watt transmitter on spacecraft

Earth   10 meter diam. antenna

20.9 Terrestrial radio links

VHF and UHF two-way radios used by cellular phones, emergency vehicles, etc. have transmitters with several watts of power but their range is limited by the curvature of the Earth to only a few miles or tens of miles. (Radio waves do not propagate through the highly conductive Earth though they do diffract slightly, so the radio horizon is somewhat beyond the optical horizon.) Mobile radios also use antennas that have gains only of the order of unity so they do not have to be pointed accurately – or even at all. Finally, in ground-to-ground radio links, the signal usually arrives from an angle near the horizon; the receiving antenna will pick up noise from the ground (thermal radiation). In these systems, then, extremely low-noise receivers are of no benefit. Broadcasting stations for FM and television also use VHF and UHF frequencies so their range is also essentially line-of-sight. Long-distance propagation in AM and short-wave broadcasting depends on reflection from the ionosphere.

20.10 The ionosphere

At altitudes above about 60 km the atmosphere is ionized by ultraviolet radiation from the Sun; electrons are stripped from the neutral particles (mostly oxygen atoms and O_2 and N_2 molecules) to produce a mixed electron and ion gas. During the day the density of this ionized gas is highest at around 250 km, the peak of the "F-region." Above the peak the ionization is less because the thinner atmosphere presents fewer particles to be ionized. Below the peak the ionization is less because the denser atmosphere exhausts the supply of ultraviolet photons; the electrons they produce quickly encounter nearby ions and recombine. At night, without sunlight, the ionization rate is zero. Recombination quickly neutralizes the ionization at the lowest altitudes, around 100 km, and depletes the F-region until sunrise.

20.10.1 Wave propagation in the ionosphere

An electromagnetic wave induces electric currents in the electron gas of the ionosphere. (The electrons, by virtue of their low mass, are accelerated by the incident wave to much higher velocities than the ions, so the ion contribution to the current is negligible.) The effective dielectric constant of an electron gas is not difficult to calculate from Maxwell's equations. (We will see that this dielectric constant becomes imaginary below a certain critical frequency which depends on the electron density; below this frequency, then, electromagnetic waves cannot propagate through the plasma.) No longer in a vacuum, we must use the general curl H equation which includes the real electric current, J, in addition to the displacement current:

$$\nabla \times \boldsymbol{H} = \boldsymbol{J} + \epsilon_0 \frac{\partial \boldsymbol{E}}{\partial t}. \tag{20.17}$$

The electrons are in rapid thermal motion, but this motion is random and contributes nothing to the current. Yet all the electrons near any given point are accelerated equally and move together to produce a net current, $\boldsymbol{J} = Ne\boldsymbol{v}$, where N is the electron density, e is the electron charge, and v is the component of velocity imparted by the electric field. We will neglect the weak $\boldsymbol{v} \times \boldsymbol{B}$ force from the magnetic field so Newton's second law of motion is just $md\boldsymbol{v}/dt = \boldsymbol{F} = e\boldsymbol{E}$. For a sinusoidal time dependence, $e^{j\omega t}$, we can write this equation of motion as $m(j\omega v) = e\boldsymbol{E}$. Substituting $\boldsymbol{J} = Ne\boldsymbol{v} = Ne^2\boldsymbol{E}/(j\omega m)$ in Equation (20.17) gives

$$\nabla \times \boldsymbol{H} = \frac{Ne^2\boldsymbol{E}}{j\omega m} + j\omega\epsilon_0\boldsymbol{E} = j\omega\epsilon_0\left(1 - \frac{Ne^2}{\epsilon_0 m\omega^2}\right)\boldsymbol{E}. \tag{20.18}$$

Note that the term in brackets is the relative dielectric constant and that it becomes negative for low frequencies, in particular for $\omega^2 < \omega_p^2$ where $\omega_p^2 = Ne^2/(\epsilon_0 m)$. This happens because the conduction current (the electron current) becomes greater than the displacement current. The total current (conduction current plus displacement current) changes sign and has the wrong polarity to source the H field of a traveling wave. This critical frequency, ω_p, is known as the *plasma frequency*. (If the local charge neutrality of an electron–ion gas is disturbed, the densities will oscillate at this frequency the way a spring and mass system oscillates at its resonant frequency.) For a wave to propagate in the plasma, the dielectric constant must be positive; only waves with frequencies lower than the plasma frequency will be reflected. The free-electron gas that gives metals their conductivity is dense enough to reflect visible light but the alkali metals (lithium, sodium, etc.) have relatively lower electron densities and are transparent in the ultraviolet.

20.10.2 Reflection of waves from the ionosphere

The reflection of radio waves is normally a process of refraction because the waves are not vertically incident on the ionosphere. As they travel obliquely upward, the dielectric constant decreases so the phase velocity increases, causing the propagation vector, which is perpendicular to the wavefront, to turn around gradually. This is shown in Figure 20.11.

For signal strength calculations over an ionospheric path the ionosphere can be considered a specular mirror and field strengths can be calculated by taking the inverse square of the total path length of the ray.

20.10.3 Daytime vs. nighttime propagation

Short-wave broadcasts from distant transmitters on the higher frequency bands are not heard at night because the ionospheric electron density is too low for

Figure 20.11. Ionospheric refraction.

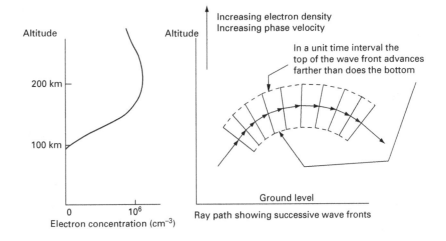

Ray path showing successive wave fronts

reflection, i.e., the waves cannot get turned around sufficiently. On the other hand, low-frequency stations, such as those in the AM broadcasting band, are received from great distances *only* at night. During the day their energy is dissipated in the lower ionosphere through collisions between the accelerated electrons and neutral atoms and molecules. Why don't higher frequency waves suffer this daytime attenuation in the lower ionosphere? Consider a low-frequency and a high-frequency wave of equal power, that is, equal field strengths. Both cause the ionospheric electrons to execute synchronized sinusoidal motion as described above. But the low-frequency wave, because its period is long, will produce higher electron velocities; we saw above that $v = Ee/(j\omega m)$. The average kinetic energy of each electron is therefore $m\langle v^2\rangle/2 = m[Ee/(\omega m)]^2/4$ where E is the amplitude of the electric field. All the electrons in the lower ionosphere suffer collisions with the neutral particles (which are present since the ionization is not 100%). The average collision leaves the electron with a random velocity, i.e., its share of the synchronized sinusoidal motion is lost. The frequency of collisions does not depend on the frequency of the electromagnetic wave (or on there being any electromagnetic wave present at all) so the rate of energy loss is inversely proportional to the square of ω, the wave frequency, and long-distance AM listeners have to wait for nighttime.

20.11 Other modes of propagation

Besides reflection by the ionosphere, there are a number of other ways that an electromagnetic wave can get around the curvature of the Earth. These include scattering from the ionized trails of meteors entering the Earth's atmosphere, scattering from irregularities in the ionosphere even when the ionosphere is otherwise not dense enough to turn the waves around by refraction, scattering from density irregularities in the neutral atmosphere (i.e., fluctuations in the index of propagation), and ducting beneath atmospheric temperature inversion layers.

Problems

Problem 20.1.   The voltage at the terminals of a receiving antenna is proportional to the E field of an incident electromagnetic wave. The "effective length" (or effective height if the antenna is vertical) is defined as the open-circuit voltage at the terminals divided by the incident E field: volts / (volts/meter) = meters. Show that the effective length is given by

$$\text{effective length} = (4RA_{\text{eff}}/Z_0)^{1/2},$$

where R, the radiation resistance, is the real part of the antenna impedance, A_{eff} is the effective area ($A_{\text{eff}} = G \cdot \lambda^2/(4\pi)$), and $Z_0 = 377$ ohms $= (\mu_0/\epsilon_0)^{1/2}$, the impedance of free space. Find the effective length of a half-wave dipole ($G = 1.6$ and $R = 73$ ohms).

Problem 20.2.   Suppose we have a 1-W transmitter connected to a dipole antenna which is aligned to provide the maximum signal strength at a distant receiving antenna. Needing more signal strength, we obtain a second, identical dipole and, using a power splitter, feed each dipole with ½W. We space the second far enough from each other so that they do not interact. We make sure that both antennas are aligned toward the receiver and we also make sure that the cables from the power splitter have equal length. At the receiving antenna, each transmitting antenna provides a field amplitude that is less than the original field by $1/\sqrt{2}$. But the two signals are in phase so the total amplitude is increased by $2/\sqrt{2} = \sqrt{2}$. Squaring this we see that the received signal strength is doubled. Have we gotten something for nothing? Could we repeat this process to increase the received power even more?

Problem 20.3.   Let the individual antennas of Problem 20.2 be AM broadcast towers with omnidirectional patterns and vertical polarization. Suppose the spacing between these antennas is $\lambda/2$. As before, they are fed symmetrically, that is, with the same power and same the phase. Find the radiation pattern in the horizontal plane: make a polar plot of the relative field strength vs. azimuth angle for a distance far from the antennas. Hint: at any observation point in the horizontal plane at a distance r from the center of the line joining the two antennas, the total voltage is the sum of the contributions from the two antennas, $e^{j\phi_1}$ and $e^{j\phi_2}$. The phases π_1 and π_2 are the phase path lengths corresponding to r_1 and r_2, the distances from the observation point to the respective antennas. These phase paths are just $2\pi r_1/\lambda$ and $2\pi r_2/\lambda$. The field strength is given by $(|e^{j\phi_1} + e^{j\phi_2}|)$.

Problem 20.4.   Consider a pair of crossed dipoles. The first dipole points in the z-direction and carries a current $\cos(\omega t)$. The second dipole points in the x-direction and carries a current $\sin(\omega t)$. Find the type of polarization and the relative power density of the radiation in the $+z$, $-z$, $+x$, $-x$, $+y$, and $-y$ directions. If you have a program like MATLAB or Mathcad, make a three-dimensional surface plot, like Figure 20.3, for which the distance from any point on the surface to the origin is proportional to the radiation power in that direction.

References

[1] Collin, R. E., *Antennas and Radiowave Propagation*, New York: McGraw-Hill, 1985.

[2] Davies, K., *Ionospheric Radio Propagation*, New York: Dover Publications, 1966.

[3] Kelley, M. C., *The Earth's Ionosphere: Plasma Physics and Electrodynamics*, New York: Academic Press, 1989.

21 Radar

The idea that radio waves could be used to detect the presence of stationary or moving objects emerged around 1900, almost as soon as radio itself. Christian Hueslmeyer, a German inventor, demonstrated an apparatus in 1904 which, when mounted on a bridge above the Rhine, rang a bell when a ship passed beneath. He used a (now) primitive spark gap RF source and coherer detector. The system may have shown only marginal potential for collision avoidance, as the German Navy demonstrated no interest. Sir Robert Watson-Watt developed meteorological radar in Britain in the 1930s and then a chain of air defense radars during World War II. In the U.S., the MIT Radiation Laboratory, set up to develop military microwave radar systems, had nearly 4000 employees between 1940 and 1945. The acronym RADAR, for Radio Detection And Ranging, has been attributed to U.S. Navy officers F. R. Furth and S. M. Tucker, who introduced it in 1940, though the term remained classified throughout the war.[1]

Today, radar goes beyond aircraft tracking to applications as diverse as space object monitoring, storm tracking, detection of clear air turbulence, and velocity measurements of speeding automobiles and baseballs. In this chapter, we look at some commonly used radars, some general system aspects of radar, and, finally, some RF components and techniques developed specifically for radar.

21.1 Some representative radar systems

Classic surveillance radar

Figure 21.1 is a block diagram of the classic radar system used to monitor air traffic. A rotating antenna sweeps continuously in azimuth while it repeatedly transmits short pulses and receives subsequent echos from targets. As each pulse is transmitted, the beam of the CRT display begins to sweep from the center in a radial direction. The detected signal from the receiver modulates the

---

[1] Butrica, A. J. *To See the Unseen: A History of Planetary Radar Astronomy*, Diane Publishing Co. 1997 (also available on the Web).

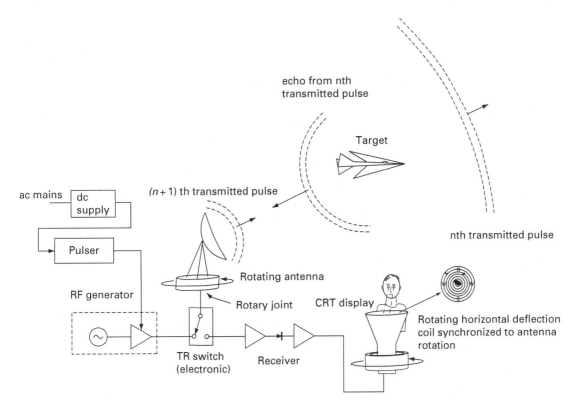

Figure 21.1. Classic surveillance radar.

intensity of the spot, rather than causing a deflection perpendicular to the sweep direction. The beam deflection coil is made to rotate continuously around the neck of the CRT, synchronized to the rotation of the antenna, so that the radial sweep direction at any instant is the azimuthal pointing direction of the antenna.

Generally there is a bright area near the center of the CRT, produced by reflections from nearby hills, buildings and towers. The receiver may incorporate an electronic gain control, slaved to the CRT sweep to compensate for the decrease in echo strength with increasing target range. Dashed lines on the figure show the spherical wavefronts for two transmitted pulses and for the reflection (echo) from one pulse, but do not indicate that the intensity of the transmitted waves is sharply concentrated in the forward direction by the high directivity of the antenna.

21.1.1 CW Doppler radar

The classic radar described above uses short pulses for good range resolution. It detects the amplitude of each returning pulse, paying no attention to the phase. Quite the opposite is the radar speed gun, which transmits cw and therefore has no range resolution, but does detect the phase to measure speed. A functional block diagram of the speed gun is shown in Figure 21.2.

Figure 21.2 Radar speed gun.

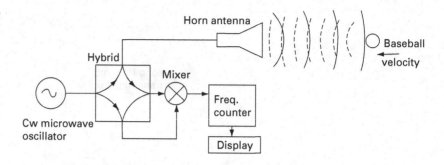

The signal from a cw microwave oscillator in the 10–20 GHz range (usually a Gunn diode negative resistance oscillator) is split to feed the horn antenna and a mixer (multiplier). The echo from the target also provides a signal to the mixer. A moving target causes a Doppler shift, so the echo has a higher frequency if the target is approaching the radar. The output from the mixer contains a signal whose frequency is equal to the Doppler shift. This frequency is counted and multiplied by the appropriate factor so that the display reads line-of-sight velocity. Note that the Doppler shift depends only on the line-of-sight velocity of the target. This radar is totally insensitive to range.

21.1.2 Chirped cw radar

The radar in Figure 21.3 differs from the radar speed gun only in that its oscillator is a "chirped" VCO, whose output frequency follows a sawtooth control voltage. Here this radar is used to determine the position of a stationary (or very nearly stationary) fluid level in a storage tank.

Figure 21.3 Radar fluid level detector.

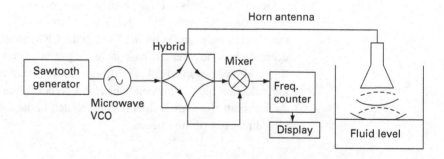

The frequency of the tone at the mixer output is given by the frequency difference of the two signals arriving at the mixer inputs. One of these signals is delayed by the round-trip time to the fluid surface, so the lower the level in the tank, the higher the output frequency of the mixer. Note that this radar cannot determine velocity; if the fluid level were changing rapidly, it would create a

Doppler shift that could not be distinguished from a change in fluid level. (Range and velocity are ambiguous for this radar.)

21.1.3 Pulse-Doppler radar

The radar of Figure 21.1 can be modified to sense velocity as well as position. The resulting *pulse-Doppler radar*, when mounted on an aircraft, can distinguish moving targets on the ground from the much stronger "ground clutter" produced by stationary objects. The earliest form of pulse-Doppler radar, called "MTI", for *moving target indicator*, used a mercury acoustic delay line as analog storage for the train of echos received after each pulse. The stored signal was subtracted from the echoes produced by the next pulse to remove the echoes from fixed targets. (This type of delay line, when configured as a recirculating shift register, was used around 1950 as the memory in some of the earliest digital computers.)

Pulse-Doppler radar really became practical with the advent of frequency-stable microwave signal sources (klystron amplifiers, rather than high-power magnetron oscillators), digital storage capability, and digital signal processing. These advances allow coherent signal averaging for increased sensitivity, as well as actual determination of target velocity. Successive received echoes, after being mixed to a low IF frequency, are stored, preserving both amplitude and phase, and then the stored echos are analyzed to see how the phase changes from one pulse to the next. A moving target causes a progressive phase change (Doppler shift) from which the line-of-sight velocity can be determined. When used as a search radar, successive received signals are stored separately for every range. Fourier analysis of the sequence of sampled IF voltages for each range determines whether there is a target at that range and, if so, determines its velocity. A problem that often arises with pulse-Doppler systems involves a range-velocity ambiguity. If target velocities are in the range $\pm v_{\max}$, the corresponding Doppler shift range is $\pm 2f\, v_{\max}/c$, where f is the radar frequency. This produces a total received bandwidth of $4f v_{\max}/c$. If we are to unambiguously determine the frequencies present in this band, the sampling rate, which is the pulse rate, must be at least $8f\, v_{\max}/c$. A high radar frequency and a high maximum velocity may dictate a pulse rate that restricts the maximum unambiguous range. (The echo from a target beyond the maximum unambiguous range arrives after the next pulse has been transmitted, as if it had been reflected from a close-in target.) This problem is discussed in Chapter 26 in the context of planetary radar astronomy.

21.2 Radar classification

Many radar systems are quite different from the classic configuration of Figure 21.1. Radar systems are conveniently classified according to several key system characteristics:

(a) Monostatic or bistatic:

A monostatic radar uses a single antenna for both transmitting and receiving, while a bistatic radar uses two separate antennas. An obvious advantage of the bistatic configuration is that it requires no TR (transmit/receive) switch. A military advantage is that only the transmitting site, which can be unmanned, is vulnerable to attack by a radiation-seeking missile. Bistatic and multistatic radar systems can also be passive, using "transmitters of opportunity," such as television broadcast transmitters.

(b) Coherent (Doppler) or incoherent:

Radars with stable frequency sources for the transmitter and the receiver L.O. can use coherent integration to extract a signal from noise. They can also use the Doppler effect to measure the line-of-sight velocity of a moving target. Incoherent radars detect the power of each pulse, making signal averaging less efficient. They cannot directly detect target velocity though they can, of course, infer velocity by observing successive changes in range.

(c) Pulse or cw:

Short pulses can distinguish closely separated targets, providing range resolution. Long pulses, whose limit is cw (continuous wave), can distinguish target velocity using the Doppler shift. Nevertheless, coherent trains of short pulses and phase-modulated long pulses can provide both range and Doppler resolution.

(d) Search or track:

Aircraft traffic monitoring is normally done with a monostatic search radar, using a rotating antenna that produces a beam pattern that is tall in elevation but narrow in azimuth. Echos from aircraft produce "blips" on a PPI (plan-position indicator) screen which is refreshed on every rotation of the antenna. A tracking radar uses servo motors to keep the antenna pointed at a single, usually fast-moving target.

(e) Mechanical or electronic scanning:

A *phased array* uses a closely spaced two-dimensional array of antenna elements, usually dipoles. Each dipole is equipped with a programmable phase shifter, allowing the array to form a beam in an arbitrary direction. If the signals received from the individual dipoles are made available to the signal processor, the processor can simultaneously form several independent receiver beams. This allows tracking of multiple targets or searching of multiple zones. Phased arrays with three or more faces of elements can be completely stationary.

(f) Detection or imaging:

Conventional radars simply detect the position and/or velocity of targets. Imaging radars make photographic-like reflectivity maps. Imaging systems include airborne side-looking (*synthetic aperture*) radars that map a strip of land parallel to the flight path and "planetary radar" (delay-Doppler) systems that image planetary objects and low-orbit artificial satellites.

21.3 Target characteristics and echo strengths

21.3.1 Radar cross-section

The strength of an echo depends on the nature of the target and on its range from the radar. In general, metal targets with dimensions commensurate with the wave length are good reflectors, especially when their geometry includes linear features. For example, airplane propellers act as dipole re-radiators. A flat metal plate produces a strong specular reflection, if perpendicular to the direction of the radar, but edge currents also produce other, less directional, echoes. Radar engineers define the *radar cross-section* (RCS) of a target, σ, as the collecting area of an object that would produce the target's observed echo strength while *isotropically* re-radiating the intercepted incident radiation. Radar cross-sections for practical targets, such as aircraft, depend on the aspect angle of the target as well as on the target's usually complicated geometry. RCS calculations for such targets are carried out using finite-element electromagnetics modeling programs. Exact expressions have long since been derived for the radar cross-sections of simple objects. A sphere, since it has no aspect angle dependence, makes a good calibration target for measuring radar sensitivity. The RCS of a metal sphere of radius a is just πa^2 (its geometric cross-sectional area) when a is greater than about $5\lambda/\pi$. But when a is less than about $\lambda/(4\pi)$, the RCS is given by the formula: $\sigma = \pi a^2 \left[9 \, (2\pi a/\lambda)^4\right]$. Note that the cross-section is proportional to a^6 and to λ^{-4} in this small-target or *Rayleigh scattering* regime.[2]

21.3.2 The radar equation

The definition of radar cross-section lets us write a simple expression for the flux density of a backscattered echo as it returns to the radar antenna

$$S_{\text{received}} = S_{\text{incident}}\,\sigma/(4\pi R^2), \tag{21.1}$$

where R is the distance to the target. This equation just says that the supposed isotropically scattered power, which is given by $S_{\text{incident}}\,\sigma$, produces a flux density determined by the area of a target-surrounding sphere whose radius is the distance to the radar. From antenna considerations discussed in Chapter 20, we can write the incident flux in terms of the transmitted power as $P_{\text{trans}}\,G/(4\pi R^2)$, where G is the gain of the antenna. Of course, this assumes the antenna is pointed directly at the target. Likewise, we can express the capture area of the antenna by $A_{\text{eff}} = \lambda^2 G/(4\pi)$. Putting this together, we obtain an expression for the received power, P_{rec}, which is just $S_{\text{incident}}\,A_{\text{eff}}$ or

---

[2] The reason the sky is not black is that the molecules that make up the atmosphere scatter direct sunlight to produce diffuse sky light. The λ^{-4} dependence of Rayleigh scattering causes sky light to be blue and sunsets to be red.

$$P_{\text{rec}} = [P_{\text{trans}}G/(4\pi R^2)][\sigma/(4\pi R^2)][\lambda^2 G/(4\pi)] = P_{\text{trans}}G^2\sigma\lambda^2/[(4\pi)^3 R^4)]. \quad (21.2)$$

This relation, however written, is known as the *radar equation*, and shows the inverse fourth-power dependence on range. For an aperture antenna, e.g., a dish antenna, the effective aperture is given by $A_{\text{eff}} = \eta_{\text{ap}} \pi(D/2)^2$ where D is the dish diameter and η_a, the aperture efficiency, is usually around 0.5. Using this, we can write the radar equation as

$$P_{\text{rec}} = P_{\text{trans}}\frac{\pi}{64}\eta_a^{\,2}\left(\frac{D}{R}\right)^4\frac{\sigma}{\lambda^2}. \quad (21.3)$$

21.3.3 Distributed targets

Targets such as aircraft subtend a solid angle much smaller than the radar beam, but distributed targets, such as rain in the atmosphere or free electrons in the ionosphere, can be much larger than the beam. In these cases the beam actually defines the extent of the target, in as much as there will be radar echoes from throughout the entire volume of the beam. When the individual scattering objects (raindrops and electrons in the above examples) are randomly distributed in space (*incoherent scattering*), the power received by the radar is the sum of the powers of the individual scatterers.[3] For this kind of target, the radar equation must be modified. We first define σ_V, the radar cross-section per unit volume of the target, to be $\sigma_V = n\sigma$, where n is the volume density of scattering particles and σ is the radar cross-section of an individual particle. For raindrops, σ is about 41% of the Rayleigh scattering cross-section given above for metal spheres. For individual electrons, $\sigma = [e^2/(4\pi\epsilon_0 m_e c^2)]$, where e and m are the charge and mass of an electron and c is the speed of light. If we extend the arguments we used for a single particle centered in the beam to include each volume element throughout the beam, we find that

$$P_{\text{rec}} = \frac{P_{\text{trans}}\lambda^2\sigma_V c\tau}{(4\pi)^2 R^2}\frac{1}{4\pi}\int G^2(\phi,\theta)\mathrm{d}\Omega. \quad (21.4)$$

In this radar equation τ is the pulse length, so $c\tau$ is the range depth of the target volume. The cross-range area of the beam is taken into account by integrating over solid angle times R^2, the square of the range. Note that, because the beam determines the target size, the return echo of a distributed target is proportional to R^{-2}, rather than R^{-4}. The last term in Equation (21.4), $(4\pi)^{-1}\int G^2 \,\mathrm{d}\Omega$, is the mean square gain, also called the *backscatter gain* of the antenna.

---

[3] The electric field of the radar echo is the sum of the field contributions from the N individual scattering particles illuminated by the beam. If the scatterers have random positions, their E-field contributions have random phases and add in the fashion of a random walk, causing the total E-field echo to be proportional to $\sqrt{N}$. The received power, proportional to the square of the received E field, is therefore proportional to N.

21.4 Pulse compression

Short pulses produce good range resolution, but shortening the pulses reduces the sensitivity of the radar unless the peak power (pulse power) can be increased to maintain the same average power. However, the peak power of any transmitter is eventually limited by voltage breakdown or other device limitations. This constraint led to the development of *pulse compression* schemes in which the transmitted pulses are modulated in such a way that the echoes can be "bunched up" by the receiver as if they had started out as short pulses. The received echo is passed through an appropriate matched filter (often a digital processor) whose output, for a point target, will be a narrow pulse.[4]

One method for pulse compression uses phase coding. The transmitted pulse is divided into N contiguous equal intervals. Each interval or *baud* is assigned a phase value of zero or 180 degrees, which will be the relative phase (and therefore polarity) of the transmitted signal during that interval. These assignments specify the code. Normally every transmitted pulse will have the same code. Figure 21.4 shows how such a code is used for pulse compression. At the receiver, the IF signal from a point target will be a replica of the transmitted pulse. This IF signal travels down a delay line. Taps along the delay line have a delay spacing equal to the baud length. Signals extracted at these taps are weighted (multiplied) by coefficients with values ± 1, whose order duplicates the code. The weighted signals are added together to form the pulse compressor output.

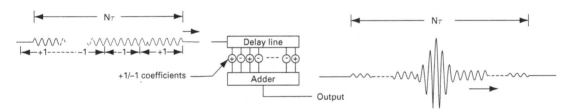

Figure 21.4. A compressor for phase-coded pulses.

Note that, for one position of the signal in the delay line, the multipliers will exactly "undo" the coding, giving the same phase to each of the N input signals to the adder. This alignment will peak at the middle of one baud, as the signal propagates down the delay line, attaining a peak amplitude N times greater than the amplitude of the uncompressed input signal. When the signal in the delay line is not in alignment with the code, there will be some output from the adder unless the code is such that the addends sum to exactly zero. There is no code with this property, though the *13-baud Barker code*, $+ + + + + - - + + - + - +$, produces

---

[4] In general, high range resolution requires that the transmitted signal have a large bandwidth. This bandwidth can be produced by transmitting either short pulses or appropriately modulated long pulses.

("sidelobe") values of only +1 or 0 when the code is not aligned and +13 when it is aligned. When longer codes are needed, it is common to use pseudorandom codes. In this case, the sidelobe levels are determined by random walk statistics; the close-in sidelobes will be less than the main lobe by a factor on the order of $\sqrt{N}$.

An analog surface acoustic wave (SAW) processor can do all the operations shown in Figure 21.4, but it is now more common to use digital processing. Typically, baseband signals are produced by mixing the IF signal, centered at ω_{IF} with $\cos(\omega_{IF}\ t)$ and $\sin(\omega_{IF}\ t)$. The resulting I and Q signals are each furnished with a tapped digital delay line/adder pulse compressor. The outputs from the two compressors can be squared and added together for immediate detection, or they can first be Doppler processed.

Pulse compression has other benefits. Because the rapidly changing code spreads the signal's spectrum, it lowers the spectral density. This gives a military radar signal an element of stealth; being less noticeable, it is less likely to provoke a hostile reaction, such as an antiradiation missile. By using different codes with low cross-correlation properties, multiple radars can share the same frequency band. This code-division multiple access property is also used in CDMA cellular telephone systems and in the GPS satellite system, where multiple users (the callers or satellites) share the same band. The GPS system also makes use of the pulse compression property to achieve the high time resolution needed for accurate range resolution (see Chapter 25).

21.5 Synthetic aperture radar

As described above, a phased-array antenna combines the signal from many individual elements to form a beam pattern. Consider a phased array directed toward a point source of cw radiation whose frequency is precisely known. The analog voltage from each element is given a programmed analog phase shift. These voltages then combine to form a vector sum. Note, however, that it is also possible to digitize the voltage (phase and amplitude) from each element and then *compute* the vector sum. If we do this, no hardware phase shifters are needed since, before summing, we can multiply each voltage, V_i by a phase factor, $e^{j\phi_i}$, to point the beam in any desired direction. This can be repeated with other sets of phase factors to form a host of beams from the one set of digitized voltages. The echo powers in these beams then form an image if there are multiple targets (assuming they all radiate at the same frequency). Finally, note that we could do all of this with only a *single* array element by moving the element successively from one position to another, recording the voltage (amplitude and phase) at each position. If the set of positions is identical to the element positions in the original phased array, we have effectively used a single small element to *synthesize* a steerable large-aperture antenna.

You can see that this all works out in a radar situation, where all the targets are illuminated by a coherent transmitter signal. The transmitter also provides a

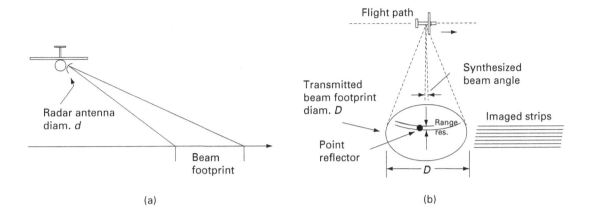

Figure 21.5. Side-looking synthetic aperture radar: (a) elevation view; (b) plan view.

phase reference for the receiver so it is practical to measure both the amplitude and the phase of the return signal at the antenna element. This is the general idea of *synthetic aperture radar* (SAR).[5] Here we assumed that the target was stationary and that the element was stationary, except while being relocated after each voltage measurement. In practice, most SAR techniques involve continuous motion of the elemental antenna, relative to the target. This continuous motion makes it possible to explain the synthesis procedure in terms of Doppler shifts. In addition, most SAR systems transmit short pulses to provide resolution in the down-range (pointing) direction, and use one-dimensional aperture synthesis to provide resolution in the cross-range direction. One of the most common SAR systems is the side-looking radar shown in Figure 21.5.

A small radar antenna of diameter d points outward and downward from the aircraft, which flies a straight-line course. The illuminated patch on the ground, called the *radar footprint*, has a large diameter, D. Since the beamwidth from the radar antenna is given by λ/d radians, we see that D is given by $D = R\lambda/d$, where R is the distance of the footprint from the aircraft. Range resolution is obtained by simply using short radar pulses. In the cross-range direction, the system synthesizes an antenna whose extent in the direction parallel to the flight path is D, since that is the dimension of the footprint that passes over any target feature. The angular size of the resulting synthesized beam is given by $\delta\theta = \lambda/D$. Substituting $D = R\lambda/d$ produces $\delta\theta = d/R$, and the cross-range resolution is therefore $R\delta\theta = d$, the diameter of the small radar antenna. It is interesting to note that the resolution is independent of λ and that making d smaller improves rather than degrades the resolution. Many range strips are observed simultaneously. You can find fascinating side-looking radar (and side-looking sonar) images by searching the internet. A "delay-Doppler" SAR system to image

---

[5] Synthetic aperture mapping is also possible when the target is not illuminated by coherent radiation, but emits its own wideband radio "noise." In this case, aperture synthesis requires at least two antenna elements. This interferometry technique, used by astronomers to make maps of the radio sky, is discussed in Chapter 26.

planetary surfaces is discussed in Chapter 26. Another interesting but difficult technique is *inverse synthetic aperture radar* (ISAR) where the aspect of the target (which might be a ship bobbing on the ocean) is variable, not known a priori, and must be inferred from the radar returns as part of the data processing.

21.6 TR switches

Monostatic radars, which transmit and receive with the same antenna, require a TR (transmit-receive) switch. In most radar applications the desired echo arrives so soon after the pulse is transmitted that the TR switch (also known as a *duplexer*) must be electronic rather than a mechanical. Here we will first look at self-duplexing radar techniques based on the use of circular polarization or circulators, then at standard TR switch circuits, and finally at RF electronic switches in general.

21.6.1 Self-duplexing radar techniques

If a radar transmits a circularly polarized signal, reflection by the target changes the sense of polarization from left-hand to right-hand or vice versa. Circular polarization can be produced by transmitting simultaneous crossed linear polarizations 90° out of phase. Figure 21.6 shows how a 90° hybrid not only produces circular polarization of one sense but also routes received circular polarization of the other sense (the return signal) into the receiver.

Note that the *x*-dipole and the *y*-dipole, together with the hybrid, are really just equivalent to two separate antennas having opposite circular polarizations. In practice, the isolation between the transmitter and receiver in this scheme is usually no better than about 30 or 40 dB, so a limiter or SPST electronic switch (a *monoplexer*) is installed at the receiver input to protect it from burnout. A waveguide version of this circuit uses a *turnstile junction*, the microwave component shown in Figure 21.7. It is classified as a six-port junction because the round waveguide supports two independent modes: *x* and *y* or RCP and LCP or any other pair of orthogonal elliptical polarizations. When two opposite

Figure 21.6. A self-duplexing radar using circular polarization.

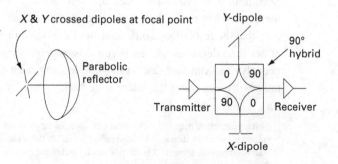

Figure 21.7. A waveguide turnstile junction combines the functions of the hybrid and crossed dipoles.

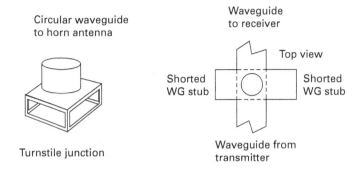

rectangular ports are fitted with shorts of appropriate lengths, the resulting four-port network is equivalent to the pair of dipoles and hybrid of Figure 21.6. The turnstile junction is described in Volumes 8 and 9 of the Rad. Lab. Series. The transmitter and receiver are connected to the remaining rectangular ports (the pair without shorts) while the antenna, usually a feed horn, is connected to the round waveguide.

A true self-duplexing circuit, shown in Figure 21.8, uses a circulator. The circulator has the property that a signal injected at the transmitter port will exit via the antenna port while a signal injected at the antenna port will exit through the receiver port. (If a signal were injected at the receiver port, it would exit through the transmitter port.) This nonreciprocal action depends on transmission through a nonreciprocal medium which, for the circulator, is a ferrite material, biased by the field of a permanent magnet. This elegant TR system is limited by available circulators to powers of tens of kilowatts.

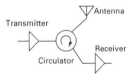

Figure 21.8. A circulator used as a TR switch.

21.6.2 TR switching devices and circuits

The classic TR circuits developed during WWII use gas discharge tubes or spark gaps and are self-activated by high-power RF on the transmission line. Lower power versions use PIN diode switches, turned on by an external bias circuit. (The radar has a timing generator providing pulses that (1) turn on the TR switch for transmitting, (2) pulse the transmitter, and (3) turn off the TR switch for receiving.) Gas discharge tube switches are usually built into a short piece of waveguide and come in two types: TR and ATR (anti-TR). The distinction is as follows: on transmit a TR device ionizes and presents a low-impedance shunt across the line. The ATR device also ionizes on transmit but it presents a low impedance in series with the line. (Reference [3], Volume 14 of the Rad. Lab. Series, is devoted mostly to these tubes.) There are two general classes of TR circuits, branch line TR switches and balanced TR switches. The former uses segments of transmission line while the latter uses hybrids.

21.6.2.1 Branch line TR switches

Figure 21.9 shows some standard branch line TR Switch circuits using TR, ATR, or both TR and ATR tubes (or PIN diode equivalents).

In the circuit of Figure 21.9(a), the switches are shown in the nonconducting (transmitter off) position. The open ATR switch is connected to the antenna by a half-wave line so it presents the same open circuit to the antenna-to-receiver line. Likewise, the open TR switch does not disturb the antenna-to-receiver line. On transmit, the TR switch places a protective short circuit at the receiver input. This short circuit is transformed by the quarter-wave line into an open circuit which does not affect the connection between the transmitter and the antenna. At high frequencies the switches contain some nonzero path lengths which form part of the half-wave or quarter-wave lines. For low-frequency designs the half-wave lines can be reduced to zero length and the quarter-wave lines can be replaced by lumped-element impedance inverters. You can see from the ways that the TR and ATR elements are used that their names are somewhat arbitrary.

Figure 21.9. Branch line TR switches.

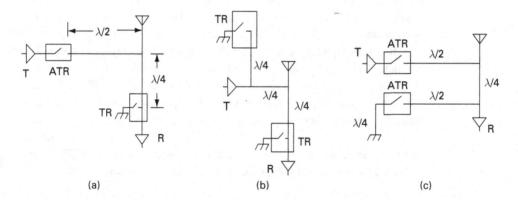

21.6.2.2 Balanced duplexers

Balanced duplexers use hybrids and can have wider bandwidths than the branch line circuits shown above (though more elaborate branch line circuits can have wider bandwidths). Both balanced duplexer circuits shown in Figure 21.10 use two 90° hybrids.

Figure 21.10. Balanced TR switches.

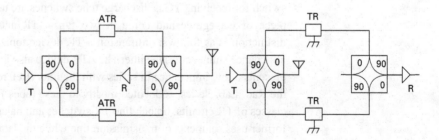

21.7 Diode switches

A single diode shunt switch circuit is shown in Figure 21.11.

Positive control voltage gives the diode a forward bias current to produce a low dynamic resistance, dV/dI. Negative control voltage turns the diode off, making its dynamic resistance very high. Let these dynamic resistance values be denoted respectively by r and R. You can verify (Problem 21.5) that the transmission values (power out/power available) for the switch of Figure 21.11 are given by:

$$\text{Isolation (forward biased state)} = 4r^2/Z_0^2$$
$$\text{Transmission (reverse biased state)} = 1/(1 + Z_0/R).$$

Note from these expressions that you could favor better isolation or lower insertion loss by transforming the line to have a larger or smaller Z_0 at the diode location. Better performance can be obtained with ladder networks analogous to multisection filters. Figure 21.12 shows how a shunt switch can be combined with a series switch to form a two-element ladder network. Isolation is improved since any signal leakage across the open series switch is shorted to ground by the closed shunt switch.

It is often more convenient to use shunt diodes than series diodes, since shunt diodes are easier to bias and to heat sink. Impedance inverters can transform series elements into shunt elements, as we saw when designing coupled

Figure 21.11. Shunt diode switch.

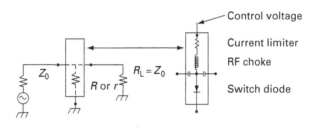

Figure 21.12. Series-shunt diode switch.

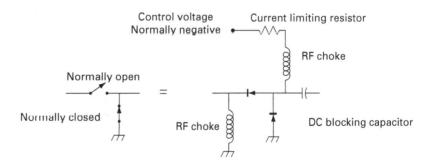

Figure 21.13. Three-section switch using only shunt sections.

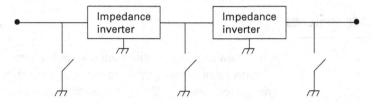

Figure 21.14. Transfer switch using quarter-wave transmission lines.

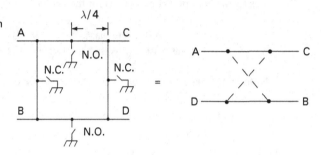

resonator filters. The switch circuit of Figure 21.13 uses two impedance inverters and three shunt switches.

The switches discussed above are all SPST switches. More complicated switches can be built up from the elementary SPST switch, but special designs can often be made such as the four-diode transfer switch shown in Figure 21.14. When the switches are in the indicated normal positions, the shorted quarter-wave lines appear as infinite-impedance shunts and do not disturb the transmission paths from A to C and from B to D. When the switches are reversed, transmission is from A to B and from C to *D*.

21.7.1 Diodes for RF switching

When ordinary diodes are used in these switching applications, the biases must be enough to keep the diode in the desired state. In particular, when the diode is off, the reverse bias voltage must be greater than the peak RF voltage and, when the diode is on, the forward bias current must be greater than the peak RF current. However, the *PIN diode*, a sandwich of p-type, intrinsic, and n-type semiconductor material, has the remarkable property that, for RF switching, the bias current and bias voltage values can be less than the corresponding RF current and voltage by perhaps an order of magnitude. The operation of the PIN diode (see reference 5) depends on having a large (small) stored charge in the intrinsic region when the diode is on (off). At high frequencies, the time between electric field reversals is much less than the transit time through the intrinsic region, determined by diffusion and drift, so the diode remains in the on (off) state. Finally, a word of caution: diodes, since they are nonlinear circuit elements, have the potential to distort a signal. In particular, they can create intermodulation products between the various signals in a complicated spectrum. For critical applications, e.g., receiver band-changing, diode switches must be turned off and on hard enough to keep any generated intermodulation products at a negligible level.

21.8 Radar pulse modulators

Any cw transmitter can generally be used as a pulse transmitter if a *pulse modulator* is added to provide the rapid turn on and turn off. Tube-type amplifiers can be operated with much higher instantaneous powers when they are pulsed. Tubes are primarily limited by their maximum anode dissipation (heat removal); the dissipation can be the result of either modest cw operation or high-power pulse operation. A cw amplifier can be converted for pulse operation by changing the output matching circuit in order to present a lower load resistance to the tube. Some tubes are available in special pulse-rated versions; they are fitted with high-emission cathodes. Gridded tubes (triodes, tetrodes, and pentodes) can be pulsed by switching the grid bias from negative, for pulse-off, to positive, for pulse-on. The negative bias keeps the tube completely turned off between pulses. Since the grid voltage and current are much smaller than the plate voltage and current, grid control requires only low-power circuitry compared to anode control. At microwave frequencies, magnetrons and klystrons replace gridded tubes. Magnetrons have no control element and therefore require high-power anode pulsers. Klystrons may or may not have a modulating anode ("mod anode") by which the beam current can be cut off. If not, they need high-power pulsers.[6]

Transistor amplifiers, unlike tube amplifiers, can make very little trade-off between duty cycle and peak power. Transistors suffer one type of breakdown or another when operated beyond their maximum continuous ratings. A high-power transistor amplifier for pulse service might differ from a cw amplifier only in that it will dissipate less heat (from the reduced duty cycle) and can therefore get by with a smaller heat sink.

No matter how an amplifier is pulsed, the power supply must furnish high-power pulses with minimum voltage droop. Duty cycles of pulsed transmitters are usually much less than unity so, in addition to at least one switching element, pulse modulators (pulsers) contain some form of energy storage element(s). The simple pulser circuit shown in Figure 21.15(a) stores energy in a capacitor.

---

[6] An air traffic control radar might have a peak power output of 2 MW and an efficiency of 50%. A klystron tube in this service could require 50 kV pulses at 80 amperes.

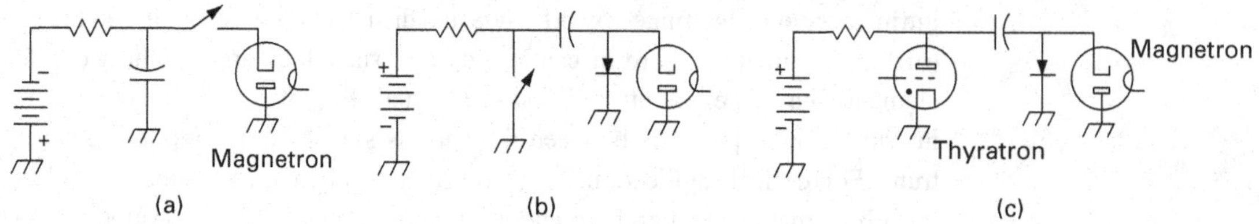

(a)   (b)   (c)

Figure 21.15. Capacitor discharge pulser.

In this circuit the tube (magnetron, klystron, or whatever) is shown as requiring negative voltage. Microwave tubes often use negative supply voltage applied to their cathodes because it is convenient to ground the external heat-dissipating anode. The version of the circuit in (b) allows one side of the switch to be grounded, which is another convenience. The diode provides a charging path for the energy storage capacitor. The circuit of Figure 21.15(c) uses a thyratron (vacuum tube version of the SCR) as the switch.

The simple capacitor discharge modulators in Figure 21.15 have several disadvantages: The voltage droops during the pulse. The droop can be reduced by increasing the size (weight, and cost) of the capacitor. Not much of the stored energy is used. Even if 10% voltage droop is permitted, only 20% of the stored energy is used for each pulse. This might be compared to a car which would not run well if the fuel tank was less than 80% full. They are expensive and heavy, a particular disadvantage for airborne equipment. Despite these disadvantages, capacitor banks are often used, as in the 430-MHZ pulse transmitter used for ionospheric research at the Arecibo Observatory, because a more efficient circuit, the line modulator discussed below, does not easily provide the flexibility needed to change the pulse width.

21.8.1 Line modulators

A length of transmission line (with the far end open) has capacitance and can therefore store electrostatic energy. When the line is discharged into a resistive load equal to its characteristic impedance, it will supply a perfect rectangular pulse rather than a drooping exponential pulse. The constant pulse amplitude during discharge is maintained by the distributed inductance of the line acting together with the distributed capacitance. In Figure 21.16(a), the line is a piece of coaxial cable, replacing the energy storage capacitor. As before, the tube is supplied with a negative pulse. A diode provides a path to recharge the line. Often the load has a higher impedance than the characteristic impedance of the line, and a pulse transformer is required.

The line supplies a pulse at half the charging voltage since, during the pulse, the charging voltage evenly divides between the load and the equivalent source resistance. The duration of the pulse is the time taken for the current to make a round trip through the line. At the end of the pulse the line is totally discharged; all the stored energy is delivered on every pulse. Figure 21.16(b) shows successive plots of the line's voltage and current distributions. In order to store more

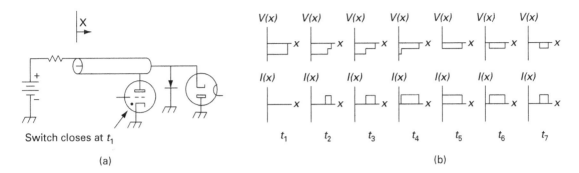

Figure 21.16. (a) Line-type modulator; (b) line voltage and currents.

Figure 21.17. Pulser using an artificial transmission line (pulse-forming network).

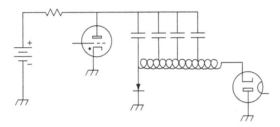

energy, it is common to use an "artificial transmission line" or pulse forming network (PFN) which is a ladder network of inductances and capacitances. A four-section network is shown in the modulator circuit of Figure 21.17.

The network looks like a lowpass filter and it is. Its cutoff frequency is given by $\omega^2 = 4/[LC]$. For frequencies well below cutoff, the network behaves like a transmission line with $Z_0 = \sqrt{L/C}$. Here L and C are in henries and farads rather than henries/meter and farads/meter as in the distributed element transmission line. The one-way time delay on this lumped line is $\sqrt{LC}$ seconds/section.

Let us consider a numerical example: Let us use four sections as in the figure above. Suppose we need a 1 microsecond pulse at 10 kV and 10 amperes. The voltage and current require that $Z_0 = \sqrt{L/C} = 1000$ ohms. We will use four sections which, for the desired one-microsecond round trip delay, requires that $8\sqrt{LC}$ be equal to 10^{-6}. These impedance and time delay equations are satisfied by $L = 125$ μH and $C = 125$ pF. We can verify that the energy stored in the line is indeed equal to the energy delivered by the pulse. The latter is just $(IV)\tau = 10 \times 10\,000 \times 10^{-6} = 0.1$ joule. The former, remembering that we must charge the line to 20 000 V, is $CV^2/2 = 4(125 \times 10^{-12}) \times 20\,000^2/2$ which is also 0.1 joule. As often happens in filter design, these are not particularly practical values; real inductors of 125 μH may well have distributed capacitances that are not negligible compared with 125 pF. We can build the line for a lower

Figure 21.18. Waveform produced by a four-section pulse forming network.

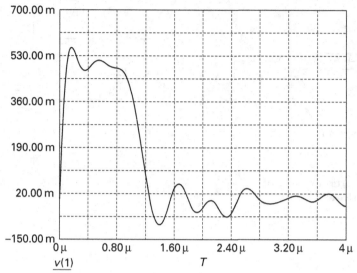

impedance and use a pulse transformer between the line and the magnetron. If we lower the line impedance to 100 ohms, the L and C values become 12.5 μH and 1250 pF, values that are more practical. Using these values, a SPICE simulation of the discharge produced the voltage waveform shown in Figure 21.18. The voltage scale is normalized, i.e., the capacitors were charged to one volt so the nominal pulse voltage is 0.5 volts. Most pulse forming networks can be tuned slightly to improve the pulse shape; the artificial transmission line is, after all, only an approximation to an ideal transmission line.

The line modulator uses all the stored energy on each pulse but, precisely because of this virtue, deserves a more sophisticated charging circuit than the resistor shown in the circuits above. Remember that when a capacitor is charged through any resistive path from empty (no energy) to $CV^2/2$, the resistor will dissipate this same amount of energy, $CV^2/2$. Here the charging resistor, no matter what value, would dissipate half the power consumed by the radar. The solution to this problem is to charge the line through an inductor instead of a resistor. Figure 21.19(a) shows the voltage waveform on a capacitor as it is *resonantly charged* through an inductor.

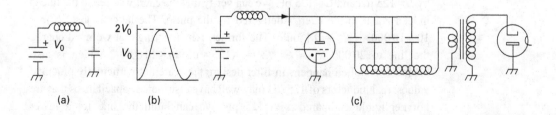

(a)   (b)   (c)

Figure 21.19. Resonant charging.

The voltage is a sinusoid, building up to a maximum of twice the supply voltage. The modulator can be triggered just as the voltage reaches this maximum. The brief pulse discharges the line and the charging curve begins anew. It would seem that the PRF is therefore determined rigidly by the charging time but, if a diode is put in series with the inductor, the charging stops at the maximum voltage and the next pulse can occur anytime. The resonantly charged modulator, with the diode and a pulse transformer, is shown in Figure 21.19(b). Note that the primary of the pulse transformer provides a charging path, eliminating the diode originally in parallel with the magnetron. Also remember that, because of the resonant charging, the supply voltage needs only to be half of the line charging voltage. Line modulators present less risk to tubes than partial-discharge capacitor modulators because there is less stored energy available when an arc occurs in the tube.

Problems

Problem 21.1. Find the radar cross-section of a flat metal plate of area A that is exactly perpendicular to the radar beam. Assume that $A \gg \lambda^2$. Hint: treat the plate as an aperture antenna with gain $4\pi A/\lambda^2$.

Problem 21.2. Find the lobe pattern produced when a point target is observed by a pulse compression radar using the 13-bit Barker code of Section 21.4. (Convolve the code with itself after padding both ends with zeros.)

Problem 21.3. Why is the side-looking radar antenna positioned to look to one side of the aircraft rather than straight down?

Problem 21.4. If we try to use the hybrid of the circular polarization duplexer as a circulator, we might consider the TR circuit shown below.

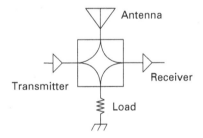

Assume the antenna, transmitter, receiver, load, and hybrid all have the same characteristic impedance. This circuit at least protects the receiver from the transmitter. What are its disadvantages (a) when transmitting, and (b) when receiving?

Problem 21.5. Verify the expressions below for $P_{out}/P_{available}$ for the circuit of Figure 21.11.

Diode state	Switch state	$P_{out}/P_{available}$
Foward biased	Isolation	$4r^2/Z_0^2$
Reversed biased	Transmission	$1/(1+Z_0/R)$

Problem 21.6. Apply your circuit analysis program (Problem 1-3) to the transfer switch circuit of Figure 21.4. Assume a 50-ohm load is connected to Port C, a 50-ohm generator is connected to Port A, and the transmission line sections have a 50-ohm characteristic impedance. Assume also that the internal switches are ideal. Find the transmission coefficient (in dB) over the frequency range from half the design frequency to twice the design frequency. Hint: the closed switches divide the circuit in two so you can ignore the bottom half.

Problem 21.7. Explain the operation of the balanced duplexers shown in Figure 21.10. What restrictions, if any, are there on the lengths of the interconnecting transmission lines?

Problem 21.8. (a) Show that when an uncharged capacitor is brought to potential V by connecting it through a resistor to a voltage source V, the energy supplied by the source is twice the energy deposited in the capacitor (CV^2 rather than $CV^2/2$).

(b) The charging efficiency in (a) is only 50%. Find the efficiency when the capacitor initially has a partial charge, i.e., when the capacitor is initially charged to a voltage αV, where $\alpha < 1$.

Problem 21.9. (a) Find the characteristic impedance of the artificial transmission line shown below. This impedance, Z_0 (which is complex), can be found by adding another LC section to the properly terminated line and noting that the new impedance must still be Z_0.

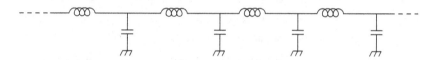

(b) Use the expression for Z_0 to show that the line has a cutoff frequency, $\omega_c = 2/\sqrt{LC}$, above which signals are reflected rather than transmitted.

Problem 21.10. Show that when $\omega \ll \omega_c$, the propagation delay per section for the artificial transmission line of Problem 21.9 is given by $\tau = \sqrt{LC}$.

Problem 21.11. (a) In a *conial scan* tracking radar, the antenna's feed horn is tilted slightly off center and mechanically rotates around the axis of the parabolic reflector. The antenna beam therefore executes a continuous tight conical scan, centered on the target. Many pulses are transmitted during the course of each scan. Draw a block diagram of circuitry to furnish x and y cross-range position error signals to the antenna's drive system in order to keep the antenna centered on the target. (b) Consider a *monopulse* tracking radar in which the antenna is effectively five antennas: one pointed directly at the target, one slightly above, one slightly below, one slightly to the left, and one slightly to the right. Draw a block diagram of circuitry to combine these five signals to provide x and y error signals.

References

[1] North, W., "High-Power Microwave-Tube Transmitters" Los Alamos Nation Laboratory, LA-12687, 1994. Available from U.S. Dept. of Energy, Office of

Scientific and Technical Information, Post Office Box 62, Oak Ridge, Tennessee 37831.

[2] Skolnik, M. I., *Radar Handbook*, 3rd edn. New York: McGraw-Hill, 2008.

[3] Smullin, L. D. and Montgomery, C. G., *Microwave Duplexers*, Rad. Lab. Series Vol. 14, New York: McGraw-Hill, 1948.

[4] Wehner, D. R., *High Resolution Radar*, Boston: Artech House, 1987.

[5] White, J. F., *Microwave Semiconductor Engineering*, New York: Van Nostrand Reinhold, 1982.

22 Digital modulation techniques

Digital modulation is both the newest and the oldest radio technique. Morse code transmissions were strictly binary, with "key down" and "key up" equivalent to multiplying the carrier by one or zero. Many modern systems also use binary keying, but the zero state is usually signaled by reversing the polarity of the signal (*binary phase-shift keying*, BPSK) or by shifting the frequency (*binary frequency-shift keying*, BFSK). This improves the probability of distinguishing zeros from ones in the presence of noise.

In this chapter we look first at some of the methods used for binary and "*m-ary*" modulation. We then see how specially shaped pulses can be used with these methods in order to avoid intersymbol interference when the pulses, dispersed in time, partially overlap at the receiver. The "8-VSB" system used for digital television in the U.S. (see Chapter 19) provides an example of pulse amplitude modulation (PAM). Finally, we discuss two newer digital modulation systems: multicarrier and spread spectrum. A glossary is provided at the end of this chapter, listing the many common abbreviations used (BPSK, BFSK, 8-VSB, PAM, etc.).

22.1 Digital modulators

Digital modulation differs from analog modulation in that only a discrete set of states (in the space of amplitudes, phases, and frequencies) is used, and that the time devoted to any state is always an integral multiple of a basic time-step. The state during this time period constitutes a transmitted "symbol," and the symbol rate is one of the parameters defining a modulation system. Figure 22.1 shows the simple and widely used binary phase shift keying (BPSK) modulation technique. Instead of turning off the carrier to indicate a zero, the carrier phase is flipped 180°. This is equivalent to multiplying the carrier signal by minus one, as shown in the figure. Binary values of 0 or 1 are handled by sending −1 or +1, respectively, to the multiplier. As shown in Chapter 8, the multiplier (mixer) produces a double-sideband suppressed-carrier RF signal.

Figure 22.1. BPSK modulator.

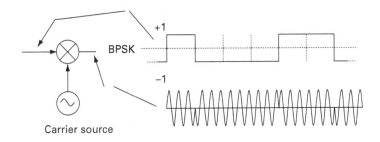

Figure 22.2. *N*-level pulse
amplitude modulation (PAM).

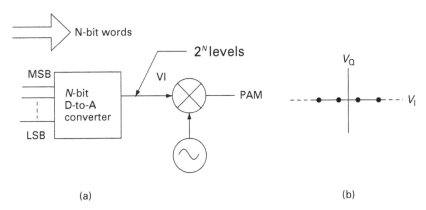

Since the BPSK signal has a nominally constant envelope, the receiver
needs a phase-sensitive or "coherent" detector. In contrast, on–off keying
(OOK) or frequency shift keying (FSK) can be detected incoherently. The
advantage in using coherent systems is one of sensitivity; for a given
situation, coherent detection provides a higher signal-to-noise ratio at the
detector output. The one-bit BPSK system can be generalized to more than
two amplitude levels. We saw in Chapter 21 that the U.S. ATSC digital
television system uses eight (2^3) modulation levels in order to send three
bits per symbol. Figure 22.2(a) shows an *N*-bit (2^N-level) pulse amplitude
(PAM) modulator. The "constellation" shown in (b) shows the signed amplitudes as discrete points on a linear scale. Note that, because the negative
amplitudes are produced by changing the phase 180°, PAM requires coherent
demodulation.

We saw in Chapter 8 that quadrature AM (QAM) systems can transmit
independent information, modulated on a *Q*-carrier which is in quadrature,
i.e., 90° out of phase, with respect to the *I* ("in-phase") carrier. This "phase
multiplexing" method is commonly used in digital communications.
Figure 22.3 shows a "4-QAM" modulator.

The binary (two-level) *I* and *Q* signals result in an *IQ* constellation
with four points. In this case, the points on the constellation have four
possible phase values, but the amplitudes are equal, since each of the four

points lies at an equal distance from the origin. You can see why this scheme is also known as QPSK (quadrature phase shift keying) and 4-PSK (four-state phase-shift keying). Figure 22.4 shows a QAM modulator with four-level modulation on both I and Q channels, producing a total of 16 states.

The constellation of output signals for this 16-QAM modulator is shown in (b). The 16 binary numbers, 0000 through 1111, could be assigned arbitrarily to the constellation points, but the assignments shown in the figure have the property that nearest neighbors differ by only one bit. As a result, in an environment with only modest noise, errors in transmission will nearly all be single-bit errors. Note that next-nearest neighboring points on the constellation differ by

Figure 22.3. (a) A 4-QAM (QPSK) digital modulator; (b) QPSK constellation.

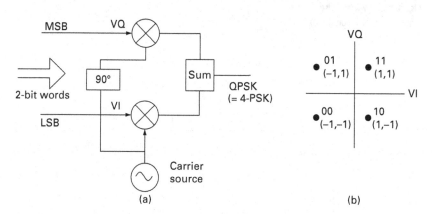

Figure 22.4. A 16-QAM digital modulator.

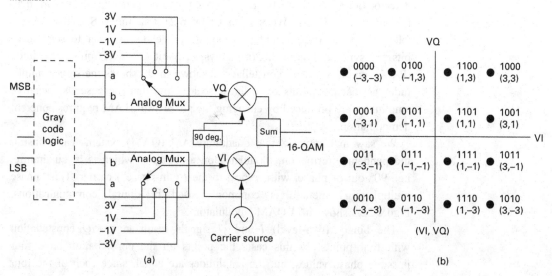

just two bits. QAM systems can also have constellations in which the points are arranged in polar fashion; a constellation with eight points arrayed on a circle in the *I-Q* plane is known as 8-PSK. Note that BPSK could just as well be called 2-PSK.

22.2 Pulse shaping

The abrupt reversals in the BPSK waveform of Figure 22.1 result in a signal with an objectionably large bandwidth. The signal is a continuous stream of rectangular RF pulses whose polarity is determined by the data. If the data bits are random, with equal probabilities of being a one or a zero, these rectangular pulses produce a power spectrum proportional to $[\sin(\pi f T)/(\pi f T)]^2$, where T is the duration of a symbol, the "baud" length. This spectrum is plotted in Figure 22.5.[1] On this graph, frequency is in units of $1/T$, and the zero is at the nominal carrier frequency.

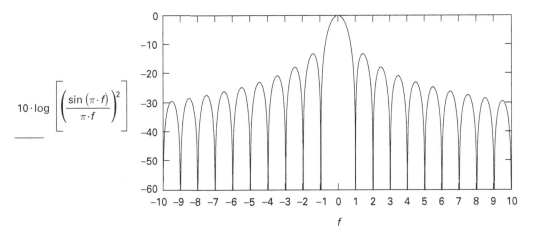

$$10 \cdot \log \left[\left(\frac{\sin (\pi \cdot f)}{\pi \cdot f} \right)^2 \right]$$

Figure 22.5. Power spectrum of a BPSK signal with random data and rectangular pulses. Frequency is in units of $1/T$, and the zero is at the nominal carrier frequency.

This wide spectrum applies to all the digital modulators discussed above, and is clearly unacceptable if other users are to use nearby frequency channels. Bandpass filtering at the transmitter, or otherwise transmitting a pulse shape whose spectrum is band-limited, solves this problem, but there is a complication – limiting the bandwith causes a widening of each pulse in the sequence so that, at the receiver, a currently arriving pulse will be contaminated by remnants or precursors of nearby pulses. This effect is known as *intersymbol interference* (ISI). However, there exists a class of pulse shapes, *Nyquist pulses*, for which (a) the bandwidth is limited and (b) there is no ISI, despite the fact that the pulses do spread into each other. One such pulse shape is $V(t) = \sin(\pi t/T)/(\pi t/T) = \text{sinc}(t/T)$, the normalized sinc function. This pulse shape is the impulse response of a

[1] The power spectrum is calculated from the signal's average autocorrelation function (see Chapter 27).

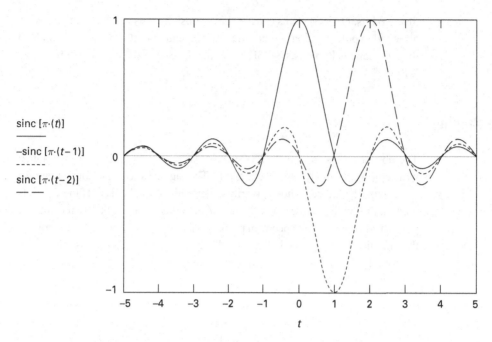

sinc [π·(t)]
_____

−sinc [π·(t− 1)]
- - - - - -

sinc [π·(t− 2)]
— —

Figure 22.6. Successive
$\sin(\pi t)/(\pi t)$ signaling pulses.

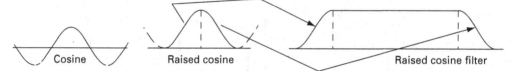

Cosine   Raised cosine   Raised cosine filter

Figure 22.7. Construction of the
raised cosine filter; $H(\omega)$ is the
voltage response.

rectangular lowpass filter with a cutoff frequency of $1/(2T)$Hz. This filter
confines all the power to a bandwidth of $1/(2T)$. Successive output pulses for
a 101 data string are plotted in Figure 22.6. The third pulse reaches a maximum
at t = 2, the instant at which we sample it. Neither of the two previous pulses has
yet died out but, at $t = 2$, both are crossing zero and thus contribute no interference to the sampled value of the desired pulse.

The same holds true for the contributions from all the other pulses, of any
amplitude and either polarity, allowing us to transmit data pulses at a rate $1/T$
without intersymbol interference.

Pulses with this property are said to be *orthogonal* to one another. While they
do invade each other's space, they do not create intersymbol interference. The
$\sin(\pi t/T)/(\pi t/T)$ pulse is an extreme case; bandwidth is minimized, but the pulse
is widely spread out in time. In practice, other orthogonal pulse shapes are
chosen, which have somewhat greater bandwidth but are only modestly spread
in time. One frequently used pulse shape is the impulse response of the *raised-
cosine* filter. Figure 22.7 shows how this filter shape is produced by replacing
the vertical sides of the rectangular filter with smooth roll-offs formed from
quarter cycles of the cosine function, raised by $1/2$ in the *y*-direction.

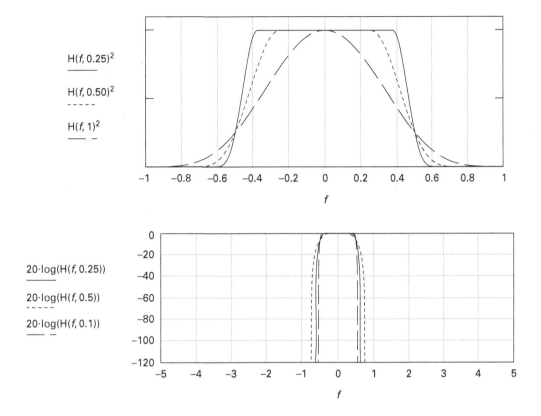

Figure 22.8. Power response, $|H(\omega)|^2$, of raised cosine filter for $\alpha = 0.25$, 0.5, and 1.0.

We can express this filter shape as

$$
H(f, \alpha) = \begin{cases} T: & 2|f|T \le 1 - \alpha \\ \dfrac{T}{2}\left[1 + \cos\left(\dfrac{\pi T}{\alpha}\left(|f| - \dfrac{1-\alpha}{2T}\right)\right)\right]: & 1 - \alpha \le 2|f|T \le 1 + \alpha \\ 0: & 2|f|T \, 1 + \alpha, \end{cases}
$$

(22.1)

where α, a roll-off factor between zero and one, determines the width of the raised cosine edge sections. Note that the $\alpha = 0$ case is just the rectangular filter discussed above. This filter's power response, $|H(f,\alpha)|^2$, for $\alpha = 0.25$, $\alpha = 0.5$ and $\alpha = 1$ (pure raised cosine with no flat center section) is plotted in Figure 22.8(a) with a linear power scale and Figure 22.8(b) with a dB power scale.

The impulse response of this filter, i.e., the Fourier transform of $H(f,\alpha)$, often called by extension a "raised-cosine pulse," is given by

$$
h(t, \alpha) = \frac{\sin(\pi t/T)}{\pi t/T} \cdot \frac{\cos(\pi \alpha t/T)}{1 - (2\alpha t/T)^2}.
$$

(22.2)

The zeros of the first term, $\sin(\pi t/T) \, / (\pi t/T)$, give this pulse shape the same orthogonality property as $\sin(\pi t/T) \, / (\pi t/T)$ alone. Figure 22.9 shows three

Figure 22.9. Successive raised-cosine signaling pulses.

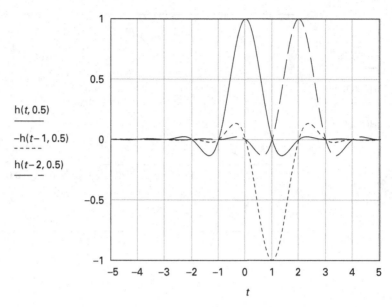

h(t, 0.5)
——————

−h(t − 1, 0.5)
- - - - -

h(t − 2, 0.5)
—— —

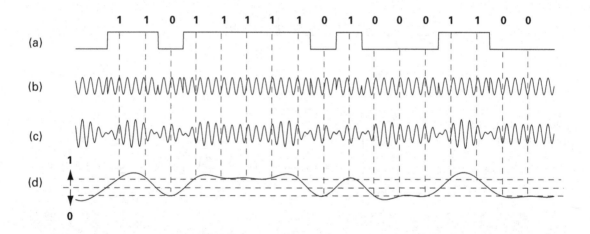

(a)

(b)

(c)

(d)

Figure 22.10. BPSK waveforms: (a) input data; (b) bi-phase modulated carrier; (c) modulated signal after raised-cosine filtering ($\alpha = 0.22$); (d) coherently detected signal.

consecutive $\alpha = 0.5$ raised cosine pulses. Note that these pulses, like the sin $(\pi t/T)/(\pi t/T)$ pulses of Figure 22.6, produce no intersymbol interference, but unlike those pulses, these are only modestly spread out in time.

Despite the bandwidth advantage, there are two penalties associated with choosing a low value for α. First, the impulse response duration of the filter is longer, requiring more computations per symbol when the filter is implemented digitally. Second, the increased ringing means that the received signal must be sampled with higher timing accuracy to avoid ISI.

The successive processes of BPSK modulation, pulse shaping, and detection are shown in Figure 22.10. The data string to be transmitted is shown in (a). In (b), the sinusoidal carrier has been multiplied by the data, producing phase

Figure 22.11. BPSK eye diagram.

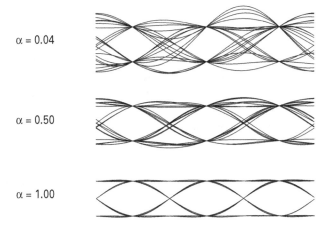

$\alpha = 0.04$

$\alpha = 0.50$

$\alpha = 1.00$

reversals. In (c), the modulated signal has been band-limited by a raised-cosine filter of $\alpha = 0.22$. In this simulation the modulation rate is high, causing obvious variations in the envelope of the filtered signal. Finally, in (d), the filtered signal has been coherently detected, i.e., multiplied by a replica of the original sine wave carrier and then lowpass filtered to remove the components at twice the carrier frequency. Note that, at the sampling instants, the sampled values are exactly the same 0's and 1's of the original data, i.e., there is no intersymbol interference. Note: for clarity in alignment with the original data, no filter delay has been added to the waveforms in (c) and (d), so they appear noncausal.

If the detected data is displayed versus time for many sweeps on a storage oscilloscope that is synchronized to the data, we obtain the well-known "eye" diagram, shown in Figure 22.11. You can see that, at the sampling instants, the value of the waveform is either zero or one; there is no intersymbol interference. Eye diagrams are presented for three values of α. You can see that, for smaller values of α (narrower bandwidths), the sampling instants must be increasingly accurate to obtain only the values zero and one, whereas larger values of α make the system more tolerant of sampling "jitter."

22.3 Root raised-cosine filter

Any receiver must contain a bandpass filter to reject unwanted signals (i.e., signals as well as noise on nearby frequencies). If the receiver filter has a rectangular shape whose bandwidth is equal to or greater than the signal bandwidth, Nyquist pulses will pass through undistorted, maintaining their orthogonality, i.e., their freedom from intersymbol interference. However, a basic result from signal processing theory (see Chapter 23) is that the overall signal-to-noise ratio of a communication link is maximized when the receiver uses a *matched filter*, which is a filter whose impulse response is proportional to the time-reversed waveform of the incoming signal pulse. Let us apply this consideration to the two pulse shapes discussed above. For the $\sin(\pi t/T)/(\pi t/T)$ signaling pulse, the matched filter is just the rectangular filter which might have been used to produce the pulse. The cascade of two rectangular filters is equivalent to a single rectangular filter, and the pulse shape is preserved. For a raised-cosine signaling pulse, however, the matched filter is the raised-cosine filter of Equation 22.1. If we put this filter at the input of the receiver, the overall response will be the *square* of the raised-cosine shape. The signaling pulses will take on the shape of a raised-cosine squared filter – a shape that is not free of intersymbol interference. To avoid this, it is common to transmit "root raised-cosine" signaling pulses, i.e., pulses whose shape is the impulse response of a filter whose amplitude response is the square root of the raised-cosine filter. A root raised-cosine filter is used at the input of the receiver as a matched filter. The pulses exiting the receiver filter will therefore have the "raised-cosine pulse" shape shown in Figure 22.9, and there will be no intersymbol interference. The frequency response of the root raised-cosine filter is given by Figure 22.8(b), if you relabel the y-scale to go from 0 to − 60 dB instead of 0 to −120 dB.

Let us summarize this section. So-called Nyquist pulses can be transmitted without intersymbol interference up to a maximum rate of B pulses per second using double-sideband AM modulation within an RF bandwidth of at least B Hz. A second carrier, at the same frequency, but phase shifted by 90°, can be used simultaneously to transmit independent pulses. When the pulses are identical except for 0° or 180° phase changes, the system is called BPSK (one carrier only) or QPSK (pulses on both I and Q carriers). Synchronous detection, i.e., phase-sensitive detection, makes this 0° and 180° phase modulation equivalent to amplitude modulation in which the sign of the signal is preserved. If the I and Q components of a QPSK signal are each allowed to take on four values, the modulation forms a 4×4 constellation and is referred to as 16-QAM. (It would be convenient if the term "AM" always referred to this kind of sign and magnitude modulation. BPSK could be called "sign modulation." Traditional AM could be called "magnitude modulation".)

Note that in the above analysis we have not needed to bother with including a phase factor in the filter amplitude response functions. These particular filter functions are purely real, except for this omitted factor, $e^{-j\omega T_{\text{delay}}}$, where T_{delay} is a frequency-independent time delay. Neglecting this factor is equivalent to setting the delay equal to zero, which causes the impulse responses to begin at non-causal negative times.

22.4 8-VSB and GMSK modulation

The U.S. digital television system, discussed in Chapter 19, modulates an I-carrier with three-bit digital data in the form of eight evenly spaced amplitude levels whose nominal values are 7, 5, 3, 1, −1, −3, −5, and −7. A bias of 1.5 is added to these levels to create a small pilot carrier (a phase reference for the receiver) resulting in five transmitted levels that are positive and three that are negative. This is a PAM system (see Figure 22.2). The resulting double-sideband suppressed carrier signal is bandpass filtered to eliminate all but a vestige of the lower sideband, reducing the bandwidth by factor of almost 2. This modulation is called 8-VSB (eight levels, vestigial sideband). Because the signal had no Q-component before filtering, the lower sideband was the mirror image of the upper sideband and contained redundant information. With a general QAM signal (using both I and Q), the lower and upper sidebands are different and both are needed. But, as a result of filtering away the lower sideband, a transmitted VSB signal acquires a Q component, making the "Q-space" unavailable for the transmission of a second television signal.

The modulation method used in GSM cellular telephones is known as *GMSK* (*Gaussian minimum shift keying*), and works as follows. A standard QPSK modulator is used, summing the signal from an I mixer with the signal from a Q mixer. But the input signals to these mixers are such as to cause the sum signal to rotate smoothly from one point on the constellation to an adjacent point during the baud time. The rotation is 90°, clockwise or counterclockwise, depending on the binary data bit. The rotation takes place at a uniform rate. Since frequency is the time derivative of phase, this modulation is actually BFSK. In addition, smoothing, derived from a Gaussian filter shape, is applied to the phase command, so that the time derivative of the frequency is continuous during the transitions between the two nominal frequencies. This system is designed to minimize bandwidth, allowing closer channel spacing to accommodate more users. Bandwidth reduction comes from using both I and Q, from changing the phase slowly and smoothly, and from allowing some intersymbol interference (which is almost all from the immediately preceding baud and can be compensated for by the receiver).

22.5 Demodulation

Synchronous demodulation is required for the BPSK and QAM systems discussed above. This requires that the receiver be able to synthesize a replica of the carrier. This carrier recovery, which takes place in the presence of data, can be done, for example, using a squaring PLL (see Chapter 12). A standard PLL suffices if a pilot carrier is transmitted. The receiver contains a matched filter, followed by a mixer whose L.O. signal is the recovered carrier. The baseband output of the mixer is sampled at the signal baud rate. The signal may include a periodic synchronizing pattern to assist the receiver in establishing the correct phase for the sampler. A digital matched filter is often used. When realized as a weighted sum of signals from a tapped delay line (a transversal FIR filter), such a filter illustrates how a matched filter is equivalent to a correlator, finding the best alignment (match) of the signal to a replica of the signaling pulse.

Sometimes a free-running L.O. will be close enough to the carrier frequency that the dephasing over a data word or packet is significantly less than $\pi/2$. In this situation, differential modulation can be used whereby, for example, a "one" is signaled by flipping the phase of a BPSK signal, while a "zero" is signaled by leaving the phase at its prior value. If both I and Q channels are implemented, the phase changes can be reliably sensed and the data can be reconstructed. In a QPSK system, both I and Q channels are always implemented and the receiver can sense differential changes between points on the modulation constellation. Differential modulation, however, results in a higher error rate than coherent demodulation.

22.6 Orthogonal frequency-division multiplexing – OFDM

OFDM is a relatively new modulation technique in which the digital data stream to be transmitted is divided into N separate parallel data substreams, each having a data rate that is $1/N$ times the data rate of the original stream. Each substream independently modulates a separate RF carrier. The modulation scheme can be any of the schemes described above, except for FSK and MSK, where the phase does not remain constant throughout the duration of a symbol. At the receiver, the N substreams are demodulated in parallel and then combined to produce the original data stream. Applications of OFDM include 802.11a , g, and n Wi-Fi modems, DAB (digital audio broadcasting) in Europe and Canada, Digital Radio Mondiale short-wave broadcasting, HD (hybrid digital) radio broadcasting on the U.S. AM and FM bands, and European DVB-T (terrestrial digital television broadcasting). The number of carriers used in these systems varies from tens to thousands.

22.6.1 Advantages of OFDM

The system is bandwidth neutral, occupying about the same bandwidth as the traditional modulation systems it replaces. Its main advantage comes from its ability to deal with multipath propagation and channel equalization. Consider the problem of equalization. The propagation channel for a digital signal should have a specific amplitude and phase response, such as the raised-cosine shape discussed above. There are several propagation effects that can distort the shape. Receiving antennas are not always flat in frequency and linear in phase, so they can modify the channel response. Constructive and destructive interference of multipath components are frequency dependent and modify the channel response. Television receivers for 8-VSB use time correlation to detect multipath components and then subtract or realign them. This process, while essential, is not simple, and is particularly difficult in a mobile environment where the multipath situation is constantly changing. At fast baud rates, new multipath components are still arriving at the time the baud ends. With the slow data rates on the individual OFDM carriers, the baud can extend far beyond the time needed for all the multipath components to arrive. Dispersion is also a classic problem in wired data links.[2] Fast DSL (*digital subscriber line*) data links over ordinary telephone circuits have to deal with cable dispersion, poorly controlled impedances, and multipath echos caused by haphazard terminations and discontinuities. An early application of OFDM was in "discrete multi-tone" DSL modems of the early 1990s. While the principles of OFDM date back to the 1960s, practical implementations had to await the advent of fast digital signal processors. Most OFDM system incorporate coding and decoding for forward error correction and are known as COFDM (coded OFDM). As in digital TV transmission, the coding usually encloses the inner modulation/demodulation "physical layer."

22.6.2 Single-frequency broadcasting networks

The essentially perfect ability of OFDM to deal with multipath propagation can be exploited in single-frequency networks, in which broadcasting transmitters, all on the same frequency, serve overlapping areas of coverage. When the identical transmitted signals are synchronized in time and frequency (which can be done using GPS satellite signals), the signals arriving at any given site are indistinguishable from multipath signals originating from a single transmitter. At any given site, the receiving antenna is aimed at the strongest signal. The weaker signals, depending on their phase, increase or decrease the signal strength. They also shift the phases of the recovered Fourier coefficients, but this is not a problem because OFDM channels can be individually equalized via an occasional "training" sequence sent by the transmitter or by using a set of pilot carriers. Single-frequency networks allow efficient use of the spectrum, since the same television or radio channels can be used in adjacent areas – something like solving the classic map coloring problem with a single color. It is assumed that 8-VSB broadcasting could also use single-frequency networking, given echo cancellers of sufficiently high performance. Same-frequency fill-in or "booster" stations with well-synchronized frequencies have been used occasionally for conventional analog AM, FM, and television, but, without echo cancellation, they produce video echos ("ghosts").

---

[2] The earliest trans-Atlantic telegraph cable, over its great length, accumulated so much dispersion that baud rate had to be slowed down to where the received signal was traced out by a chart recorder. Likewise, early overland telegraph lines were too dispersive to support baseband telephony. The first solution, series loading coils, was a product of the then recent theory of waves on transmission lines, which showed that the reactance per unit length had to be increased, relative to the resistance per unit length.

22.6.3 Implementation of OFDM

The key to practical implementation of OFDM is the discrete Fourier transform (DFT) and its inverse (IDFT), which, in practice are implemented using FFT algorithms. At the transmitter, the data is arranged as a stream of data blocks, each containing N pairs of digital numbers. Each pair of numbers can be considered as the real and imaginary parts of a complex number. Let the complex data numbers in a given block be denoted as D_n, for $n = 0$ to $N-1$.

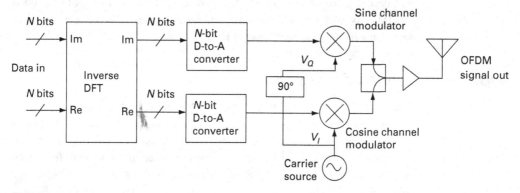

Figure 22.12. OFDM transmitter block diagram.

Figure 22.12, a block diagram for an OFDM transmitter, shows that these numbers are first inverse Fourier-transformed as follows:

$$V_m = \sum_{n=0}^{N-1} D_n e^{j2\pi mn/N}. \tag{22.3}$$

The resulting set of N complex numbers, V_m, for $m = 0$ to $N - 1$, is read out serially at a rate N/T, where T must be no greater than the spacing between blocks of input data. If V_0 appears at $t = 0$, we can express the output numbers as a function of time:

$$V(t_m) = \sum_{n=0}^{N-1} D_n e^{j2\pi nt_m/T} = \sum_{n=0}^{N-1} D_n e^{j\omega_n t_m}, \tag{22.4}$$

where ω_n is $2\pi n/T$. Note that this is just the sum of N carriers at equally spaced frequencies, ω_n, and that each carrier is multiplied (amplitude modulated) by a coefficient, D_n.

This set of N modulated carriers is converted from baseband to a QAM signal at RF by using a "cosine mixer" for the real part and a "sine mixer" for the imaginary part. Of course the real and imaginary outputs of the IDFT must be converted to analog voltages before they are applied to the mixers. These (suppressed) carriers are often called subcarriers, although they are not part of the sideband structure of some main carrier. Figure 22.13 shows an OFDM

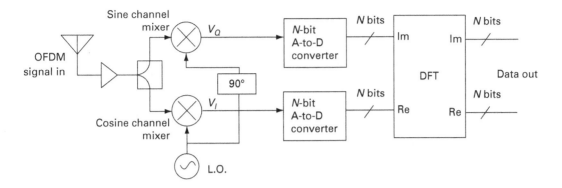

Figure 22.13. OFDM receiver block diagram.

receiver. Note that it is a mirror image of the transmitter. The operations done by the transmitter are undone in the reverse order by the receiver. I and Q samples from the cosine and sine mixers are digitized and combined to from the sequence of complex numbers, V_m, which is Fourier transformed to recover the date (the complex amplitudes of the subcarriers),

$$\frac{1}{N}\sum_{m=0}^{N-1} V_m\, e^{-j2\pi nm/N} = D_n, \tag{22.5}$$

and we see that the data has indeed been recovered from the separate carriers. Thus the inverse Fourier transform effectively acts as N carrier oscillators and N mixers to create N modulated carriers while the Fourier transform at the receiver acts as N L.O.'s and N mixers to detect the (complex) amplitude of each carrier. During the interval T, the n-th carrier has constant amplitude; it is an RF pulse with a rectangular envelope. The Fourier transform of this pulse has the shape $\sin(\Delta\omega T/2)/(\Delta\omega T/2)$, where $\Delta\omega$ is the offset from ω_n. The shape plotted in Figure 22.5. We see that the spectrum for any one of the carriers has considerable overlap with the spectra of nearby carriers. Nevertheless, at the nominal frequency of any one of the carriers, the power spectrum of any other carrier is crossing through zero. For this reason, the carriers are said to be orthogonal in the frequency domain, analogous to Nyquist pulses in the time domain, whose maximum value occurs at an instant when the voltage from any earlier pulse is crossing zero.

22.6.4 OFDM guard interval

The key to dealing with multipath in OFDM is the use of a guard interval after every T-second transmission of the group of multicarrier signals. To see that this is necessary, consider the situation shown in Figure 22.14, where no guard interval is used. In the figure, the data is advancing in time toward the right. Ideally we would receive N signal samples from a

Figure 22.14. OFDM without a guard interval.

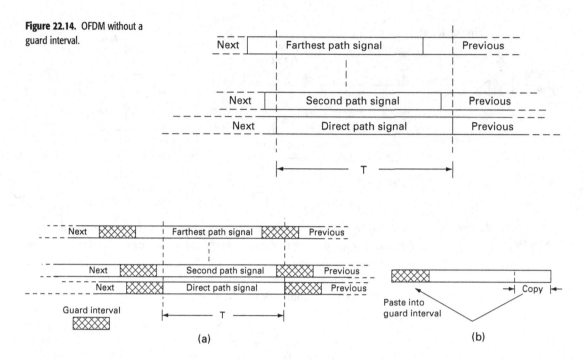

Figure 22.15. OFDM with a guard interval.

direct signal during the time interval 0 to t and then apply the Fourier transform to this set of numbers. But signals arriving on paths other than the direct path will arrive somewhat later, as shown in the figure, where the signal from the farthest path arrives late by about $0.2T$. A set of voltage samples taken from $t = 0$ to T will be the sum of the signal voltages from the various paths so the first 20% of the samples will be contaminated by data from the previous signal interval, arriving via the farthest path and the first 5% or so will be contaminated by the previous signal arriving over all the delayed paths.

Now let us add a guard interval of length αT, as shown in Figure 22.15(a). The contributions from the delayed paths during the interval $t = 0$ to T will now be from whatever has been placed in the guard intervals. You might think of zeros as a possible choice; at least they contain no data from the previous signal. However, they would compromise the orthogonality property. A much better scheme is to extend the data cyclically into the guard intervals, as shown in Figure 22.15(b). Data from the first part of the sequence is pasted into the guard interval so that when the guard interval advances into the sampler, the data sequence for the data interval begins over again. The result is that each multipath signal is properly demodulated over a complete interval and orthogonality is maintained. The overall system works as follows. Blocks of $2N$ real numbers are considered as blocks of N complex numbers. Each arriving block is subjected to an inverse DFT, whose output is also a block of N complex numbers. These output blocks are cyclically extended from length N to length $N(1+\alpha)$.

The numbers are converted to analog voltages and fed to a QAM modulator at a rate $N(1+\alpha)/T(1+\alpha) = N/T$. The receiver takes a burst of N samples every $T(1+\alpha)$ seconds, nominally aligned with the frame of data arriving over the direct path. Sampling bursts are separated by a time interval αT. The frames of N samples are Fourier-transformed to yield the complex amplitude of each carrier. This set of amplitudes is the original data. Many methods are possible to get the proper framing at the receiver and to regenerate the carrier for coherent detection. DAB audio broadcasting uses differential PSK modulation in order to tolerate carrier phase errors. DVB-T television, however, uses PSK with coherent detection.

22.6.5 Disadvantages of OFDM

One disadvantage of OFDM is that the signal has a high dynamic range, requiring that the transmitter have excellent linearity. (The same is true for any transmitter that transmits the sum of many independent signals, e.g., a cellular telephone base station transmitter.) Another potential problem is use in a mobile environment where different multipath signal components will generally have different Doppler shifts, as when the receiver is moving away from the transmitter but is moving toward a reflective object (maybe a building or tower) which provides a strong additional signal. The spacing between OFDM subcarriers would have to be considerably greater than the Doppler spread of the multipath signals. But, if there are appreciable differences in the time delays of the multipath components, the carriers must be closely spaced. This conflict is less severe at low radio frequencies, since Doppler shifts are proportional to frequency.

22.7 Spread-spectrum and CDMA

Best known now for its application in CDMA cellular telephony, spread spectrum was originally developed to provide secure communications links by using signals spread out in frequency and, therefore, reduced in spectral amplitude below the noise level of common intercept receivers. Besides being hard to detect, the signals cannot be demodulated without finding the long code word used in the spreading process. In addition, spreading makes the signals hard to jam.

22.7.1 Direct sequence spreading

Figure 22.16 shows how a rapidly changing binary pseudorandom code sequence multiplies a data stream to produce a signal whose bandwidth is determined by the "chip"[3] length of the code, rather than by the relatively slow data baud.

---

[3] Code elements, unlike signal elements, carry no information and are called "chips" rather than "bauds."

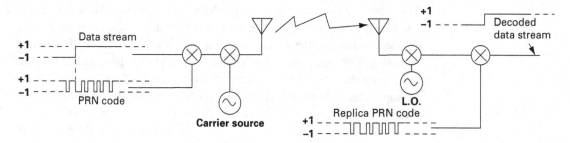

Figure 22.16. Pseudonoise spread spectrum.

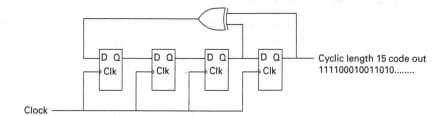

Figure 22.17. Four-stage maximum length shift register code generator.

This system is known as a *direct-sequence* spread spectrum (DSSS). We saw the same circuitry in Chapter 21, where its purpose was to produce a wide radar pulse with modest power that could be decoded (decompressed) to form a narrow high-power pulse. The purpose there was to get around the peak power limitation of the transmitter, but the system also provides the radar with a signal that is harder to intercept and jam. We will encounter the circuit again in Chapter 25, where its use in GPS satellites provides the high range resolution of pulse compression radar as well as multi-access – allowing all the satellites to transmit on the same frequency.

The pseudonoise or "direct sequence" spreading used in Figure 22.16 often uses *maximum length shift register sequences*, which are pseudorandom codes of length $2^N - 1$ generated by a circuit made of an N-stage shift register and a two (or more)-input modulus 2 adder. Figure 22.17 shows a four-stage generator and its 15-chip sequence. For this four-stage generator, only two feedback taps are needed, so the modulus-2 adder is a simple exclusive-OR gate. You can see that, for this kind of code generator, the shift register will never contain all zeros, as that state would then repeat endlessly. All other states do occur, so the length of the sequence is 2^N-1. A sequence of length $2^{32}-1$ can be generated by a 32-stage shift register with taps at stages 32, 31, 30, and 10. These taps can be added with a cascade of three ex-OR gates. With a 1-MHz clock, this generator would run for more than an hour before the sequence starts to repeat. For a given value of N, there are usually several sets of taps, each of which will yield a different maximum length sequence. The key property of the maximum length sequences is that they are noise-like. There is essentially zero correlation between different codes or between a code and a time-shifted version of itself.

The power spectrum of the signal transmitted by this system has a $[\sin(\pi fT)/(\pi fT)]^2$ shape, where ω is the offset from the carrier frequency and T is the chip length. Thus, the bandwidth is about $1/T$, independent of the code length. Longer codes, however, provide better security against interception. This system uses coherent detection; there must be provision for the receiver L.O. to acquire the correct phase (and, therefore, frequency) and for the replica code generator to acquire the phase for alignment with the incoming signal.

Wide frequency spreading requires a short chip time which, in turn, requires a high-precision time alignment of the replica code. Another spreading technique, *frequency hopping*, requires less alignment precision.

22.7.2 Frequency-hopping spread spectrum (FHSS)

In this system, the transmitter carrier frequency "hops" pseudorandomly from one frequency to another throughout the band. The timing of the hops is usually regular, though it, too, could be pseudorandom. An advantage of the hopping scheme is that the bandwidth is determined primarily by the assignment of hopping frequencies and not by the chip time – the time between hops. A convenient scheme is to hop to a new frequency for each data baud. These systems usually use incoherent detection since, until the advent of the direct digital synthesizer (DDS), it was difficult to achieve phase-coherent changes in frequency. One modulation scheme uses frequency shift keying. At the transmitter, the FSK signal is up-converted to the RF band by a frequency hopping carrier generator. At the receiver, a synchronized hopping L.O. converts the signal down to a constant IF frequency. The FSK IF signal can then be demodulated by an ordinary incoherent FM discriminator.

22.7.3 CDMA in cellular telephone systems

By itself, a direct sequence spread spectrum link is as power-efficient as a narrowband link, but deliberately has low spectral efficiency (data rate/RF bandwidth). However, the low cross-correlation property of the codes makes it possible for many users to have spread spectrum links operating in the same band. To each user, the signals from the others appear as random noise. This is known as *code-division multiple access* (CDMA). One advantage of CDMA is that no synchronization is required between users, in contrast to conventional time division and frequency multiplex systems where users must occupy scheduled frequencies and time slots. Another advantage is that the CDMA system quality degrades gracefully; as users are added, each user sees the background noise increase somewhat and experiences a lower signal-to-noise ratio. In time division and frequency division multiplexing, the maximum number of users is strictly limited by the number of frequency and time slots. For the same reason, in a CDMA cellular telephone system, adjacent cells can use the same band. Signals from adjacent cells will be, on average, weaker, and contribute only modestly to the background noise. Moreover, since a mobile user moving from one cell to another stays on the same frequency, there is a "soft handoff" from one base station to the next, rather than a momentary disconnection while new time or frequency slots are allocated.

Problems

Problem 22.1. Show that the impulse response of the filter function of Equation (22.1) is indeed the pulse shape of Equation (22.2).

Problem 22.2. Show that the filter whose frequency response is the *square* of that of the raised-cosine filter will produce intersymbol interference, i.e., show that such a filter, when excited by delta function at $t=0$, will have an impulse response that is not identically zero at $t=T$.

Problem 22.3. Consider the simple single-frequency "network" for traditional AM broadcasting in which a single "booster" transmitter site is located at the fringe of the coverage area of the main transmitter. Describe what a listener, midway between the sites, could experience if (a) the carrier frequencies differ by 400 Hz; (b) the carrier frequencies differ by 0.1 Hz; (c) the carriers are locked to the same frequency.

Problem 22.4. Write a program to simulate the pseudonoise code generator of Figure 22.15 but use 10 stages, rather than four. One input to the ex-OR gate is the output of the tenth stage. Find which stage should be connected to the other input of this gate to produce a sequence of length 1023.

Hint: Start with the shift register in the all-ones state. Count the clock pulses needed to bring it back to this state. Do this for different taps until you find one that requires 1023 pulses.

Glossary

ATSC	Advanced Television Systems Committee: developed the ATV standard for US digital television broadcasting.
BER	Bit Error Rate: probability that a received "one" is deemed a zero or vice versa
BFSK	Binary Frequency-Shift Keying.
BPSK	Binary Phase-Shift Keying.

CDMA   Code Division Multiple Access: spread spectrum modulation that allows multiple users to share the same nominal frequency.

COFDM   Coded Orthogonal Frequency Division Multiplexing: multicarrier modulation system suited to multipath propagation environments. Includes coding for forward error correction.

DAB   Digital Audio Broadcasting standard used in Europe and Canada.

DDS   Direct Digital Synthesizer: frequency synthesizer with phase-continuous frequency transitions (see Chapter 12).

DFT   Discrete Fourier Transform, equivalent to a matrix multiplication that converts a set of N complex numbers to another set of N complex numbers.

DSL   Digital Subscriber Line: Telephone company high-speed digital service

DTV   Digital Television

DSSS   Direct Sequence Spread Spectrum: frequency spreading done via multiplication with pseudorandom ± 1 sequence.

DVB-T   Digital Video Broadcasting – Terrestrial: European over-the-air digital television.

FEC   Forward Error Correction

FDM   Frequency Division Multiplexing: users are assigned different frequencies

FHSS   Frequency Hopping Spread Spectrum: transmitted signal "hops" rapidly between pseudorandomly chosen frequencies.

FIR   Finite Impulse Response: FIR filters are often based on a weighted sum of the signals from a tapped delay line.

GMSK   Gaussian Minimum Shift Keying: a particular frequency shift keying with smooth transitions between ones and zeros.

GPS   Global Positioning System: satellite navigation system.

GSM   Global System for Mobile Communications: the dominant cellular phone system.

HD radio Hybrid Digital radio: a transition U.S. radio broadcasting standard in which COFDM digital signals share frequency channel assignments with AM and FM signals.

IDFT   Inverse Discrete Fourier Transform: equivalent to a matrix multiplication that inverts (undoes) a discrete Fourier transform.

ISI   Intersymbol Interference: self-interference caused by spreading of nearby pulses.

PAM   Pulse amplitude modulation

OFDM   Orthogonal Frequency Division Multiplexing: multi-carrier modulation system suited to multipath propagation environments.

OOK   On–Off Keying

QAM   Quadrature Amplitude Modulation: system in which two independent baseband signals are transmitted at the same frequency. One is up-converted to an "in-phase" carrier $\cos(2\pi f_{RF} t)$, while the other is up-converted to a "quadrature" carrier $\sin(2\pi f_{RF} t)$.

QPSK   Quadrature Phase Shift Keying: QAM using BPSK on both I and Q carriers.

TDM   Time Division Multiplexing: users on a single frequency channel have assigned time slots.

VSB   Vestigial Sideband: The remaining vestige of the lower sideband of an AM signal after asymmetric bandpass filtering.

References

[1] Federal Communications Commission *DTV Report on COFDM and 8-VSB Performance*, Office of Engineering and Technology, September 30, 1999.
[2] IEEE Std 802.11a-1999(R2003) (Supplement to IEEE Std 802.11–1999) [Adopted by ISO/IEC and redesignated as ISO/IEC 8802-11:1999/Amd 1:2000(E)]. This standard describes the Wi-Fi COFDM format.
[3] Lahti, B. P., *Modern Digital and Analog Communication Systems*, 3rd edn, Oxford: Oxford University Press, 1998.
[4] Proakis, J. B., *Digital Communications*, 4th edn, New York: McGraw-Hill, 2000.
[5] Schulze, H. and Lüders, C., *Theory and Applications of OFDM and CDMA Wideband Wireless Communications*, New York: John Wiley, 2005.

23 Modulation, noise, and information

In this chapter we examine how noise degrades the accuracy of digital data transmission and the fidelity of analog transmission. We begin with an explanation of matched filtering, a subject which came up in Chapter 22. We show that, for binary data links using matched filtering and coherent detection, the probability of error depends only on the noise level at the input of the receiver and on the energy, but not the shape, of the transmitted pulses. We look at two example systems: BPSK (binary phase-shift keying) with coherent detection and OOK (on–off keying) with envelope detection. The error rates and channel capacities (maximum error-free data rates when using forward error correction coding) are calculated and compared with Shannon's expression for the capacity of a band-limited channel. Finally, traditional AM and FM are examined with respect to their noise characteristics.

23.1 Matched filtering

We stated in Chapter 22 that, in the presence of noise, the post-detection signal-to-noise ratio is maximized when the predetection bandpass shape of the receiver is that of a *matched filter*.

A matched filter is one whose impulse response is proportional to the time-reversed waveform of the incoming signal pulse, as will be shown below. For example, if the input signal is a pulse whose shape is the same as the symmetric impulse response of a root raised-cosine filter, then the receiver should use a root raised-cosine filter or an equivalent cascade of filters. Sometimes we deal with complicated pulses, such as the biphase coded pulses used in pulse compression radar. In these cases, it is common to use a front-end bandpass filter which is a matched filter for the individual subpulses, followed by coherent detection and then a decoder which undoes the plus/minus phase coding. Here the cascade of the front-end filter and the decoder is equivalent to a matched filter for the coded pulses. Note that coherent detection is down-conversion to baseband, a linear operation. We can think of the entire matched filter as the actual detector, whose output samples are the received symbols.

Let D denote the output of the receiver's filter (which we have just defined as the "detector" output) and let $h(t)$ be the impulse response of this filter, a real function. For a signal pulse unaccompanied by noise, the output from the filter is given by

$$D_S = \int\limits_{-\infty}^{\infty} h(t) V_S(-t) dt, \tag{23.1}$$

where we have assumed the pulse position to be such that the filter output should be sampled at $t = 0$. The output signal power, in units of volts^2, is given by $P_S = D_S{}^2$. The noise output of the filter can be written in terms of N_0, the standard one-sided noise density, in units of volts^2/Hz, as

$$P_N = \frac{N_0}{2} \int\limits_{-\infty}^{\infty} |H(\omega)|^2 d\omega, \tag{23.2}$$

where $H(\omega)$ is the filter transfer function, i.e., the Fourier transform of $h(t)$. Note that we have assumed that N_0 is a constant (white noise). Using Parseval's theorem, we can write P_N in terms of $h(t)$:

$$P_N = \frac{N_0}{2} \int\limits_{-\infty}^{\infty} h(t)^2 dt. \tag{23.3}$$

With these expressions for P_S and P_N, we can write the output signal-to-noise ratio as

$$\text{SNR} = \frac{P_S}{P_N} = \frac{\left(\int\limits_{-\infty}^{\infty} h(t) V_S(-t) dt \right)^2}{\frac{N_0}{2} \int\limits_{-\infty}^{\infty} h(t)^2 dt}. \tag{23.4}$$

At this point, we invoke Schwarz's inequality[1] which gives us

$$\text{SNR} \le \frac{\int\limits_{-\infty}^{\infty} h(t)^2 dt \int\limits_{-\infty}^{\infty} V_S(-t)^2 dt}{\frac{N_0}{2} \int\limits_{-\infty}^{\infty} h(t)^2 dt} = \frac{1}{N_0} \int\limits_{-\infty}^{\infty} V_S(-t)^2 dt = \frac{1}{\frac{N_0}{2}} \int\limits_{-\infty}^{\infty} V_S(t)^2 dt. \tag{23.5}$$

---

[1] Schwarz's inequality is written as $|\int f(x)g(x)dx|^2 \le \int |f(x)|^2 dx \int |g(x)|^2 dx$. This can be seen by considering the integral of fg to be the dot product of a multidimensional vector, while the integral $\int |f(x)|^2 dx$ is the length of the vector f and $\int |g(x)|^2 dx$ is the length of the vector g. For two vectors A and B, we know that $|A \cdot B|^2 \le |A \cdot A||B \cdot B|$.

Looking at Equation (23.4), we see that if $h(t) = \beta V_S(-t)$, where β is any constant, then the SNR will be equal to its maximum possible value, the right-hand expression in Equation (23.5). This is the matched filter. Note that, with a matched filter, the signal-to-noise ratio for a pulse is independent of the pulse shape and is simply the time integral of $V_S(t)^2$, i.e., the energy of the pulse, $\mathscr{E}_S$, divided by the noise spectral density N_0. (N_0 has units of volts^2/Hz, which is the same as volts^2 sec.) Thus, when a matched filter is used,

$$\text{SNR} = \frac{1}{N_0/2} \int_{-\infty}^{\infty} V_S(t)^2 dt = \frac{2\mathscr{E}_s}{N_0}. \qquad (23.6)$$

23.2 Analysis of a BPSK link

Coherent detection makes the BPSK link especially easy to analyze. The signal at the output of the matched filter is just the sum of the signal voltage and the noise voltage, which we will assume to have a Gaussian distribution, characteristic of thermal noise. The detected signal voltage will be either A or $-A$, so the Gaussian noise distribution will be centered at either A or $-A$, as shown in Figure 23.1. The one/zero decision threshold will, of course, be set at $V = 0$ for this symmetric situation.

The probability distribution for a zero (the solid curve) is the normal Gaussian distribution function, $p_0 = (2\pi\sigma^2)^{-1/2} \exp[-(V+A)^2/(2\sigma^2)]$. By inspection of the figure we can write

$$p_e = \int_{0}^{\infty} \frac{e^{-(V+A)^2/2\sigma^2}}{\sqrt{2\pi\sigma^2}} dU = \int_{A}^{\infty} \frac{e^{-V^2/2\sigma^2}}{\sqrt{2\pi\sigma^2}} dU = \frac{1}{\sqrt{2\pi}} \int_{\frac{A}{\sigma}}^{\infty} e^{-u^2/2} dU.$$

$$(23.7)$$

Now let us relate this probability of error to the signal-to-noise ratio. We saw in the previous section that, at the output of a matched, the ratio of the square of the signal portion of the sampled output to the average square of the noise portion is $2\mathscr{E}_S/N_0$, where $\mathscr{E}_S$ is the energy of the pulse, i.e., the time integral of the V_S^2, and

Figure 23.1. Gaussian voltage probability distributions for received BPSK ones and zeros for $A = 1.5\sigma$ (SNR $= 1.5^2$). The shaded area in the figure is the probability p_e of a transmission error, i.e., the probability that a received zero will be interpreted as a one or that a received one will be interpreted as a zero.

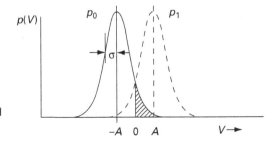

N_0 is the noise power density. Therefore $A^2/\sigma^2 = 2\mathscr{E}_S/N_0$, and we can rewrite Equation (23.7) as

$$p_e = \frac{1}{\sqrt{2\pi}} \int_{\sqrt{2\ \mathscr{E}_S/N_0}}^{\infty} e^{\frac{-u^2}{2}} du. \tag{23.8}$$

We have been implicitly working at baseband, i.e., assuming that the signal (and the noise) have been converted down from an RF carrier frequency, ω_0, through multiplying the modulated RF signal by a sine wave in phase with the (normally suppressed) carrier. This, as we have seen, constitutes synchronous detection. In this common situation, the matched filter is a baseband filter.

It is interesting to note that we could alternately have used a matched filter at the RF (or some IF) frequency. In this case, the output of the filter is an RF sine wave, multiplied by the baseband pulse. We can sample this directly, but the sampling must be very precise, so that the samples are taken at points that, with a dc pulse, would be at the peaks of the RF sine wave. The SNR will be the same as for the synchronous conversion to baseband because even though the RF bandwidth is twice the baseband bandwidth and contains twice as much total noise power, the synchronous sampling will respond to only half this noise, e.g., the "cosine" component.

If symbols are arriving at the rate $1/T$ in the minimum baseband bandwidth of $1/(2T)$ that eliminates intersymbol interference (see Chapter 22), then $2\mathscr{E}_S/N_0 = (\mathscr{E}_S/T)/(N_0/[2T]) = P_S/P_N = SNR$ the signal-to-noise ratio. This probability of error from Equation 23.8 is plotted vs. SNR in Figure 23.2.

This figure displays the well-known "dropout" effect in digital communications systems. If, for example, the input SNR drops from ten to five, a factor of only 2, the error rate increases by a factor of 100. The probability of error for the BPSK distributions shown in Figure 23.1, where $A/\sigma = 1.5$, is 0.067.

Figure 23.2 Envelope probability distributions $P(E, A, \sigma)$, for $\sigma = \sqrt{2}$ and $A = 0$ or $1.5\sqrt{2}$.

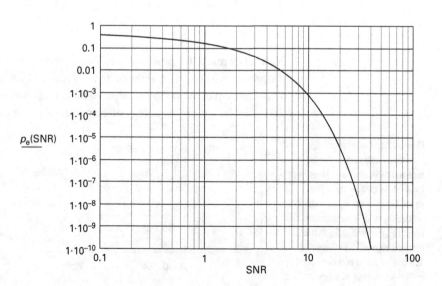

23.3 On–off keying with envelope detection

The BPSK system discussed above, using coherent detection, is suited for data transmission under low signal conditions. (When there is ample signal power, multilevel "M-ary" modulation can obviously transmit data at a faster bit rate.) But let us now consider simple on–off keying with envelope, i.e., noncoherent detection. Before we can calculate the probability of error, we must find the distribution function of the output voltage from the envelope detector.

23.3.1 Envelope probability distributions

As we have seen, the IF signal, containing random noise, is a random variable with Gaussian probability distribution. The presence of a signal offsets the Gaussian curve. The output of the envelope detector is also a random variable. Let us find the probability distribution of the envelope, first without the presence of a signal. Figure 23.3(a) shows the in-phase and quadrature components, N_I and N_Q, of the IF noise voltage. The length of their vector sum is E, the envelope voltage.

Since the I and Q noise components are uncorrelated, their joint probability distribution is the product of their individual Gaussian distributions.

$$p(V_I, V_Q) = \frac{e^{-V_I^2/2\sigma^2}}{\sqrt{2\pi\sigma^2}} \times \frac{e^{-V_Q^2/2\sigma^2}}{\sqrt{2\pi\sigma^2}} = \frac{e^{-(V_I^2 + V_Q^2)/2\sigma^2}}{2\pi\sigma^2} = \frac{e^{-E^2/2\sigma^2}}{2\pi\sigma^2}.$$

(23.9)

To obtain the distribution in terms of E and θ, we note that

$$p(E, \theta)E\mathrm{d}\theta\mathrm{d}E = \frac{e^{-E^2/2\sigma^2}}{2\pi\sigma^2} E\mathrm{d}\theta\mathrm{d}E,$$

(23.10)

from which we identify

Figure 23.3 (a) I and Q noise phasors alone; (b) noise phasors together with a sine wave of amplitude A.

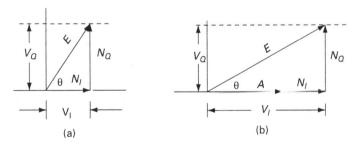

(a)   (b)

$$p(E, \theta) = \frac{E e^{-E^2/2\sigma^2}}{2\pi\sigma^2}. \qquad (23.11)$$

Integrating over θ, from 0 to 2π, we find $p(E)$, the envelope probability distribution:

$$p(E) = \int_0^{2\pi} P(E, \theta)\mathrm{d}\theta = \frac{E e^{-E^2/2\sigma^2}}{\sigma^2}. \qquad (23.12)$$

This function is known as the Rayleigh probability distribution. Note: for both the I and the Q noise components, the variance is σ^2. Therefore, the total noise power, which is the expectation of E^2, is $2\sigma^2$.

Now we must find the envelope distribution function when the IF signal is the superposition of noise plus a sinusoidal carrier. This is shown in Figure 23.3(b). Again, both noise voltages, N_I and N_Q, have Gaussian distributions, but now the distribution for N_I is centered at A, the amplitude of the carrier, so we can write

$$\begin{aligned} p(V_I, V_Q) &= \left[\sqrt{2\pi\sigma^2}\right]^{-2} e^{[-(V_I - A)^2 + V_Q^2]/2\sigma^2} \\ &= (2\pi\sigma^2)^{-1} e^{[-(E\cos(\theta) - A)^2 + E\sin(\theta)^2]/2\sigma^2}. \end{aligned} \qquad (23.13)$$

Expanding the argument of the exponential on the right-hand side we have

$$p(V_I, V_Q) = (2\pi\sigma^2)^{-1} e^{-(E^2 + A^2)/2\sigma^2} e^{-2AE\cos(\theta)/2\sigma^2}. \qquad (23.14)$$

Following the steps used in the carrier-free case, and integrating over θ, we find

$$\begin{aligned} p(E, A, \sigma) &= \frac{E e^{-(E^2 + A^2)/2\sigma^2}}{\sigma^2} \frac{1}{2\pi} \int_0^{2\pi} e^{-AE\cos(\theta)/\sigma^2}\mathrm{d}\theta \\ &= \frac{E e^{-(E^2 + A^2)/2\sigma^2}}{\sigma^2} I_0(AE/\sigma^2), \end{aligned} \qquad (23.15)$$

where I_0 is the modified Bessel function of order zero. This function, $p(E, A, \sigma)$, known as the *Rician* distribution, is plotted in Figure 23.4 for $A = 0$ (the envelope distribution of noise alone) and for $A = 1.5\sqrt{2}$, for the same noise power and average transmitted power (assuming equal probabilities for transmitted ones and zeros) used for the BPSK example of Figure 23.1.

Let us calculate the probability p_e that a bit is received incorrectly when we have (arbitrarily) set the decision threshold at the intersection of the ON and OFF envelope distribution functions, which we will denote as E_t. We will also assume that we transmit, on average, equal numbers of ones and zeros. The probability of error is, therefore, $\frac{1}{2}p_{e0} + \frac{1}{2}p_{e1}$, where p_{e0} is the probability that a transmitted zero is received incorrectly and p_{e1} is the probability that a transmitted one is received incorrectly or

Figure 23.4 Probability distribution $p(E,A,\sigma)$, for the envelope of a signal that is the sum of a sine wave of amplitude $A = 1.5\sqrt{2}$ plus noise with $\sigma = 2$.

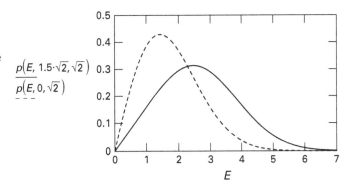

$$p_{\text{eOOK}}(A,\sigma) = 1/2 \int\limits_{E_t}^{\infty} p_0(E,A,\sigma)\mathrm{d}E + 1/2 \int\limits_{0}^{E_t} p_1(E,A,\sigma)\mathrm{d}E. \quad (23.16)$$

Evaluating P_{eOOK} for $A = 1.5\sqrt{2}$ and $\sigma = \sqrt{2}$ for comparison with the BPSK example of Figure 23.1 (same average transmitted power, but twice the noise power (quadrature as well as in-phase components), yields $P_{\text{eOOK}} = 0.34$ for a threshold $E_t = 2.3$, compared to only .067 for P_{eBPSK}.

Channel capacity

We have seen how to calculate the expected bit error rate as a function of the signal and noise powers for two example situations, coherently-detected PSK and envelope-detected on–off keying. We also know that data error rates can be reduced by expanding (encoding) the data so that it contains redundancies.[2] Of course, the net data rate slows when redundant bits are transmitted. These redundancies may be crude, such as transmitting a packet of data several times, or elegant, such as the nested combination of block codes and convolutional codes used in data links and digital broadcasting systems. Given a particular communications link with an optimal encoding scheme, it is obvious to ask how many bits must be transmitted, on average, for each data bit, if we are willing to tolerate a certain average data error rate. One might expect that, as the error tolerance is reduced to zero, the optimal transmission efficiency would also go to zero, i.e., that an infinite number of bits would have to be transmitted for each recovered data bit, as indicated by the crossed out curve in Figure 23.5. Shannon's remarkable "noisy channel theorem" showed that this is not so. For any given communications link, there must exist coding schemes that will permit data transmission with an arbitrarily low error tolerance while still

---

[2] Making the data redundant enough so that errors can be detected and corrected at the receiver is known as forward error correction (FEC). In reverse error correction, the receiver can only detect errors. To make corrections, it must request repeat transmissions.

Figure 23.5 Transmission efficiency vs. tolerated output error rate when using optimum channel encoding.

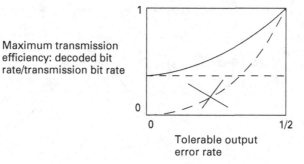

achieving a non-zero efficiency. This situation is shown by the solid curve Figure 23.5, which shows the efficiency for a link using an optimum error coding scheme.

The maximum efficiency is unity if we are willing to tolerate an output error rate of 1/2, but this is the point at which no net information is transmitted since a one is randomly declared a one or a zero and vice versa. But as the error rate tolerance is lowered to zero, the maximum efficiency approaches an asymptotic *non*zero value. This asymptotic maximum efficiency value is called the *channel capacity* (bits/bit) and, when multiplied by the bit rate, in bits/sec, the result, in units of bits/sec, is a rate and is also referred to as the channel capacity. (Context usually resolves any confusion between the two.) The ususal statement of Shannon's theorem is that it is possible to transmit data at a rate equal to or less than the channel capacity, with an arbitrarily low data error rate. While Shannon's work shows that optimum codes must exist, it does not show how to construct then. However, it does show how to calculate the channel capacity for a given link, the highest standard against which we can judge the efficiency of any coding scheme proposed for the link. If the arbitrarily low data error rate is set too low, the encoded data must be in the form of extremely long blocks. Fortunately, acceptably low error rates can be achieved with acceptably short data blocks.

Binary symmetric channel

The coherent BPSK link discussed above is an example of a *binary symmetric channel*. The transmitted symbols are either one or zero, and every received signal is deemed either a one or a zero. From the symmetry of Figure 23.1, it is obvious that the probability p_e that a transmitted one will be received in error as a zero is equal to the probability that a transmitted zero will be received in error as a one, as indicated in Figure 23.6. This figure, a sort of probability flow graph for a binary channel, is equivalent to a 2×2 matrix whose coefficients are probabilities. The binary symmetric channel corresponds to a symmetric channel matrix.

Figure 23.6. Binary symmetric channel: p_e is the probability that a symbol x_i is wrongly received as a symbol y_j.

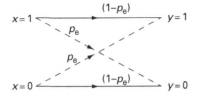

The channel capacity for this link, which depends only on p_e and on p_i, the probability that the i-th transmitted bit is a one, is given by

$$C(p_e, p_i) = p_e \log_2(p_e) + (1 - p_e) \log_2(1 - p_e)$$
$$-p_i \log_2(p_i) - (1 - p_i) \log_2(1 - p_i). \qquad (23.17)$$

This expression is always greatest for $p_i = 1/2$ (equal probability of transmitted ones and zeros). Figure 23.7 shows C(p_e, 1/2) plotted versus p_e.

Figure 23.7 Channel capacity of a binary symmetric channel vs. transmission error rate, assuming ones and zeros are transmitted with equal probability.

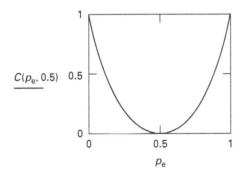

Note that the channel capacity falls to zero at $p_e = 1/2$, where the receiver is equally likely to interpret a transmitted one as a zero and vice versa. Then C rises again to unity for $p_e = 1$, where the receiver always mistakes a one for a zero and zero for a one. In this case, the message is transmitted faithfully, assuming the receiver realizes that it should complement the bits. (It takes only a single transmitted bit to resolve this ambiguity.)

The reader will find excellent treatments of information theory, channel capacity, and coding in the texts listed at the end of this chapter but the origin of Equation (23.17) warrants at least a brief discussion, in the limited context of a binary channel. The discussion should at least serve to show that, while the concepts are subtle, the mathematics is simple. In the binary channel, the transmitter sends a symbol which is denoted either x_1 (a one) or x_2 (a zero). The *information* in a transmitted symbol is defined as $I(x_i) = \log_2 (p(x_i)^{-1})$, where $p(x_i)$ is the probability of x_i. Note that if $p(x_i)$ is nearly one, the information contained in x_i is nearly zero. If the transmitted symbols are almost always ones (or zeros), very little information is transmitted by sending a one (or zero), as we would have expected the one (or the zero). But if $p(x_i)$ is a very small number, the occurrence of x_i is highly informative, as we would not have expected it. A second definition is that of *mutual information*,

$I(x_i; y_j) = \log_2[p(x_i|y_j)|p(x_i)]$, where $p(x_i|y_j)$ is the conditional probability that x_i was transmitted, given that y_j was received. Note that while the distribution of transmitted symbols is generally flat (uniform), the distribution of $I(x_i; y_j)$ for a given j is not uniform, but peaked around the value of i most likely to have resulted in y_j.

The expected value of the mutual information is the sum of the $I(x_i; y_j)$'s, weighted by their probabilities, i.e.,

$$\langle I(x_i; y_j)\rangle = \sum_i \sum_j p(x_i, y_j) I(x_i; y_j), \qquad (23.18)$$

where $p(x_i, y_j)$ is the joint probability of x_i and y_j, i.e., $p(x_i\, y_j) = p(x_i)\, p(y_j|x_i)$. For the binary symmetric channel discussed above, $p(y_j, x_i) = p_e$ for $i \neq j$ and $(1-p_e)$ for $i = j$. The channel capacity is given by the maximum value of $\langle I(x_i; y_j)\rangle$ with respect to the set of values $p(x_i)$, by

$$C = \max\langle I(x_i; y_j)\rangle \text{ w.r.t } \{p(x_i)\}. \qquad (23.19)$$

For the binary symmetric channel, the maximum occurs when equal numbers of ones and zeros are transmitted, i.e., when $p(x_1) = p(x_2) = 1/2$. Evaluation of Equation (23.19) yields Equation (23.17), the expression for the channel capacity. Note that the channel capacity is based on the relative frequency of transmitted ones and zeros and on the values of $p(y_j\,|\,x_i)$, which is the conditional probability that the receiver produces y_j when the transmitter sent x_i. However, to calculate the values of $I(x_i; y_j)$, we need to know $p(x_i\,|\,y_j)$, the conditional probability that when the receiver produces y_j, the transmitter had sent x_i. The two conditional probabilities are related through Bayes' theorem: $p(x_i, y_j) = p(x_i)\, p(y_j\,|\,x_i) = p(y_j)\, p(x_i\,|\,y_j)$.

Channel capacity of the BPSK and OOK example channels

For the BPSK example shown in Figure 23.1, with $A/\sigma = 1.5$, we found the probability of error to be $p_e = .067$. Using this value in Equation 23.17, the channel capacity is 0.645 data bits/transmitted bit. For the comparable asymmetric OOK channel of Figure 23.4 the two integrals in Equation 23.16 are, respectively, the probability that a transmitted zero is received as a one and vice versa. If we calculate the channel capacity using Equation 23.19, we find that it maximizes at about $C = 0.08$, if the threshold is set at $E_t = 2.5$ and the proportion of transmitted zeros is 52%. In addition to low channel capacity, another disadvantage of the OOK system is that the threshold must be changed when either the signal or noise level changes.

Channel capacity of a bandpass channel

Shannon also presented a formula for the *maximum* channel capacity for a band-limited channel with added Gaussian white noise – the channel most amenable to analysis, and a channel often encountered in practice, for example, in spacecraft telemetry links. This formula states that the maximum channel capacity is given by

$$C = B \log_2[1 + S/N] = B \log_2[1 + S/(BN_0)] \qquad (23.20)$$

where S is the received signal power, N is the noise power which is equal (for white noise) to BN_0, where B is the channel bandwidth, and N_0 is the noise spectral density at the receiver in Watts/Hz. The figure of merit for any particular modulation scheme is how close its channel capacity approaches this ideal maximum channel capacity. To see that Equation 23.20 is reasonable note that, when S/N is high, it is essentially the practical number of quantization levels, n, and $\log_2(n)$ converts this into bits. The bandwidth factor, B, will be the symbol rate or close to it.

For illustration, suppose the signal power available from a spacecraft is 10^{-16} Watts, contained within a bandwidth of 100Hz, and the noise density at the receiver input is 4×10^{-19} Watts/Hz. With these numbers, Equation 23.20 produces $C = 181$, which is the maximum rate (bits/sec) at which information could be transmitted over the channel at an arbitrarily small error rate, if the modulation scheme and signal coding are optimum.

Let us look at our example BPSK channel in the light of Equation 23.20. In that example, the SNR at the output of the detector was $(A/\sigma)^2 = 1.5^2$. If we are using a matched filter, we saw that $(A/\sigma)^2$ be equal to $2\mathcal{E}/N_0$. Let us assume we transmit bits at a rate of $1/T$ and use receiver bandwidth of $1/(2T)$. The received power is $S = \mathcal{E}/T$ and the noise at the receiver is $N_0/(2T)$, from which we have $S/N = 2\mathcal{E}/N_0$. Using Equation 23.20, we see that the maximum channel capacity will be $C = 1/(2T)^{-1} \log_2(1 + 2\mathcal{E}/N_0) = .850/T$ bits/sec. But, from the bit error probability, a function of A/σ, we had calculated the actual channel capacity of this BPSK link to be .645 bits/bit. If we multiply this by the bit rate, $1/T$, the channel capacity becomes $.645/T$ bits/second. This BPSK link, therefore, has $.645/.850 = 76\%$ of the maximum possible channel capacity.

For much larger values of S/N, we would find that, while the ideal channel capacity increases, our BPSK channel capacity saturates at the value $1/T$. This is simply due to staying with two-level binary modulation, as opposed to N-level PAM modulation.

Noise in analog FM and AM systems

Let us first look at the noise produced at the output of an FM receiver. The instantaneous received signal voltage, V_{SIG}, is accompanied by noise, V_N, as shown in Figure 23.8.

This is the same as the diagram in Figure 23.3(b), except that now we are interested in the phase angle rather than the magnitude of the vector sum of the signal plus noise. This "phase noise," ϕ_N, will cause noise in the detected output of an FM or PM detector. Clearly the angle ϕ_N becomes smaller if the signal strength is increased. But there is another way to defeat the noise. The signal-to-noise ratio at the detector output depends on the ratio of the signal's modulation phase excursions to the phase noise. If the modulation level is increased, even without increasing the signal strength, the output signal-to-noise ratio will be improved.

Figure 23.8. Signal and noise voltages.

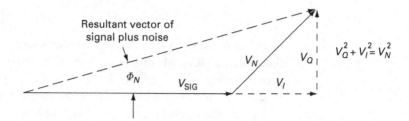

If, for example, the rms phase noise is 1/10 radian and the modulation index is 1 radian, the phase SNR is 100. If the deviation is increased to produce a modulation index of 5 radians, the phase SNR increases to 2500. The improvement has been obtained not by increasing the signal power but by increasing the signal bandwidth. In the case of amplitude modulation, the SNR depends on the noise modulating the length of the vector $V_{\text{SIG}} + V_N$. Since V_N is fixed in any given situation, the only way to improve the SNR in AM is to increase V_{SIG}, i.e., increase the transmitted power.

Analysis of the SNR improvement in FM

A quantitative analysis of the SNR improvement in FM is simpler if we take the noise to be the background noise produced by the detector when the signal is unmodulated, i.e., the total power in the hiss coming from the loudspeaker or other output device. For the signal we will take an audio sine wave with 100% modulation (maximum deviation). We can represent the noise, V_N, by in-phase (I) and quadrature (Q) noise components, V_I and V_Q where $V_I^2 + V_Q^2 = V_N^2$ as shown in Figure 23.8. Both V_Q, and V_I are phasors rotating at ω_0, the frequency of the unmodulated carrier. Their amplitudes are random and independent. The I component of the noise is the most effective in causing amplitude fluctuations and therefore contributes noise in AM demodulation. But it is mostly the Q component, since it is perpendicular to V_{SIG}, that causes angle fluctuations and therefore contributes noise in the FM demodulation. For $V_N \ll V_{\text{SIG}}$, the instantaneous angle noise, $\phi_N(t)$, is just $V_Q(t)/V_{\text{SIG}}$ radians. Since V_{SIG}, the carrier amplitude, is constant, the power spectrum of ϕ_n (call it S_ϕ) is proportional to the power spectrum of V_Q. The spectral distribution of V_Q can be assumed uniform (white) so S_ϕ is also uniform. The integral of S_ϕ over the IF band gives the mean square phase fluctuation, $\langle(\phi_n(t))^2\rangle$, so we can write $S_\phi = \langle(\phi_n(t))^2\rangle/B_{\text{IF}} = \langle V^2{}_Q(t)\rangle/(V^2{}_{\text{SIG}}B_{\text{IF}}) = \langle V^2{}_N(t)/2\rangle/(V^2{}_{\text{SIG}}\, B_{\text{IF}})$ where B_{IF}, the IF bandwidth, is twice the maximum deviation. An FM demodulator produces an output spectral density proportional to the time derivative of the phase, $\langle(d\phi_n(t)/dt)^2\rangle$. The spectral density of the noise in the (one-sided) audio band at the detector output is therefore given by $2\omega_a{}^2 S_\phi(t)$, and the total noise power is the integral of this spectral density over the output bandwidth of the detector, i.e., the audio band (0 to B_a radians):

$$\text{Output noise power} = \int_0^{B_a} 2\omega^2 S_\phi d\omega = S_\phi \int_0^{B_a} 2\omega^2 d\omega = \frac{2}{3} S_\phi' B_a^3 \tag{23.21}$$

$$= \frac{B_a^3}{6 k_{osc} A_{MAX}} \frac{\langle V_N^2 \rangle}{V_{SIG}^2}.$$

The maximum amplitude of the sine-wave signal at the output of the detector is $k_{osc} A_{MAX}$, the maximum deviation, so the maximum signal power is just $P_{SIG} = k_{osc}^2 A_{MAX}^2 / 2$. Taking the noise power from Equation (23.21), the signal-to-noise ratio at the detector output is:

$$\text{Maximum output SNR} = 3 \left(\frac{k_{osc} A_{MAX}}{B_a} \right)^3 \frac{V_{SIG}^2}{\langle V_N^2 \rangle}. \tag{23.22}$$

Note that the output SNR improves as the cube of the ratio of the maximum deviation to the full audio bandwidth.

Output signal-to-noise ratio for an AM signal with the same carrier power

Let us consider an AM system with the same carrier power and the same audio bandwidth, in order to compare its output SNR to that of the FM system. Again the modulation will be a single sine-wave tone and the amplitude of the carrier will be V_{SIG}. At 100% modulation, the amplitude of the sine wave modulation envelope will also be V_{SIG} so the audio signal power at the AM detector output will be $V_{SIG}^2 / 2$. For the AM system, the IF bandwidth needs to be only $2B$, twice the audio bandwidth (wide enough to accommodate the highest audio frequency but, to minimize noise, no wider). The noise voltage at the detector output will be V_I', the in-phase component of V_N', where the primes distinguish the noise voltages in the AM IF band from the noise voltages in the wider FM IF band. The noise power will be given by $\langle V_I'^2 \rangle = \langle V_N'^2 \rangle / 2$. The SNR at the AM detector output is therefore

$$\text{AM output SNR} = \frac{V_{SIG}^2}{\langle V_N'^2 \rangle}. \tag{23.23}$$

Comparison of noise, FM vs. AM under strong signal conditions

All that remains in order to compare the FM and AM systems is to note that the ratio of the IF noise powers is just the ratio of the IF bandwidths, i.e.,

$$\frac{\langle V_N^2 \rangle}{\langle V_N'^2 \rangle} = \frac{2 k A_{max}}{2 B_a} = \frac{k A_{max}}{B_a}. \tag{23.24}$$

Using Equations (23.22)–(23.24), we can write

$$\frac{\text{FM SNR}}{\text{AM SNR}} = \frac{3 k_{osc}^2 A_{MAX}^2}{B_a^2} = 3 (\text{Deviation ratio})^2. \tag{23.25}$$

In FM broadcasting the deviation ratio is about five so the output SNR for a maximum-amplitude audio tone will be higher with FM than with AM by a factor of $3 \cdot 5^2 = 75$ or about 19 dB. The FM sound in television uses a deviation ratio only 1/3 as large as that of FM broadcasting so the signal-to-noise improvement is correspondingly less. Remember that the above analysis is only for the situations of high signal-to-noise. When the signal is comparable to or lower than the noise, the phase is almost totally determined by the noise and the FM system is useless.

FM, AM and channel capacity

The improvement in a signal-to-noise ratio that is possible with wideband FM is an example of increasing the channel capacity of a communications channel by increasing the bandwidth.

Note that S/N_O has units of bandwidth and can be considered a "natural" bandwidth for a given N_0 and S. N_0 is determined by the noise added along the channel, such as atmospheric noise and noise added by the receiver. The signal power, S, is determined by the transmitter power, transmitter and receiver antenna gains, and propagation loss. Channel capacity, from Equation (23.20), vs. bandwidth is plotted in Figure 28.9. Both are normalized to S/N_0.

Note that, for $b > 1$, the channel capacity has essentially reached an asymptotic value of 1.44 S/N_0. If we are below the knee of the curve, we can increase the channel capacity significantly at no cost in transmitter power by (somehow) using more bandwidth. We have seen that FM broadcasting does just this. On the other hand, Equation (23.20) shows that it is expensive (and ultimately impractical) to increase channel capacity by increasing power since the log term increases slowly.

Figure 23.9. Channel capacity vs. bandwidth (both are normalized to S/N_0, the "natural bandwidth").

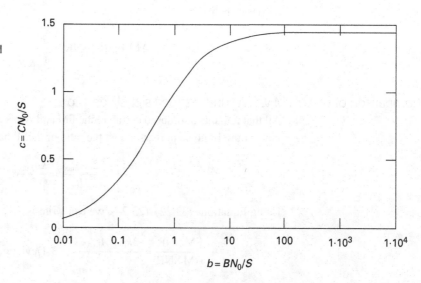

In AM, the bandwidth is fixed by the highest modulation frequency; the total bandwidth in standard full-carrier DSB AM is twice the highest modulation frequency. When a weak AM station is received, the signal-to-noise ratio is low enough to put us beyond the knee of the channel capacity vs. bandwidth curve and there would be no gain in going to a modulation scheme that increases the bandwidth. But if the station is strong, we are probably far below the knee of the curve. In this case, changing to FM modulation (without changing transmitter power) can bring us up the curve where the higher channel capacity allows a higher signal-to-noise ratio.

Problems

Problem 23.1. Show that the differential equation $x^2y'' + xy' - x^2y = 0$ (modified Bessel's equation of order zero) is satisfied by the function

$$I_0(x) = \frac{1}{2\pi} \int\limits_0^{2\pi} e^{x\cos(\theta)} d\theta.$$

Problem 23.2. Consider an asymmetric binary channel in which the probability that a one is correctly received is 0.9 but the probability that a zero is correctly received is 1/2. (A transmitted zero is equally likely to be declared a zero or a one.) Suggest a simple coding method to transmit data reliably through this channel.

Problem 23.3. Consider the situation where a rectangular pulse of length T is transmitted, but the receiver instead of using a $\sin(1/2\ \omega T)/(1/2\ \omega T)$ matched filter, uses a rectangular filter of bandwidth β/T. The noise at the input is Gaussian. Find the value of β that maximizes the SNR at the output of the filter. Compare this to the value that would have been obtained if the matched filter had been used.

References

[1] Lahti, B. P., *Modern Digital and Analog Communication Systems*, 3rd edn, Oxford: Oxford University Press, 1998.
[2] Proakis, J. B., *Digital Communications* – 4th edn, New York: McGraw-Hill, 2000.

24 Amplifier and oscillator noise analysis

A fundamental limitation in amplifiers and oscillators is the internally generated random noise whose physical causes are shot noise in semiconductors and thermal noise (Johnson noise) in resistors. Amplifier noise is a practical concern whenever this noise level is comparable to the signal being amplified. Oscillator noise, characterized as phase or frequency fluctuations, adds noise to FM systems, limits how closely carriers may be spaced in frequency division multiplexing (FDM), and limits the length of signal averaging times in coherent communication systems. In this chapter we discuss the way transistor noise is characterized, treating the transistor itself as a two-port network, and how these characteristics are taken into account to design low-noise circuits.

24.1 Amplifier noise analysis

In Chapter 17 we quoted a formula showing how the noise figure of any active two-port device depends on the impedance of the signal source:

$$F = F_{\min} + (R_n/G_s)|Y_s - Y_{opt}|^2. \tag{24.1}$$

Here $F_{\min}$ is the minimum noise figure; $Y_s = G_s + jB_s$ is the source admittance, $Y_{opt} = G_{opt} + jB_{opt}$ is the source admittance that produces $F_{\min}$, and R_n is a real parameter known as the *noise resistance*. Equation (24.1) applies as well to audio and instrumentation amplifiers as to radio and microwave amplifiers, as long as the four noise parameters of the device, $F_{\min}$, R_n, G_{opt}, and B_{opt}, are known for the frequency in question. While Equation (24.1) requires some effort to derive, its use is simple, at least with RF (narrowband) amplifiers.

24.1.1 Noise matching

To design a low-noise RF amplifier, we get out the data sheet for the transistor, find the value of Y_{opt} for the desired frequency, and design an input network to convert the impedance of the intended source (probably 50 ohms) into $1/Y_{opt}$.

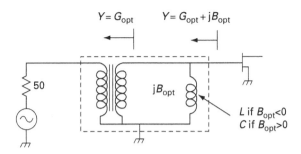

Figure 24.1 shows how a transformer and a reactor can form the matching network. (Of course there are many equivalent networks.)

With this input network, the transistor is said to be "noise matched," because the noise figure of the overall amplifier will be $F_{\min}$. The input network that provides noise matching generally does not provide impedance matching, i.e., it does not convert the source impedance into the complex conjugate of the amplifier's input impedance. The resulting power mismatch means we will have less signal at the amplifier output than with a conjugate impedance match. But converting the source impedance into $1/Y_{opt}$, rather than into $1/Y^{*}_{in}$, will reduce the transistor noise more than the mismatch reduces the signal power. The net effect is to maximize the signal-to-noise ratio, i.e., provide the minimum possible noise figure.

(Note: In some cases simultaneous noise matching and impedance matching can be achieved or at least approached. If the device has a nonzero reverse transfer coefficient, the amplifier's input impedance is not fully determined by the input network. Depending on the transistor characteristics, a designer might find an output network to make the input impedance of the amplifier approach Z_0 while using an input network that provides noise matching. The use of feedback can provide additional degrees of freedom with which to pursue simultaneous matching.)

24.1.2 Equivalent circuits for noisy two-port networks

Equation 24.1 was derived by Rothe and Dahlke [4] who pointed out that a noisy two-port network (transistor, amplifier, etc.) can be modeled, for small signals, as an ordinary passive linear two-port with two external frequency-dependent noise sources. These noise sources can be series voltage generators at the input and output, shunt current generators at the input and output, a current source at one end and a voltage source at the other end, or, finally, both a voltage source and a current source at the same end. Figure 24.2 shows these equivalent circuits.

Note that when a current generator and a voltage generator appear on the same side they can be interchanged without changing their parameters, so Figures 24.2(e) and (f) (or g and h) are actually the same circuit.

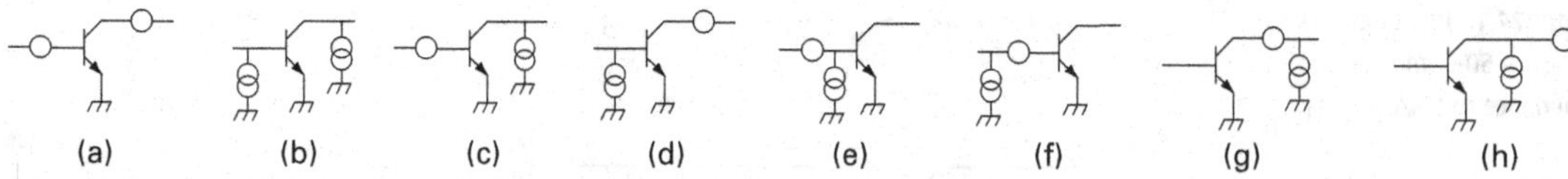

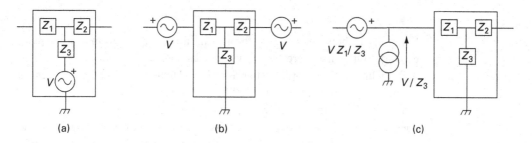

Figure 24.2. Equivalent representations of a noisy transistor as a noiseless transistor with two external noise generators.

(a) (b) (c) (d) (e) (f) (g) (h)

Figure 24.3. (a) An arbitrary two-port network with an internal voltage source; (b) and (c) networks equivalent to (a), but with external sources.

(a) (b) (c)

Note: Replacing internal sources with equivalent external sources is a sort of generalization of Thévenin's theorem and Norton's theorem, from one-port networks to two-port networks. Figure 24.3(a) shows an arbitrary two-port network with an internal voltage generator. Both (b) and (c) are equivalent to (a). In this example, the network is passive, but it can just as well be active, containing dependent voltage and current sources. When the internal source is a sine-wave source, the external generators are specified by complex amplitudes. When the internal source is a noise source, the external sources are specified by their rms values and the correlation between them. For example, if the external sources are a voltage source and a current source, the correlation is the complex number $\langle iv^* \rangle$, the time average of the product of the complex amplitude i and the complex conjugate of the complex amplitude v. When a two-port has multiple internal noise sources,[1] the resulting external sources can be consolidated into two external sources.

24.1.3 Noise figure of the equivalent circuit

Figure 24.2(e) (or 24.2f) is the easiest equivalent circuit to analyze for noise figure. Since both noise sources are placed at the input, the transfer characteristics (Z, Y, or S parameters) of the two-port will not come into play. In Figure 24.4, the circuit to be analyzed, the amplifier block can represent a transistor, a tube, or a complete amplifier. The two external noise sources, v and i, are equivalent to the device's internal sources of noise, whatever their number and physical nature. The other current source, i_T, represents the thermal

---

[1] A circuit model for a transistor usually contains several noise sources.

Figure 24.4. Signal source and noisy amplifier – equivalent circuit.

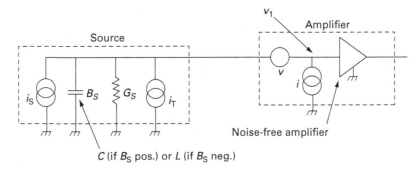

C (if B_S pos.) or L (if B_S neg.)

noise of the source, i.e., $\langle i_\mathrm{T} \rangle^2 = 4kTG_\mathrm{s}$. Since there are four noise parameters to be determined, we need more than just $\langle i^2 \rangle$ and $\langle v^2 \rangle$. The other two parameters come from the fact that i and v are generally not independent but are correlated. The part of i that is correlated with v defines the "correlation admittance," Y_c, through the relation $i_\mathrm{c} = Y_\mathrm{c}\, v$. Thus

$$\langle iv^* \rangle = \langle (i_\mathrm{c} + i_\mathrm{u})v^* \rangle = \langle i_\mathrm{c}v^* \rangle = \langle Y_\mathrm{c}|v|^2 \rangle = Y_\mathrm{c}\langle |v|^2 \rangle (G_\mathrm{c} + jB_\mathrm{c})\langle |v|^2 \rangle. \quad (24.2)$$

To find an expression for the noise figure, we must find the voltage v_1 at the input of the device and use the definition of noise figure to find the ratio of the average of $\langle |v_1|^2 \rangle$ to the part of $\langle |v_1|^2 \rangle$ attributable to the source. The result is independent of the input impedance of the device since, using superposition, we can find v_1, considering the noise sources one at a time. They all have the same effective source impedance, $1/Y_\mathrm{s}$. For convenience, then, we can take the device to have an infinite input impedance. It follows immediately that

$$v_1 = (i + i_\mathrm{T})/Y_\mathrm{s} + v. \quad (24.3)$$

To get the equivalent input *power* we use the average squared magnitude of v_1:

$$\langle |v_1|^2 \rangle = ((\langle |i|^2 \rangle + \langle |i_\mathrm{T}|^2 \rangle)/|Y_\mathrm{s}|^2 + \langle |v|^2 \rangle + (\langle iv^* \rangle/Y_\mathrm{s}) + (\langle i^*v \rangle/Y_\mathrm{s}^*). \quad (24.4)$$

Here we have used the fact that there is no correlation between i and i_T or between i_T and v. The noise figure is given by this total power divided by the part due to the noise current of the source:

$$F = \frac{((\langle |i|^2 \rangle + \langle |i_\mathrm{T}|^2 \rangle)/|Y_\mathrm{s}|^2 + \langle |v|^2 \rangle + (\langle iv^* \rangle/Y_\mathrm{s} + \langle i^*v \rangle/Y_\mathrm{s}^*)}{\langle |i_\mathrm{T}|^2 \rangle/|Y_\mathrm{s}|^2} \quad (24.5)$$

or

$$F = 1 + \langle |i|^2 \rangle/\langle |i_\mathrm{T}|^2 \rangle + (\langle |v|^2 \rangle/\langle |i_\mathrm{T}|^2 \rangle)|Y_\mathrm{s}|^2(1 + Y_\mathrm{c}/Y_\mathrm{s} + Y_\mathrm{c}^*/Y_\mathrm{s}^*). \quad (24.6)$$

The noise current of the source is given by $\langle |i_\mathrm{T}|^2 \rangle = 4kT_0G_\mathrm{s}$. It is convenient now to introduce the terms "noise conductance," g_n, and "noise resistance," R_n, which are defined by

$$\langle |i|^2 \rangle = 4kT_0 g_n \quad \text{and} \quad \langle |v|^2 \rangle = 4kT_0 R_n. \tag{24.7}$$

(The lower-case g_n follows the original notation of Rothe and Dahlke who reserved G_n for the part of i that is correlated with v.) In terms of these parameters we have

$$F = 1 + (g_n/G_s) + (R_n/G_s)|Y_s|^2(1 + Y_c/Y_s + Y_c^*/Y_s^*). \tag{24.8}$$

Replacing Y_s by $G_s + jB_s$ and Y_c by $G_c + jB_c$, this becomes

$$F = 1 + (g_n/G_s) + (R_n/G_s)(G_s^2 + B_s^2) + 2(R_n/G_s)(G_c G_s + B_c B_s). \tag{24.9}$$

Differentiating with respect to B_s and setting the derivative to zero, we find the optimum source susceptance,

$$B_{\text{opt}} = -B_c. \tag{24.10}$$

Differentiating next with respect to G_s and setting that derivative equal to zero gives the optimum source conductance:

$$G_{\text{opt}} = \sqrt{g_n/R_n - B_c^2}. \tag{24.11}$$

After a few more steps we find

$$F = 1 + 2R_n(G_c + G_{\text{opt}}) + (R_n/G_s)([G_s - G_{\text{opt}}]^2 + [B_s - B_{\text{opt}}]^2), \tag{24.12}$$

which is the desired relation, the Rothe–Dahlke formula. The first term on the right is just T_{min}. Summarizing these relations between the equivalent external noise generators and the standard noise parameters we have

$$R_n = \frac{\langle |v|^2 \rangle}{4kT_0} \tag{24.13a}$$

$$B_{\text{opt}} = \frac{-\text{Im}\langle iv^* \rangle}{\langle v^2 \rangle} \tag{24.13b}$$

$$G_{\text{opt}} = \sqrt{\frac{\langle |i|^2 \rangle}{\langle |v|^2 \rangle} - \left(\frac{\text{Im}\langle iv^* \rangle}{\langle |v|^2 \rangle} \right)^2} \tag{24.13c}$$

$$F_{\text{min}} = 1 + \frac{2\langle |v|^2 \rangle}{4kT_0} \left(\frac{\text{Re}\langle iv^* \rangle}{\langle |v|^2 \rangle} + G_{opt} \right). \tag{24.13d}$$

The inverse relations are

$$\langle |v|^2 \rangle = 4kT_0 R_n \tag{24.14a}$$

$$\langle |i|^2 \rangle = 4kT_0(G_{\text{opt}}^2 + B_{\text{opt}}^2) \tag{24.14b}$$

$$\text{Im}\langle iv^* \rangle = -B_{\text{opt}}\langle |v|^2 \rangle \qquad (24.14c)$$

$$\text{Re}\langle iv^* \rangle = 4kT_0 R_n \left(\frac{F_{\text{min}} - 1}{2R_n} - G_{\text{opt}} \right). \qquad (24.14d)$$

24.1.4 Devices in parallel

Noise matching allows us to realize the best noise figure, F_{min}, from a transistor. We saw earlier that multiple devices in series (an amplifier cascade) result in an unavoidable increase in noise figure. What about amplifiers in parallel? Identical transistors might be literally paralleled, emitter-to-emitter, base-to-base, and collector-to-collector. Or amplifiers might have their inputs directly connected and their outputs combined with a hybrid coupler. The result for two devices in parallel (see Problem 24.4) is that the resulting amplifier, compared to the individual amplifiers, has half as much noise resistance, twice as much noise conductance, and twice as much correlation admittance. This leaves F_{min} unchanged. It does, however, double Y_{opt}, making it easier to noise match to a low-impedance source, and can be a useful technique, for example, at very low frequencies where transformer matching is impractical.

24.1.5 Noise measure

We pointed out earlier that an amplifier with a good noise figure is of little value if its gain is low (a piece of wire has a perfect noise figure, unity, but has no gain). The application of negative feedback may lower the noise figure of a given transistor but at the same time it will reduce the gain. To recover the lost gain we can cascade two or more amplifiers. We saw in Chapter 17 the expression for F_∞, the noise figure of a cascade of identical amplifiers:

$$F_\infty - 1 = \frac{F - 1}{1 - 1/G}. \qquad (24.15)$$

This quantity, called *noise measure*, is a better figure of merit than noise figure in that a low-gain amplifier scores low, even if its noise figure is good. The optimum input network should, therefore, minimize F_∞ rather than F. But the gain, G, depends also on the output network. Haus and Adler [1] showed that the *minimum noise measure* is a fundamental invariant when the gain, G, is taken to be the "available gain" of the device (the maximum gain that can be obtained by varying the output match while leaving the input untouched). Minimum noise measure, like minimum noise figure, is therefore determined by the input network. Haus and Adler proved that the minimum noise measure of a device is left unchanged when the device is embedded in an arbitrary network of lossless reactances. It had been noticed earlier that the noise figure of a device could sometimes be lowered by using feedback. Haus and Adlers' results showed that any improvement in noise figure must be accompanied by a decrease in gain such that the minimum noise measure remains constant. A corollary is that changing the orientation of the device (common-emitter, common-collector, common-base) does not change the minimum noise measure. Circuit techniques, then, cannot produce a breakthrough low-noise amplifier. Any standard configuration can have the best possible noise performance. Circuit techniques, however, can improve bandwidth and stability. If we know the characteristics of a given device, we can calculate the minimum noise measure (best possible noise performance) at every frequency. There is, however, no theory available that tells us how well we can do over a given frequency band. R. N. Fano showed how well impedance matching can be done over a given band but no one has done the same for noise matching.

24.2 Oscillator noise

The frequency stability of an oscillator is determined by changes in ambient temperature and humidity, mechanical vibration, component aging, power supply variations, load variations, and finally, random noise generated by the circuit components – primarily the active element (transistor, tube, op-amp, etc). Except for noise, these effects are systematic, i.e., nonrandom, and can, in principle, be eliminated by using constant-temperature chambers, thermal compensation schemes, and shock mounting. But even when the systematic variations have been reduced to an acceptable level, random variations remain and determine the power spectrum of an oscillator (called "line shape" in spectroscopy). An ideal oscillator would have a power spectrum that is a delta function, i.e., an infinitely narrow unmodulated carrier. Noise broadens the spectrum of any real oscillator. In some applications the concentration of power is important; a specification might state, for example, that at least 90% of the power must be contained in a band within 0.01 Hz of the nominal frequency. More often it is only necessary that the spectral density beyond some distance from the center frequency be very low. In such a case it might be specified that "the noise power in a 1-Hz bandwidth centered 1 kHz from the carrier must be at least 60 dB below the total power." Spectral energy outside the narrow band containing most of the power is referred to as "sideband noise."

In a superheterodyne receiver the noise sideband power of the local oscillator can mix with an incoming signal that is nominally outside the IF passband to produce noise that falls within the passband, as shown in Figure 24.5. An undesired signal that is translated to a frequency outside the IF passband will normally be rejected by the IF bandpass filter. But a noisy L.O. will add noise sidebands to this signal that extend into the bandpass. If the undesired signal is near the bandpass and if it is much stronger than the desired signal, this noise can override the desired signal.

Figure 24.5. Translation of phase noise into the passband of a receiver.

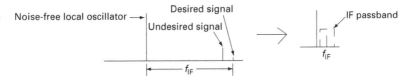

(a) Only the desired signal falls within the IF passband

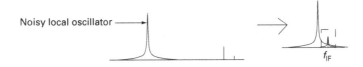

(b) Oscillator noise sidebands on the undesired signal fall within the IF passband

A similar situation occurs in MTI (moving target indicator) radar where the Doppler effect is used to distinguish moving objects from background "clutter" (the ground and stationary objects). Moving targets can be identified as spectral components shifted away from the frequency of the strong clutter echo. But, in as much as the transmitter's master oscillator or the receiver's local oscillator is noisy, moving targets with small velocities will be buried in the broadened clutter return. Good subclutter visibility therefore requires low-noise oscillators.

24.2.1 Power spectrum of a linear oscillator

In the simplest analysis, an oscillator is considered to be a linear system with just enough positive feedback to sustain oscillation. We have already used this approach to design a feedback oscillator. Here we will present a simple analysis [2, 3] that predicts the shape of the spectral line. This shape does not depend on the circuit configuration so here we will consider the simple series-mode oscillator presented in Chapter 11. This circuit, shown in Figure 24.6, uses a noninverting voltage amplifier having voltage gain, A (independent of frequency), infinite input impedance, and zero output impedance.

Figure 24.6. Prototype oscillator for phase noise analysis.

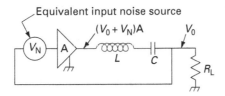

We will put an equivalent generator at the input to account for the noise produced by the amplifier. As long as the amplifier is linear, we need only consider the amplifier's noise spectrum in the vicinity of the oscillation frequency. The feedback voltage is taken from the load resistor forming the bottom leg of a voltage divider whose top leg is the series LC circuit. By inspection of Figure 24.6, we can write the loop equation:

$$V_0 = \frac{(V_0 + V_N)AR}{R + jX} \tag{24.16}$$

where A is the voltage gain of the amplifier and $X = \omega L - 1/(\omega C)$. Note that V_0 and V_N are spectral densities, i.e., their units are volts $/\sqrt{Hz}$. Solving for V_0 we have

$$V_0 = \frac{V_N AR}{R(1 - A) + jX}. \tag{24.17}$$

Expanding X around the resonant frequency, we let $\omega = \omega_0 + \omega'$, where $\omega_0^{-2} = LC$. The reactance, X, then simplifies to $X \approx 2\omega' L$ and Equation (24.17) becomes

$$V_0(\omega') = \frac{-V_N A}{(A - 1) - 2j\omega' L/R}. \tag{24.18}$$

The value of A is determined by the condition $\int |V_0^2(\omega')|/R \, d\omega' = P$, where P is the total output power:

$$P = \frac{1}{R} \int V_0^2(\omega') d\omega' = \frac{1}{R} \int_{-\infty}^{\infty} \frac{A^2 V_N^2 d\omega'}{(A - 1)^2 + 4\omega'^2 L^2/R^2} = \frac{\pi A^2 V_N^2}{2(A - 1)L}. \tag{24.19}$$

Solving for $(A - 1)$ we find

$$\frac{A - 1}{A^2} = \frac{\pi V_N^2}{2PL}. \tag{24.20}$$

Note that $A - 1$ can be very small; the voltage gain is only slightly greater than unity. Using Equation (24.20) and the fact that A is approximately unity, Equation (24.18) gives the power spectrum:

$$S(\omega) = \frac{|V_0(\omega')|^2}{R} = \frac{4P^2 Q^2 R^2}{\pi^2 V_N^2 \omega_0^2} \frac{1}{1 + (\omega'/\Delta)^2} \tag{24.21}$$

where Δ, the half-width of the line, is given by

$$\Delta = \frac{\pi V_N^2 \omega_0^2}{4PQ^2 R}. \tag{24.22}$$

The $1/(1 + x^2)$ "Lorenztian" line shape of Equation (24.21) gives the bell-shaped curve shown in Figure 24.7.

Figure 24.7. Lorentzian line shape.

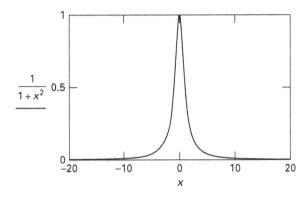

$\dfrac{1}{1+x^2}$

24.2.2 Sideband shape

As explained earlier, it is often important that the far-out sideband noise be low compared to the carrier power. If we assume $\omega' \gg \Delta$ then Equations (24.21) and (24.22) reduce to

$$\frac{S(\omega')}{P} = \frac{\omega_0^2 V_N^2}{\omega'^2 4 Q^2 RP} \ \text{Hz}^{-1}, \tag{24.23}$$

and we see that the far-out sideband noise falls off as $1/\omega'^2$.

Phase noise

So far, the oscillator noise has the same character as if it had been produced by amplifying white noise and passing it through a very narrow filter. The oscillator output voltage is equivalent to the sum of a noiseless phasor, i.e., the carrier, and a random noise phasor, as shown in Figure 24.8.

Figure 24.8. Vector diagram showing oscillator noise.

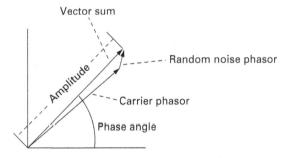

When the noise component is parallel to the carrier, the vector sum has the correct phase but an altered amplitude. When the noise is perpendicular to the carrier, the result is mostly just a phase error. The signal can be passed through an amplitude limiter to remove the amplitude fluctuations. This leaves only the perpendicular noise components, i.e., the phase noise. The parallel and perpendicular noise signals are uncorrelated, so if the amplitude variations are removed, the total noise power is cut in half. The symbol $\mathscr{L}(\omega')$ is often used to represent the phase noise power in a 1-Hz bandwidth at an offset of ω' from the nominal carrier frequency:

$$\mathscr{L}(\omega') = \frac{\omega_0^2 V_N^2}{\omega'^2 8 Q^2 RP} \text{ Hz}^{-1}. \tag{24.24}$$

We can express the noise power $V_N^2(\omega')/R$ as FkT where F is the noise figure of the amplifier, k is Boltzmann's constant, and T is the reference temperature, 290 K:

$$\mathscr{L}(\omega') = \omega_0^2 \frac{FkT}{\omega'^2 8 Q^2 P} \text{ Hz}^{-1}. \tag{24.25}$$

This result shows that best performance is obtained by using a low-noise device, by running it at high power, and by maximizing the loaded Q (by using a high-Q resonator with light loading).

24.3 Effect of nonlinearity

The analysis given above actually gives a lower limit for the noise. Very close to the carrier, oscillators depart from the predicted Lorenztian line shape and, as the offset frequency approaches the carrier, phase noise increases faster than ω'^{-2}. This additional noise is usually attributed to nonlinearity in the active device. We have made the assumption that the device is linear but some non-linearity *must* be present in any actual oscillator or its amplitude would either decay or increase without limit. The nonlinearity can translate baseband $1/f$ noise up to the radio frequencies.

Problems

Problem 24.1. A transistor with noise parameters, F_{min}, G_{opt}, B_{opt}, and R_n, is provided with a noise matching network to produce an amplifier with $F'_{min} = F_{min}$, $B'_{opt} = 0$, and $G'_{opt} = 1/50$. What is the value of R'_n?

Problem 24.2. Can a transistor be fitted with an input network such that the correlation admittance, Y_c, is zero for the overall amplifier?

Problem 24.3. The portion of the noise current, i_c, that is correlated with the noise voltage cannot exceed the total noise current, i.e., $\langle |i_c|^2 \rangle \leq \langle |i|^2 \rangle$. Use this to prove the following inequality:

$$F_{min} - 1 \leq 4R_n G_{opt}.$$

Problem 24.4. Show that when two identical transistors are paralleled, the minimum noise figure of the combination is the same as the minimum noise figure of the individual transistors.

Problem 24.5. A standard amplifier for use in 50-ohm systems has a gain G and a noise figure F. This amplifier is combined with a 50-ohm directional coupler to form the feedback amplifier shown below. The 180° coupler has a power gain of α in its "main channel" and therefore a power gains of $1-\alpha$ in the adjacent arms. The 180° path of the coupler provides the negative feedback. (Assume that the amplifier has zero phase shift.)

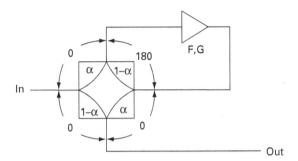

(a) Find the gain, G', of the feedback amplifier.
(b) Find the noise figure, F', of the feedback amplifier.
(c) Combine the results of (a) and (b) to show that the noise measure of the feedback amplifier is the same as the noise measure of the imbedded amplifier, i.e., show that

$$\frac{F'-1}{1-1/G'} = \frac{F-1}{1-1/G}.$$

This problem provides a simple demonstration of noise measure remaining invariant when an amplifier is imbedded in a lossless network. The negative feedback reduces the noise figure but it also reduces the gain and the noise measure stays the same. (What would positive feedback do – short of causing oscillation?)

Problem 24.6. It is certainly possible (and common) for an amplifier to have a noise temperature less than its physical temperature. Is it possible to build an *electronic cold load*, i.e., an active one-port circuit whose impedance is resistive but which generates less noise power than a resistor at ambient temperature?

Problem 24.7. A differential amplifier, e.g., an op-amp, is a three-port device (output and two inputs). Find an equivalent circuit which represents an op-amp as a noiseless device with three external noise sources. Arrange for these noise sources to be on the input side of the amplifier.

Problem 24.8. Which of the following statements best describes the operation of an oscillator?
(a) Noise is necessary to get an oscillator started. Once running, the oscillation is self-sustaining and noise plays no role.
(b) The output of an oscillator consists of amplified noise; the very narrow line shape derives from a resonator whose high loaded Q is effectively increased by the circuit's positive feedback.

Problem 24.9. Discuss the consequences (sound or picture quality) of a noisy oscillator used as a local oscillator in (a) an AM receiver; (b) an FM receiver; (c) an NTSC television receiver; (d) an ATSC television receiver.

Problem 24.10. Sketch a design for an oscillator that somehow regulates its own gain in order to stay as linear as possible, i.e., runs with just enough gain to maintain oscillation.

References

[1] Haus, H. A. and Adler, R. B., Optimum noise performance of linear amplifiers, *Proc. IRE*, Vol **46**, pp. 1517–1533, 1957.

[2] Leeson, D. B., *A simple model of feedback oscillator noise spectrum*, Proc. IEEE, Vol. **54**, pp. 329–330, February 1966.

[3] Robins, W. P., *Phase Noise in Signal Sources*, London: Peter Peregrinus on behalf of the Institution of Electrical Engineers, London, 1984.

[4] Rothe, H. and Dahlke, W., *Theory of noisy fourpoles*, Proc. IRE, Vol. **44**, pp. 811–818, June 1956.

[5] Vendelin, G. D., Pavio, A. M. and Rohde, U. L. *Microwave Circuit Design Using Linear and Nonlinear Techniques*, New York: John Wiley 1990.

25 The GPS navigation system

The basic idea of GPS navigation is simple. A GPS receiver simultaneously monitors ("tracks") the signals from four GPS satellites. Using a highly accurate onboard clock, each satellite effectively stamps its signals with the time at which they are transmitted. The receiver has an internal clock and can, therefore, determine the apparent travel time of the signals from each satellite. Let τ_i denote the travel time of the signal from the i-th satellite. The receiver contains a database and can look up the position of the i-th satellite, x_i, y_i, and z_i, corresponding to the time of transmission. Let x, y, and z represent the (unknown) position of the receiver. For each of the four satellites, *distance = rate × time*, so we can immediately write four equations:

$$
\begin{aligned}
(x - x_1)^2 + (y - y_1)^2 + (z - z_1)^2 &= c^2(\tau_1 - \Delta)^2 \\
(x - x_2)^2 + (y - y_2)^2 + (z - z_2)^2 &= c^2(\tau_2 - \Delta)^2 \\
(x - x_3)^2 + (y - y_3)^2 + (z - z_3)^2 &= c^2(\tau_3 - \Delta)^2 \\
(x - x_4)^2 + (y - y_4)^2 + (z - z_4)^2 &= c^2(\tau_4 - \Delta)^2,
\end{aligned}
\tag{25.1}
$$

where c is the speed of light and Δ is the offset (error) in the receiver's clock (the same error that made determination of longitude a challenge for early navigators). Because there are four equations and four unknowns, a unique solution can be found for x, y, z, and Δ, thus determining both position and time from scratch. (The receiver's clock error does not matter.) While these equations are nonlinear, the ample processing power available, even in hand-held receivers, makes the solution process easy. An obvious method is to begin with an estimated position, e.g., the last determined position, and then linearize and solve Equations (25.1) in the vicinity of this position.

25.1 System description

The GPS system is operated by the U.S. Department of Defense. In 1993 the system began running a full complement of 24 active satellites, with four satellites

Figure 25.1. (a) Block IIa GPS
satellite, 1816 kg; (b) GPS
satellite orbits.

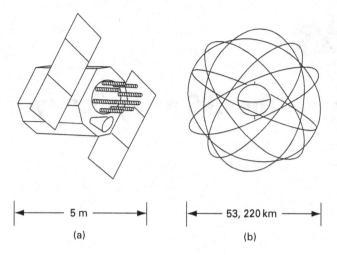

|← 5 m →| |← 53, 220 km →|

(a)   (b)

in each of six orbital planes, inclined at 55° with respect to the equatorial plane.
The satellites were in 12-hour circular Earth-centric orbits, 20 200 km (3.17 Earth
radii) above the surface of the Earth as shown in Figure 25.1.

At any point on the Earth, from six to twelve satellites will be above the
horizon at any time. As the number of active satellites was increased to 31, the
constellation was modified to provide better availability in the event of satellite
failures. Orbits are adjusted once a year. The satellites have had operational
lifetimes of approximately 10 years.

25.2 GPS broadcast format and time encoding

The system uses two L-band frequencies, 1575.42 MHz ("L1") and
1227.6 MHz ("L2"). Each satellite transmits on both of these frequencies
using CDMA (code division multiple access) modulation. The L1 signal contains a "navigation message," described below, which is transmitted repeatedly
as a continuous 50-Hz bit stream. Before transmission, this slow bit stream is
exclusive-ORed (modulus-2 added) with a one-bit pseudorandom noise (PRN)
sequence (the CDMA code) having a high clock rate. This produces a fast bit
stream, spread out in bandwidth to approximately the clock rate ("chip" rate) of
the PRN code. The L2 signal may also be modulated by the data stream, but is
usually modulated only by the PRN code, equivalent to sending a data stream of
all 1's. Each satellite is assigned a unique PRN code. In order to receive data
from the i-th satellite the receiver converts the incoming signal L1 to baseband
and multiplies it with a replica of that satellite's PRN code. When the replica,
whose binary levels are ±1, is precisely positioned (phased), in order to decode
or "strip" the code from the signal, it reproduces the original 50-Hz bit stream,
as shown in Figure 25.2.

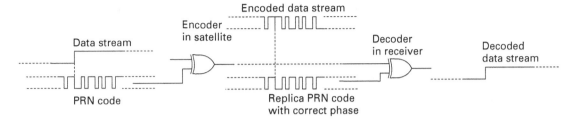

Figure 25.2. CDMA coding (spreading) and decoding (de-spreading).

If the phase of the replica code is shifted off by as little as one chip, the decoder output will be as random and as spread in frequency as its input. This decoding process is the way the receiver selects the desired signal. Signals from other GPS satellites will also be superposed in the received signal, but they have different PRN codes which, when multiplied by the selected code, average to zero for any value of phase shift. Thus, other signals appear as added noise. This added noise is insignificant, in that the received signal is already less than the thermal background noise by some 20–30 dB.[1]

The 50-Hz data streams contain the GPS time,[2] orbital corrections, satellite clock correction, propagation delay effects, and satellite condition, as well as an almanac of long-term satellite data and an auxiliary service, the current difference between GPS time and UTC time. But the precise time of transmission is inferred not from the data but from the phase of the PRN code. We saw in Chapter 21 that just such codes are used for pulse compression in radar to achieve high time resolution (and hence high range resolution) without using narrow high-power pulses. When PRN codes are correlated (convolved) with themselves, a sharp peak is formed, allowing precise time resolution. Thus, PRN coding does double duty in the GPS system; it provides multiple access (many satellites using the same frequency) and it also provides pulse compression for timing accuracy. The *P-code* (Precision), a high rate code transmitted on both L1 and L2, has a seven-day period. Therefore, a receiver that has successfully delayed the output of its replica code generator to match the incoming signal, i.e., to acquire the signal, can use that delay value to estimate the signal travel time. The P-code was originally intended only for military use.[3] The other transmitted code is the *C/A-code* (Coarse ranging and Acquisition). This code, always present, along with the P-code, on L1, has a 1.023-MHz chip rate vs. 10.23 MHz for the P-code. It therefore nominally provides only one-tenth the range resolution of the P-code. The C/A code is intended for civilian use; consumer GPS receivers use only the C/A code on L1. The C/A-code repeats every millisecond. This makes acquisition easier, but introduces a

---

[1] Assuming a receiving antenna with a gain of 3dB, the received power of a GPS signal is $-160\,\text{dBW}(10^{-16}$ watts).

[2] GPS time increases continuously, unlike Universal Coordinated Time (UTC) which is occasionally adjusted by a leap second. GPS and UTC coincided at the start of Jan. 6, 1980, UTC.

[3] Usually the P-code is referred to as the *P(Y)*-code because the published P-code can be replaced by a classified *Y-code* for military security.

1-msec ambiguity in the time, which must be resolved using time information transmitted in the data stream.

25.3 GPS satellite transmitter

Figure 25.3 is a functional block diagram of the transmitter aboard a GPS satellite and shows the signals that modulate L1 and L2. All signals are derived from an onboard 10.23-MHz atomic frequency standard. Newer satellites carry two rubidium and two cesium frequency standards. A data formatter produces the continuous one-bit data stream at a 50-Hz baud rate (20 msec/baud). This stream is exclusive-ORed separately with the C/A code and the P-code to create two-bit streams, with respective (main lobe) bandwidths of 1.023 MHz and

Figure 25.3. GPS satellite transmitter block diagram.

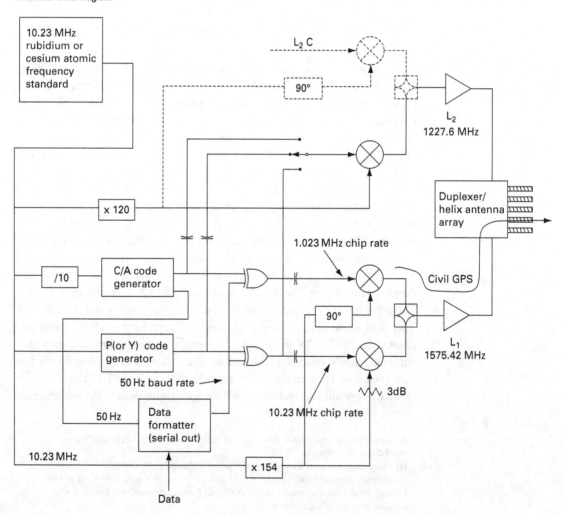

10.23 MHz. The streams are offset to be symmetric about zero volts (indicated by a coupling capacitor on the diagram) and then multiplied by sine waves at 154×10.23 MHz $= 1575.42$ MHz, one shifted 90° with respect to the other. These products are summed to form the L1 signal. The output of each multiplier is a double-sideband suppressed carrier (DSSC) signal. The sum of these two signals, as we have seen, is known as quadrature AM modulation (QAM).[4] The modulating signal fed to each mixer has only two values, $+V$ or $-V$, which determine the polarity of the mixer output, so these mixers can also be called BPSK (binary 0 and 180° phase-shift keying) modulators. The L1 C/A "Civil GPS" signal path is indicated on the diagram.

The original L2 signal is produced by a single mixer, whose signal input can set by the ground control station to be the C/A code, the P-code, or the exclusive-OR of the data and the P-code. Normally the P-code is selected and the L2 signal provides dual-frequency operation for ionospheric delay correction (described below). Newer satellites also provide an "L2C" civil L2 signal, as shown in dotted lines. This signal will provide civil users with dual-frequency capability for high-precision surveying.

25.4 Signal tracking

To see how tracking can be done, we will first consider the case of the L2 signal carrying only the long PRN code and no navigation data. We will then describe the somewhat more complicated case of tracking a signal containing the navigation data, exclusively-ORed with the short C/A code.

25.4.1 Tracking the L2 signal

Figure 25.4 is a block diagram of a hypothetical receiver set up to track this signal. The IF band, which is 20.46 MHz wide to contain the P-coded signal, is established by a bandpass filter which precedes the mixer but follows the low-noise amplifier (LNA), in order that the filter loss not degrade the receiver's noise figure.

Surface acoustic wave (SAW) filters are common in this very small fractional bandwidth application. A 20-MHz lowpass filter (LPF), probably not extending down quite all the way to dc, further defines the bandpass and eliminates the sum frequency component from the mixer. The IF signal is sampled at 41 MHz

---

[4] QAM is normally used to transmit two independent signals which will be independently recovered. In this case, however, the two signals have been made independent by the coding, so they could be algebraically added, with the sum feeding a single mixer. This would introduce "self noise," but a GPS receiver is dominated by thermal noise, so the system performance would not suffer. The QAM system has the advantage of producing a constant-amplitude signal which does not require a linear power amplifier.

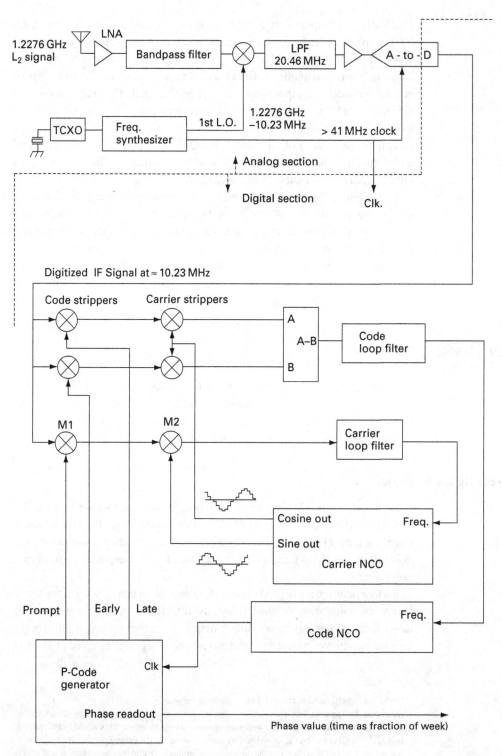

Figure 25.4. A GPS navigation
receiver tracking the L2 signal.

(Nyquist sampling at twice the signal bandwidth to avoid aliasing). From this point on, the receiver uses all-digital processing.

The digital portion of the block diagram includes the blocks necessary to track the L2 signal from one satellite. This entire section would be duplicated n times in a receiver that receives signals simultaneously from n satellites. The receiver synchronously (coherently) detects the signal by stripping off the PRN code and also mixing down to dc (stripping off the carrier). These operations can be done in either order. Once these stripping operations are working properly, the phase of the replica code will be aligned with the code on the incoming signal, allowing the time of transmission to be inferred, as described above. The L.O. used to strip the carrier is the *carrier NCO*, a numerically controlled oscillator in a digital phase lock loop. To convert the signal to precisely zero frequency, the receiver must tune the carrier NCO to compensate for Doppler shift, caused by line-sight-velocity, and by inevitable error in the first L.O. frequency.

Mixer M1 (a digital multiplier) multiplies the digital IF signal by a correctly phased replica of the ±1 PRN code for the desired satellite. This strips the code from the signal, leaving just the $\approx$ 10.23 MHz carrier signal. This carrier signal feeds mixer M2, as the reference signal for the PLL. Two additional code stripper mixers are provided. One is fed by the replica code, advanced by half a chip (*early* code), while the other is fed by the replica code, delayed by half a chip (*late* code). The outputs of these code strippers are mixed down to dc. The carrier NCO is part of a conventional PLL, so its phase is in quadrature with the phase of the reference. Therefore, a version shifted by 90° (the "cosine" output) is provided to synchronously demodulate the code-stripped early and late signals. When the code is in perfect alignment with the satellite signal, the demodulated early and late signals will have equal average values, half the value they would have if they had been on time ("prompt"). But if the alignment shifts slightly, the demodulated early and late signals will be different. A subtractor forms this difference and uses it to increase or decrease the frequency of the *code NCO*, which is the time base for the code generator. Thus, the code loop is also a conventional PLL.

Note that the two loops, while they can run independently, are related. Suppose that, while both loops are tracking a satellite signal, the line-of-sight velocity suddenly increases. The increase in Doppler shift causes the carrier loop to shift the L.O. by an amount $\Delta\omega/\omega_0 = \Delta v/c$. Since range = velocity × time, we can expect that the code tracking loop will shift the code NCO by the same fractional amount. In fact, the code loop can be assisted by properly scaling the carrier NCO frequency control value and adding it directly to the code NCO's frequency command. The scaling is chosen to make the fractional frequency shifts equal. Using this *assisted tracking* technique, the bandwidth of the code loop can be narrowed, for more accurate tracking. Note that it does not matter that the carrier NCO frequency is determined partly by frequency error in the first L.O., because the code NCO loop will pump up a bias to counteract this non-Doppler component.

25.4.2 Tracking the C/A code on the L1 signal

Figure 25.5 shows a hypothetical consumer GPS receiver tracking the L1 short C/A code-plus-data signal on L1. The RF section is tuned to L1, and, for the slower chip rate code, the IF bandwidth is now just 2.046 MHz. The presence of the 50-Hz data stream will cause the polarity of the carrier stripper outputs to change sign, according to whether a given data bit is a one or a zero. This will defeat the operation of the conventional PLLs used above for a signal with no data. A remedy for this is to use loops that detect the signals before subtracting them to form the loop error signals.

Consider the loop controlling the code NCO. The detected early and late signals are lowpass filtered, i.e., time averaged, over an interval substantially shorter than one data baud. Then they are squared (detected) and the difference of these power values forms the loop's error signal. The hypothetical carrier loop shown here works in the same way. Two carrier stripping mixers are fed quadrature L.O. signals ($-45°$ and $+45°$). When the loop is locked, these mixers will produce equal powers. The carrier NCO also provides a $0°$ output, in phase with the original carrier, which feeds mixer M2 which strips the carrier to produce the data stream. The $0°$ carrier is also used for the code loop.

Since the C/A code repeats every millisecond, rather than every week, the phase of the code generator provides only the LSBs of the time. The MSBs come from data contained in the navigation message. The 50-baud data is divided into 300-bit subframes, so a new subframe begins every six seconds. Every subframe contains a TOW (time of week) word, which is the GPS elapsed time, measured in units of six seconds (6000 msec) from the start of the current week, that corresponds to the transmission of the first bit in the next subframe. A distinctive header word identifies the beginning of each subframe. (Every subframe also contains the PRN code i.d. number that identifies the satellite[5].) The receiver has a millisecond counter synchronized to the code generator, i.e., the counter is clocked at the each time the code cycles through its starting point. The receiver's control block checks to verify that the MSBs of the millisecond counter are correct in relation to the TOW word. If incorrect, the control block enables the counter to synchronously set itself to the TOW × 6000 at the beginning of the next 1-msec C/A code cycle. The complete time value from the receiver is, therefore, the time, rounded to milliseconds, from the counter plus the code generator phase, in fractions of a millisecond with submicrosecond precision.

25.5 Acquisition

From the discussion of tracking, you can see that neither loop can lock unless the other is also locked; they track simultaneously or not at all. Acquisition of

---

[5] Reference [1] contains a detailed format for a complete 12.5-minute navigation message, made of 25 frames, each having five subframes made of ten 30-bit words.

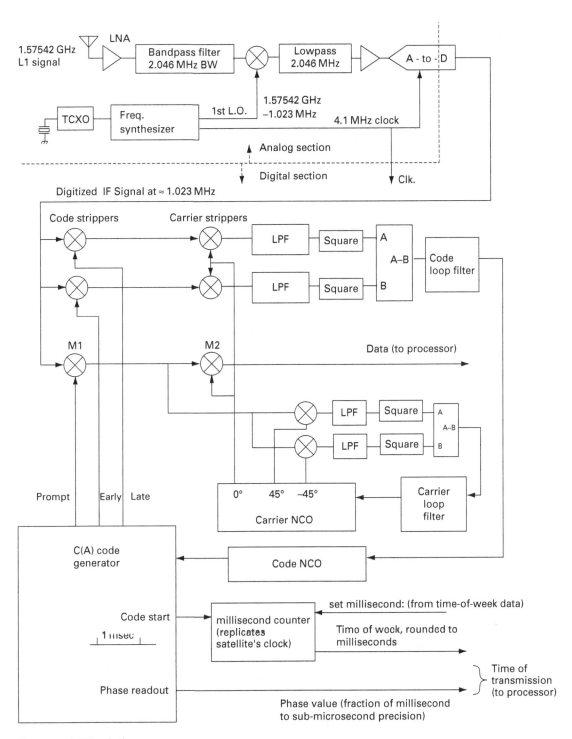

Figure 25.5. A GPS navigation receiver tracking the L1 C/A signal.

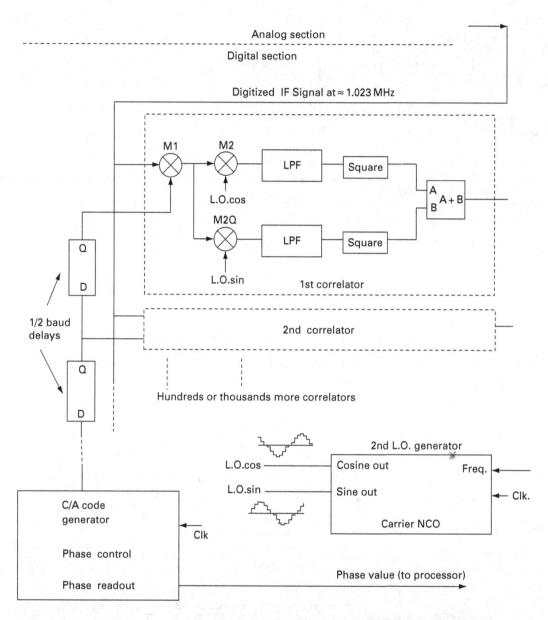

Figure 25.6. GPS navigation receiver in acquisition mode.

the signal therefore requires that the receiver perform a two-dimensional search over both frequency space and code phase. Enough digital processing power is usually available so that the search can be done in a parallel fashion. Figure 25.6 shows a setup for acquisition processing.

One search process is as follows: Mixers (multipliers) M1, M2 and M2Q are provided for every code phase in increments of half a baud. Successively delayed versions of the replica code are fed to each of the M1 mixers. The M2 and M2Q mixers are all fed from a common L.O. generator. The L.O. frequency is stepped in equal frequency increments. At each frequency, the M2 mixer output signals are averaged (lowpass filtered) over an interval, a "dwell time," and the search ends when the detected power (the square of the sum of the two mixer outputs) exceeds a detection threshold. At that point, the L.O. frequency and carrier phase are close enough for a handover to the tracking mode. Longer dwell times require smaller frequency steps, so it takes more time to search for a weak signal. The changing data bits do not spoil the correlation as long as the dwell times are short compared to the data baud length (unless a dwell happens to straddle a data transition). The circuitry or processing resources for one Doppler/code phase cell is called a correlator. With enough correlators running in parallel, the GPS receiver can simultaneously search multiple frequency shifts, multiple code delays, and multiple satellites. Once the signals have been acquired, relatively little processing power is required for tracking.

The C/A signal is easier to acquire than the P-signal, as it has a higher signal-to-noise ratio and a shorter code (less space to search). Receivers that use the P-code usually acquire the C/A code as a first step in acquiring the P-code. (Hence the designation "C/A:" Civil and Acquisition.)

25.6 Ionospheric delay

GPS signals, at 1.5 GHz and 1.2 GHz, interact only slightly with the Earth's lower atmosphere, but the interaction with the ionosphere contributes the largest term to the measurement error budget. We saw in Chapter 20 that the ionosphere's free-electron gas, produced by ionizing ultraviolet solar radiation, changes the dielectric constant from ϵ_0, its vacuum value, to become

$$\epsilon = \epsilon_0 \left(1 - \frac{Ne^2}{\epsilon_0 m \omega^2} \right) = \epsilon_0 \epsilon_r, \tag{25.2}$$

where N is the electron density, e is the electronic charge, m is the electron mass, and ϵ_r is the relative dielectric constant (relative to ϵ_0). The phase velocity of the signal is given by $v_{phase} = \omega/k = (\mu_0 \epsilon)^{-1/2}$. From Equation (25.2) we see that, in the ionosphere, ϵ is less than ϵ_0. Therefore, v_{phase} is greater than the speed of light and the wavelength, $\lambda = 2\pi/k$, is longer than its vacuum value. Information, however travels at the group velocity,[6] $v_{group} = d\omega/dk$, which, as you can verify, turns out here to be

$$v_{group} = \frac{c^2}{v_{phase}} = c\sqrt{1 - Ne^2/(\epsilon_0 m \omega^2)}. \tag{25.3}$$

---

[6] Group velocity: A narrowband signal around $\omega = \omega_0$ can be represented by a Fourier integral as $V(t) = \int F(\omega') e^{j(\omega_0 + \omega')t} d\omega'$, where the function $F(\omega')$ is concentrated around zero. When this signal is launched as a wave into a dispersive medium where the wave number, k, can be approximated as $k \approx k_0 + \omega' dk/d\omega$, the wave at t, z becomes $V(t,z) = e^{j(\omega_0 t - k_0 z)} \int F(\omega') e^{j\omega'(t - z\, dk/d\omega)} d\omega'$. Except for a phase factor (the first term on the right), $V(t,z)$ is identical to $V(t=0, z=0)$ when $z = t d\omega/dk = t v_{group}$.

Thus v_{group} is less than c, and the ionosphere increases the satellite-to-ground propagation delay. The propagation time is just $\int (v_{\text{group}})^{-1}\, ds$ over the signal path or

$$T_{\text{prop}} = c^{-1} \int \left[1 - \frac{Ne^2}{(\epsilon_0 m \omega^2)}\right]^{-1/2} ds \approx c^{-1} \int \left[1 + \frac{Ne^2}{2\epsilon_0 m \omega^2}\right] ds, \quad (25.4)$$

where we have assumed that the second term in the brackets is much less than unity. We see that the additional delay caused by the ionosphere is given by

$$\Delta T_{Iono} = \approx c^{-1} \int \frac{Ne^2 ds}{2\epsilon_0 m \omega^2} = \frac{c r_e}{2\omega^2} \int N ds = \frac{c r_e \text{TEC}}{2\omega^2}, \quad (25.5)$$

where $r_e = e^2/(\epsilon_0\, m\, c^2) = 2.82 \times 10^{-15}$ m (the "classical electron radius"), and TEC $= \int N\, ds$, the total electron content, i.e., the number of electrons in a one square meter column along the signal path.

Under worst case conditions (the ionosphere enhanced by a solar storm and the satellite near the horizon), the value of TEC might reach 10^{19}, in which case ΔT_{iono} would be 43 nsec, corresponding to a position error of 14 m. Normally the ionospheric error would be less than one-tenth of this.

If the receiver is able to measure the delay on both frequencies, L1 and L2, the delay caused by the ionosphere can be calculated and subtracted from the measured delay. From Equation (25.5), the difference between the measured propagation delays will be $T_{\text{L2}} - T_{\text{L1}} = \frac{1}{2} c\, r_e\, \text{TEC}\, (1/\omega_2{}^2 - 1/\omega_1{}^2)$. This provides the value of TEC, from which we can calculate the true delay,

$$T = T_{L1} - \frac{\omega_2^2 (T_{L2} - T_{L1})}{2(\omega_2^2 - \omega_1^2)}. \quad (25.6)$$

25.7 Differential GPS

Over a limited area, GPS receivers tracking the same satellites will be subject to essentially the same systematic errors – satellite clock errors, ephemeris errors, ionospheric delay, and even intentional systematic errors. ("Selective availability" limited the accuracy for civil users by introducing pseudorandom errors in the satellite clocks which could be corrected only by authorized users. However, this was discontinued in 2000.) Thus, over the limited area, individual determinations of absolute position will be biased by the same amount, and calculated differential positions between them will be accurate. The U.S. Coast Guard maintains differential GPS reference stations at well-surveyed positions near harbors. They continuously broadcast the current correction to their GPS reading so that users, nominally aboard ships, can apply that correction to their own GPS reading. This system requires a means of transmission – dedicated beacon transmitters or piggybacking data on subcarriers of local FM stations. It also requires that the users be able to receive these signals and have GPS receivers equipped to apply the corrections.

In an even more accurate form of differential GPS used for precise surveying, the phase of the carrier itself is used to get extremely fine time resolution. This interferometric technique can produce accuracy better than 1 cm. This carrier phase tracking mode requires dual-frequency receivers for ionospheric correction. The maximum practical distance from the reference station is about 30 km.

25.8 Augmented GPS

The Federal Aviation Administration (FAA) has set up a large-scale "wide area augmentation system" (WAAS) with about 25 ground reference stations spread through the U.S. Each of these stations monitors all the GPS satellites within view, on both frequencies, L1 and L2. The data are consolidated at a central location where correction data are generated. The measured TEC values are used to create a model ionosphere in the form of a set of vertical TEC values, from which the TEC for an arbitrary path can be estimated. Correction data is also produced for the satellite clock and ephemerides errors. These data are sent to two geostationary satellites, which broadcast them back down to GPS users. The down-link is on the GPS L1 frequency, and provides the same signal strength as the GPS signals. The WAAS signals are spread with the same kind of C/A codes used by GPS, but with different PRN sequences. This allows a GPS receiver, equipped with suitable firmware, to receive correction data from a WAAS satellite as if it were just another GPS satellite.[7] The WAAS data is transmitted at a net rate of 250 bits/sec. Convolutional encoding for forward error correction is used at a rate of 1/2, so the data stream is 500 bits/sec. In practice, the use of WAAS data reduces C/A code ("Standard Positioning Service") GPS position errors to about 2 m.

25.9 Improvements to GPS

Replacement GPS satellites include the new L2C (Civil) signal, whose signal path is shown at the top of Figure 25.3. This signal is actually two time-multiplexed signals, a medium-length code, CM, and a long-length code, CL. The CL code is transmitted without data. It can, therefore, be tracked with the system shown in Figure 25.4, which uses conventional phase lock loops which provide better tracking than the loops used in Figure 25.5, for a signal containing the data stream. The CL2 signal also provides dual-frequency operation so that civil GPS receivers can correct for ionospheric delay. The CM code contains the data, but convolutionally encoded for forward error correction.

---

[7] The WAAS satellites contain atomic frequency standards and can, in fact, be used as additional GPS navigation satellites.

This slows the net data rate to 25 bits/sec. Both the CM and CL codes have a chip rate of 511.5 kHz. The CM code repeats every 20 msec, while the CL code repeats every 1.5 sec. The L2C multiplexes CL with CM+Data by alternating between them every 1/1.023 microseconds.

25.10 Other satellite navigation systems

The Russian *Glonass* system and the European *Galileo* system (scheduled for operation in 2013) are very similar to GPS. Both are L-band systems, using PRN coding for time resolution. The Glonass system, however, does not use the PRN coding for multiple access. Instead, each satellite transmits on its own assigned frequency, which requires that receivers be tunable. The Chinese *Beidou* system began with two geostationary satellites, but is expected to eventually have five geostationary and 30 medium Earth orbit satellites. All of these systems were preceded by the interesting *Transit* or NAVSAT system, which operated from 1964 until 1991. A ship at sea could locate its position within 200 m by analyzing the curve of Doppler shift vs. time from a single satellite pass. With only five satellites, a given user could get a position fix only once every several hours.

Problems

Problem 25.1. The nominal GPS signal power is -160 dBW (10^{-16} W) at the output of the receiver antenna (at the L1 frequency assuming the antenna gain is 3 dB). Assume an antenna temperature of 290 K, since the low directivity of the antenna causes ground noise to dominate. (a) Assume that the receiver noise figure is 4 dB and calculate the signal-to-noise ratio for a bandwidth of 2.046 MHz (the nominal bandwidth of the C/A signal). (b) Use the Shannon formula, $C = B \log_2 (1+S/N)$, to calculate the channel capacity in bits/sec. Compare this with the 50-Hz data rate of the navigation signal.

Problem 25.2. Estimate the gain of the GPS satellite antenna, assuming the beam angle just encloses the Earth. Then use the information given in Problem 25.1 to estimate the power of the satellite transmitter.

Problem 25.3. "Selective Availability" (SA), a now-discontinued GPS security feature, degraded the navigation accuracy for civil users. Suggest some ways to implement this feature.

Problem 25.4. Describe how you would use off-the-shelf RF components, a deep-memory sampling oscilloscope, and a PC to make an "off-line" detection of a GPS satellite signal.

Problem 25.5. Devise an algorithm and write a computer program to solve the set of equations (25.1) for x, y, z, and Δ.

Problem 25.6. Derive an expression for the Doppler shift of the GPS signal, assuming the orbit passes directly overhead.

References

[1] Navstar global positioning system interface specification IS-GPS-200, Revision D, 7 December 2004 Navstar GPS Space Segment/Navigation User Interfaces. This 193-page document provides a full description of the GPS satellite signals including navigation message formats, code generator details, and algorithms to make corrections using the data in the navigation message. You can find this document in PDF format by searching for "IS-GPS-200."

[2] Global positioning system standard positioning service Performance standard October 2001, Assistant secretary of defense for command, control, communications, and intelligence. This document provides general information on the system architecture and performance. To find this document in PDF format, search on "GPS SPS performance."

[3] Kaplan, E. D. and Hegarty, C. J. Ed., *Understanding GPS* 2nd edn; Boston: Artech House, 2006.

[4] Misra, P. and Enge, P., *Global Positioning System*, Ganga-Jamuna Press 2001.

[5] Prasad, R. and Ruieri, M., *Applied Satellite Navigation using GPS, Galileo and Augmentation Systems*; Boston: Artech House, 2005.

26 Radio and radar astronomy

Radio astronomy was discovered accidentally in 1931 by Karl Jansky, a physicist at Bell Telephone Laboratories. Jansky had been assigned to identify the sources of noise encountered in a newly installed transatlantic short-wave radiotelephone service. Using a directional receiving antenna on 20.5 MHz, he observed that one component of the noise, a wideband hiss, had a diurnal variation that reached a maximum intensity on average four minutes earlier each day. Jansky knew that the stars advance in just this way (in siderial time) and deduced that the source of the hiss must be outside the solar system. His observations showed that this "cosmic noise" came from the galactic plane and was strongest from the direction of the galactic center (in the constellation Sagittarius).

After Jansky, the second pioneer of radio astronomy was a radio engineer, Grote Reber, who in 1937, built a 9-m (30-ft) parabolic reflector beside his house in Wheaton, Illinois. This was maybe the first modern dish antenna. Reber began his observations using a receiver at 3 GHz, which pushed the high-frequency state of the art, because he assumed that cosmic radio noise was the low-frequency tail of the thermally generated radiation spectrum from white-hot stars. The intensity of this radiation would increase as the square of the frequency, so using the highest practical frequency would make detection easier and would also make his antenna more directive. Detecting nothing at 3 GHz, he worked his way down finally to 160 MHz, where he was able to make contour maps of the cosmic noise intensity. The radiation he and Jansky observed is now known to be *synchrotron radiation*, caused by the centripetal acceleration of fast, i.e., non-thermal, electrons spiraling in a magnetic field. By the end of World War II, the Sun (an ordinary thermal source under low sunspot conditions) had been detected at microwave frequencies. After the war, a previously-predicted spectral line at 21 cm (1420 MHz) was quickly detected. This famous neutral hydrogen line corresponds to the energy difference between the parallel and antiparallel orientations of the magnetic moment of the nucleus (the proton) with respect to the magnetic moment of the spinning electron. Many radio telescopes have been built in the decades following the war, the largest being the

305-m (1000-ft) diameter dish built by Cornell University at Arecibo, Puerto Rico. Discoveries in radio astronomy include some one hundred atomic recombination and molecular lines, pulsars, natural masers, and the isotropic 3K blackbody cosmic background radiation, a remnant left from the Big Bang.

26.1 Radiometry

Most of the radio sources found in nature emit wideband noise; their radiation comes from a great number of individual radiators whose contributions add randomly to produce *Gaussian noise*. (A histogram of voltage samples from the antenna terminals forms a Gaussian curve centered on zero.) Since such a signal is itself a random process, we can only measure its average properties. The most important of these is the average of the square of the voltage, i.e., the power. If we take, say, several n-minute averages of the power, these averages will be scattered around the true average. If n is made larger, the averages will be distributed more tightly about the true average. It might not take long to measure the receiver output power to a precision of, say, 10% or even 1%. But that is almost never long enough to measure the power of a radio source to the same precision or, for that matter, even to detect a source. The problem is that the power from the source is masked by other sources of noise including receiver noise, antenna noise, and cosmic background noise. When an astronomer is trying to detect a source in a certain direction, the first step is to average the power received from that direction, the on-source direction. The next step is to measure the power from a nearby, but off-source, direction. Finally, the latter "off" power is subtracted from the former "on" power. For weak sources (most sources), these powers are almost identical and might correspond to a system temperature[1] of, say, 100 K. Yet the astronomer may need to detect a source that raises the system temperature by only 10 mK. This requires that the "on" and the "off" powers both be measured to an accuracy of, say, $3.3/\sqrt{2} = 2.3$ mK for a 3-sigma detection. The fractional accuracy of the "on" and "off" power measurements must therefore be $0.0023/100 = 2.3 \cdot 10^{-5}$ or about one part in 50 000 (see Problem 26.1).

If we average N samples of the squared voltage, the relative standard deviation, $\delta P/P$, will be $\sqrt{2/N}$ if the voltage has a Gaussian distribution and if all the samples are independent (see Problem 26.2). A signal from a channel of bandwidth, B, can furnish $2B$ independent samples every second. Integrating for a time T we can therefore collect $2BT$ independent samples and the relative standard deviation of the power measurement will therefore be

[1] A *system temperature* of 100 K, for example, means that the equivalent noise power at the receiver input is the same as the noise power from a resistor at 100 K. This equivalent noise power is the sum of the actual noise power (sky noise from the main lobe, ground noise picked up from back-lobes, some thermal noise if the antenna is lossy, plus a contribution representing the noise generated in the receiver.

$$\delta P/P = (BT)^{-1/2}. \tag{26.1}$$

Usually a radio astronomer expresses power in terms of antenna temperature and would write $\delta T/T_{\text{system}}$ rather than $\delta P/P$. Equation (26.1) is commonly known as the "radiometer equation." Note that sensitivity increases with bandwidth, contrary to many communication situations where increasing the bandwidth just increases the noise and decreases the signal-to-noise ratio. We arrived at Equation (26.1) by considering digital (discrete) signal processing. It is also common to use an analog square-law detector followed by an analog lowpass filter, which does the averaging. When the lowpass filter is just a simple RC integrator, an analog signal analysis shows that the radiometer equation can be written as

$$\delta T/T_{\text{sys}} = (2BRC)^{-1/2}. \tag{26.2}$$

26.2 Spectrometry

Many sources produce "colored" noise rather than white noise, i.e., the flux density from these sources varies with frequency. Often these variations reveal characteristic shapes of atomic or molecular lines. Instead of just measuring the total power, the signal is divided into adjacent frequency "bins" and the power in each bin is measured. When a spectrum of bandwidth B is divided into n frequency bins, each of bandwidth B/n, the radiometer equation shows that the integration time must be increased by n. It is therefore especially important in radio astronomy, where weak sources may require many hours of integration, to measure the n individual spectra simultaneously rather than sequentially. Simultaneous analysis is done with multichannel radiometers ("multiplex" spectrum analyzers). The simplest multiplex spectrometer is just a bank of n filters, each followed by its own square-law detector and averager. Such an instrument is called a filterbank, but might better be called a radiometer bank. Today, most radio spectrometry is done digitally, often using digital *autocorrelators* (see Chapter 27).

26.3 Interferometry

The classic single-dish radio telescope such as Reber's backyard dish or the Arecibo dish has an angular resolution which is diffraction limited; the angular size of the beam (between half-power points) is about λ/D radians where D is the diameter of the dish. Such a telescope can make maps by scanning the vicinity of a source and doing radiometric averages at each point, but the resolution of the map is limited to λ/D. Better resolution requires a larger antenna. Interferometry uses more than one antenna to form a beam whose size corresponds to the *spacing* between the antennas rather than their diameters. The simplest interferometer has just two elements. The beam formed by the two antennas together has a series of narrow lobes in one dimension. If a point source moves across these lobes, the radiometer will trace out the lobes, just as a point source moving across the beam of a single dish antenna traces out the beam pattern. (Of course the interferometer's fine lobe structure is multiplied by the broader beam pattern of the individual antennas.) The multilobed pattern of a two-element interferometer is unsuitable for the intensity mapping described above but it is often used to set an upper limit on the angular size of a source.

26.3.1 Imaging interferometry

By using data from multiple antennas, it is possible to synthesize a beam that is small in both dimensions, i.e., a beam that would correspond to a filled-aperture antenna whose diameter is the size of the interferometer array. The VLA (Very Large Array) at Socorro, New Mexico, is an example of such a system. Signals from 27 antennas are transmitted up to 21 km through optical fibers (originally through low-loss circular waveguides) to the central processing laboratory. The VLBA (Very Long Baseline Array) has 10 antennas with spacings as large as the distance from Hawaii to St. Croix in the Virgin Islands. Its signals, together with timing information, are recorded on magnetic media at each station and sent, physically or electronically, to New Mexico for after-the-fact combining and processing. Imaging interferometers (phased arrays) work essentially as follows. Suppose we have an array of antennas such as shown in Figure 26.1. Assume the antennas are all pointed in the same direction,

Figure 26.1. An interferometer antenna array.

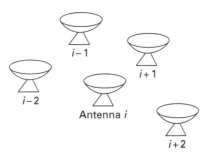

say up. If the voltages from the antennas are all added together in phase, squared, and averaged, the result is a narrow beam, pointing straight up. This process can be written as follows:

$$P_0 = \left\langle \sum_n V_n^2 \right\rangle = \sum_{n,m} \langle V_n V_m^* \rangle. \tag{26.3}$$

Now suppose we add the voltages, multiplied by a progressive phase shift, to form a beam tilted slightly off the vertical. If we do this addition, followed by squaring and averaging, we get the *k*-th beam:

$$P_k = \left\langle \sum_{n,m} V_n \mathrm{e}^{\mathrm{j}\phi_{n,k}} V^* \mathrm{e}^{-\mathrm{j}\phi_{m,k}} \right\rangle = \sum_{n,m} \mathrm{e}^{\mathrm{i}(\phi_{n,k}-\phi_{m,k})} \langle V_n V_m^* \rangle. \qquad (26.4)$$

This is the general case; for the straight-up beam, $\phi_{n,k}$ and $\phi_{m,k}$ are both zero. The important thing to notice is that we can multiply the signals from every possible antenna pair and average the products first, independent of what beam we wish to form. After averaging these pairs we can *then* form each beam by weighting the averaged products with the appropriate phase factors and performing the double sums. Note also that the averaged products do not have to be measured simultaneously; we could even use one fixed antenna and one mobile antenna and measure the average product for one baseline after another. The VLA has 27 antennas so it can measure $(27 \times 26)/2$ baselines simultaneously. But, as the Earth turns, these baselines become different baselines so, over time, it is possible to collect averaged products from a huge number of baselines. This set of baselines is virtually the same as all the baselines one could form between pairs of points on a dish with a radius of 21 km, and the final synthesized beam can be as sharp and clean as if it had come from a filled dish. An array can thus have the resolution of an enormous antenna even if it does not have the corresponding sensitivity (collecting area). On the other hand, because the individual antenna elements have relatively wide beams, an array has a relatively wide field of view. Radio astronomers hope to have, by the year 2020, a super large array, the SKA (see reference 5) which would have a total collecting area of one square kilometer and provide baselines as long as 3000 km.

26.4 Radar astronomy

Ionized meteor trails produced echoes on WWII radars, but echoes from the Moon were not observed until after the war. In January 1946, a U.S. Army Signal Corps group headed by John H. DeWitt, Jr. observed real-time Moon echos using modified WWII radar equipment delivering 3 kW of power at 110 MHz to a 64-element dipole antenna array. In Hungary, one month later, Zoltán Bay used similar equipment, together with electrolytic cells ("coulombmeters") to do signal integration. By the end of his experiment, the on-target cell had accumulated significantly more gas than the off-target cell. Many large radars were built in the late 1950s, and radar echo detections from the Sun, Venus, Mercury and Mars were made in the 1960s. Radar was used to measure the rotation rates of Venus and Mercury. Saturn's rings were detected in the 1970s, as were the large moons of Jupiter, many asteroids and several comets.

26.4.1 The Moon

The Moon's distance from the Earth, R, is 3.8×10^5 km, and it has a *radar cross-section*, σ, of about 6.6×10^5 km^2. This means that the power density of the echo's SR, in watts per square meter, received back on the radar antenna will be given by $\sigma/(4\pi R^2)$ times the power density incident on the target, S_{inc}. (Note that this defines *radar cross-section* as the collecting area of an equivalent target that isotropically scatters the intercepted incident power.) The incident power density is just $S_{inc} = P_T G/(4\pi R^2)$, where P_T is the transmitter power and G is the antenna gain. The power into the receiver, P_R, is equal to $S_R A_{eff}$ where A_{eff} is the effective collecting area of the antenna. We have seen earlier that G is equal to $4\pi A_{eff}/\lambda^2$. The Moon's radar cross-section is about 7% of its geometric cross-section, i.e., it is 7% as reflective as a metal sphere of the same size. An echo from the Moon will therefore produce a power at the receiver of

$$P_R = P_T \frac{4(\pi A_{eff}/\lambda^2)}{4\pi R^2} \frac{\sigma}{4\pi R^2} A_{eff} = \frac{P_T A_{eff}^2 \sigma}{4\pi R^4 \lambda^2}. \tag{26.5}$$

Suppose we have a radar with a 1-kW transmitter and a modest 5-m diameter dish antenna with an aperture efficiency of 50%. The effective area of this antenna is therefore half of its geometric cross-section or $A_{eff} = 0.5 \times \pi (5/2)^2 = 9.8\,\text{m}^2$. It is an experimental fact that σ for a planet is approximately independent of wavelength, so we see from Equation (26.5) that the received power is inversely proportional to λ^2 (because of the fundamental antenna relation between gain and effective area). Suppose that the wavelength is 30 cm, i.e., the frequency is 1 GHz. With this radar system, the power received from the Moon would be

$$P_R = \frac{1000\,\text{W}(9.8\,\text{m}^2)^2 6.6 \times 10^{11}\,\text{m}^2}{4\pi(3.8 \times 10^8\,\text{m})^4 (0.30\,\text{m})^2} = 2.7 \times 10^{-18}\,\text{W}. \tag{26.6}$$

Whether this amount of power is easy to detect or not depends on the noise it competes with. Let us assume that the noise from the antenna (sky noise from background cosmic radio sources plus some "spill-over" noise from the surrounding ground) has an equivalent noise temperature, $T_{ant} = 50$ K. Let us also assume that our receiver has an equivalent noise temperature, $T_{rcvr} = 35$ K. The total equivalent input temperature is therefore given by $T_{system} = 85$ K and the noise power will be $kT_{sys}B$. If we make the bandwidth, B, very small we decrease the noise power and hence detection becomes easier. But if we make B very small we will need a very accurate prediction of the Doppler shift caused by the relative motion of the Earth and Moon in order to tune the return echo into the passband of the filter. Finally, the rotation of the Moon causes a Doppler *broadening* of the return echo; if we make B less than 25 Hz we will begin to exclude some of the broadened signal. Let us compromise and use a bandwidth of 100 Hz. The signal-to-noise ratio at the receiver output will then be given by

$$S/N = \frac{P_R}{kT_{sys}B} = \frac{2.7 \times 10^{-18}\,\text{W}}{1.38 \times 10^{-23}\,\text{W/Hz/K} \times 85\,\text{K} \times 100\,\text{Hz}} = 23.0. \tag{26.7}$$

This SNR of 23 is large enough that the signal would be visible on an oscilloscope connected at the intermediate frequency (IF) voltage output; no signal averaging would be needed. The modest radar system assumed here could be assembled for a few thousand U.S. dollars.

26.4.2 Venus

Venus is not nearly so easy to detect, being some 280 times more distant than the Moon. Its radar cross-section is about 20 times that of the Moon, so with the radar system described above, the SNR of the return echo would be lower by a factor $280^4/20$ or 307 million! This requires a *much* larger radar. The Arecibo radar, however, has an effective antenna diameter of 200 meters and an average power of 1 MW at 2.38 GHz. It would out-perform our Moon example radar by a factor of $(10^6/10^3)$ $(2.38/1)^2$ $(200^2/5^2)^2$ or 1.4×10^{10}. Thus, with the same assumed bandwidth and system temperature, the Venus echo from the Arecibo radar would have an SNR of 23 $(1.4 \times 10^{10})/(307 \times 10^6) = 1090$. This is overkill for a simple detection but is needed for high-resolution mapping.

26.4.3 Delay-Doppler mapping

Except for the Moon, planetary targets have angular sizes much smaller than the radar beam. Nevertheless, images of photographic quality can be made by a technique known as delay-Doppler mapping. This method uses short pulses (or pulse compression) to obtain adequate range resolution. The relative motion between the radar and the target provides resolution in the transverse direction via the Doppler effect. This was the method used to get the first surface images of Venus, whose cloud cover kept its features hidden to optical telescopes. The technique is essentially the same as side-looking or synthetic aperture radar by which photographic-like images of ground targets are made from an airborne radar with only a small dish antenna. Figure 26.2 shows how delay-Doppler mapping works. This is a view of the planet as seen from the radar. When a short pulse is transmitted, the first echo to return comes from the front cap or sub-Earth point. At subsequent times the echo signal

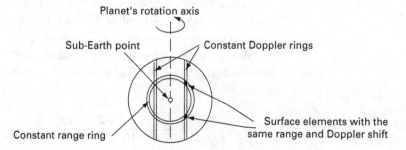

Figure 26.2. Geometry for delay-Doppler radar imaging.

corresponds to loci of equal range that are rings on the planetary surface, centered on the sub-Earth point. Because the planet is rotating, surface points on the right-hand side are moving away from the radar and their echoes are Doppler-shifted to lower frequencies. Likewise, points on the left-hand side are up-shifted. The magnitude of the of Doppler shift is proportional to the cross-range distance, i.e., the distance in the direction perpendicular to both the planet's axis of rotation and the line between the radar and the planet. Loci of constant Doppler shift are rings parallel to the rotation axis of the planet and to the line from the planet to the radar.

The two shaded surface elements have the same range and the same Doppler shift. This fundamental ambiguity can be removed, at least for Venus maps made at Arecibo, by tilting the radar beam slightly to illuminate only the Northern or only the Southern hemisphere of Venus. Obviously this technique requires the narrow beam provided by a very large antenna. Data taking is done as follows: a series of IF voltage samples is read and stored separately as each return pulse arrives. After many pulses have been received, we have a time series for each range ring. Each series is Fourier analyzed to get the Doppler spectrum. The magnitude of a given Doppler component corresponds to the reflectivity of the surface element at the intersection of the range ring and the Doppler ring.

26.4.3.1 Overspreading

The standard delay-Doppler[2] method will not work when the planet is too large and/or spins too fast. This is because the Nyquist sampling theorem requires that the (complex) sampling rate be at least equal to the bandwidth of the signal so that high-frequency Doppler components do not "alias" and appear as lower frequency components (the stroboscope effect). This establishes a minimum pulse repetition frequency (PRF). But if the depth of the planet is too large, even the minimum PRF will result in having more than one pulse on the planet at a time, causing the echoes from different ranges to "fold" together.

The bandwidth of the return echo (twice the highest Doppler shift) is given by $B = 2f_{radar} 2v_{max}/c = 4f_{rad}r_{planer} \Omega /c$ where r_{planet} is the radius of the planet, Ω is its apparent angular velocity,[3] f_{radar} is the frequency of the radar, and α is the angle between the radar beam from the Earth and the planet's rotation axis. Therefore, the Nyquist sampling requirement is $f_{sample} > 4f_{radar} r_{planet} \Omega/c$. The condition to avoid range folding is $1/f_{sample} > 2r_{planet}/c$. Multiplying these two inequalities gives us $1 \geq 8 f_{radar} r^2_{planet} \Omega/c^2$, which imposes an upper limit on the frequency of the radar:

---

[2] Note that delay-Doppler mapping of space objects is the same technique as pulse-Doppler radar, discussed in Chapter 21. But while the planet presents a "target" in every range-Doppler cell, a pulse-Doppler surveillance radar is likely to have a target (often an airplane) in only a single range-Doppler cell.

[3] The *apparent* angular velocity of the planet corrects for the planet's tilt and the relative translational (orbital) motion of the Earth and the planet.

$$f_{\text{radar}} < \frac{c^2}{8r^2\Omega}.$$
(26.8)

Using Equation (26.8), the radar frequency should be less than about 1.5 GHz for Venus and less than about 1.8 GHz for Mercury. These are practical frequencies for conventional microwave radar technology. For Mars, however, the radar frequency would have to be lower than 14 MHz, because of the high rotation rate. Equation (26.5) shows that such a low frequency would require an extremely large antenna. To make matters worse, sky noise and atmospheric noise at 14 Mz are both much higher than at microwave frequencies.

While standard delay-Doppler does not work for Mars and other overspread targets, Mars images were made in 1988 using the VLA in New Mexico. During the observations, Mars was illuminated with a 10-GHz signal from the JPL NASA Solar System Radar transmitter at Goldstone, California. The resolution of these images was only about 100 km.

A modified delay-Doppler technique, developed around 1986 for radar probing of the ionosphere, was first used at Arecibo in 1990 to solve the overspreading problem for Mars. This "long code" technique is used with the 2.38-GHz transmitter, which puts out a continuous wave (cw) rather than pulsed power, but is biphase-modulated with a long pseudorandom code – for Mars, a sequence of 10-µs bauds. The code elements shift the RF phase by zero or 180°, equivalent to leaving the signal unchanged or changing its polarity. The length of the sequence must be greater than 0.33 seconds (the length of a coherent integration or "look"), though the sequence used at Arecibo repeats only every 3054 hours and can be considered totally random. To see how this technique works, imagine first a simple point target. The echo from such an object would be a delayed version of the code. At the receiver we sample the return echo at the baud rate and multiply successive samples of the echo signal by successive bauds of an identical "replica" code to undo the echo's phase reversals. When the replica code has the correct alignment, it undoes the phase reversals and produces a cw signal. We can separately multiply the echo samples by the replica code for all alignments. Identifying the alignment that produces a cw signal gives us the distance (range) to the point target. If the point target has a velocity component parallel to the radar beam, the recovered cw signal will be Doppler shifted and we can determine the velocity component. Next, imagine that the echo comes from *two* point reflectors, separated in time by at least one baud (1500 m in range). Again multiplying the echo by the replica code, we would find two alignments that produce cw signals and could thereby deduce the range and line-of-sight velocity of each target. Note, however, that the "decoded" signal from either target contains white noise that is the signal from the other target, randomized because the replica code is not aligned with its echo. This "self-noise" has power equal to that of the desired signal, assuming the targets have equal radar cross-sections. We can consider the range rings on a planet to be a set of N such targets, separated in range delay by the

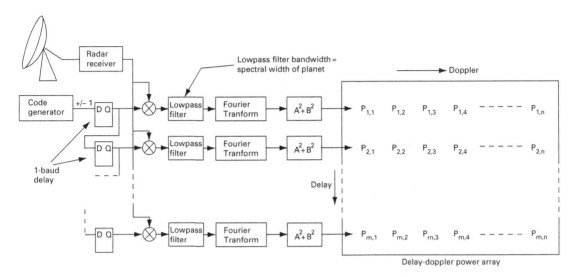

Figure 26.3. Long-code delay-Doppler method.

baud length. (Each range ring has a Doppler spectrum.) The self-noise voltage is therefore the sum of N voltages, which add as a random walk. The delay time depth of Mars is 23 ms so, for a baud length of 10 μs, $N = 2300$. If we assume that the signals from the rings are approximately equal in strength, the self-noise power will be N times the signal power. Here, this factor can be reduced to 2300 $(7.5/100) = 177$ by lowpass filtering the data sequences before doing the Doppler analysis, since the self-noise is spread out over a much larger bandwidth (100 kHz) than the Mars echo bandwidth (7.5 kHz). This still appears disastrous, until we remember that the self-noise is in addition to "real" noise, i.e., cosmic noise, antenna noise, and receiver noise. Under the worst weak-signal conditions, the self-noise is insignificant compared to the real noise so, while the system performance will be poor, the self-noise is not to blame. Under the best conditions, the self-noise dominates the real noise and it will determine the integration time needed to pull the signal out of the total noise. When the long-code technique is used with the Arecibo S-band radar system to observe Mars, the effective signal-to-noise ratio is decreased by a factor of 5.2 or 1.7, depending on the polarization of the echo.[4] Figure 26.3 shows a signal processing arrangement that produces the decoded signals for each range, does a Fourier transform on each decoded range signal, and stores the resulting delay-Doppler data points in an array.

---

[4] The radar transmits a signal with circular polarization. Smooth portions of the planetary surface create specular (mirror-like) reflections, which reverse the sense of the circular polarization. Rough portions of the surface create multipath reflections, which cause some of the echo power (usually less than half) to return with the same circular sense as the transmitted signal. The radar is equipped to receive both polarizations.

Problems

Problem 26.1.   Use the radiometer equation to find how much time is needed to make a 3-sigma detection of a 10 mK radio source if the radio telescope's system temperature is 100 K. Assume the predetection bandwidth, B, is 10 MHz. A "3-sigma detection" requires that the 0.01 K contribution from the source be $3\sqrt{2}$ times the δT fluctuation predicted by the radiometer equation. (The factor $\sqrt{2}$ takes account of the increased fluctuation when the "off" is subtracted from the "on", assuming an equal time T is spent measuring each.)

Problem 26.2.   Verify that if a noise power estimator is defined as the average of N independent samples of the squared noise voltage, the standard deviation of this estimator will be $\sigma^2\sqrt{2/N}$. Assume that the probability distribution of the noise voltage is Gaussian with zero mean and variance σ^2. Hint: for a zero-mean Gaussian distribution, the expected value of V^2 is σ^2 and the expected value of V^4 is $3\sigma^4$.

Problem 26.3.   Pulsars (rotating neutron stars) are radio sources that turn on and off with very regular periods ranging from about 2 milliseconds to about 2 seconds. Given a radio telescope at your disposal, how would you go about searching for pulsars, i.e., what kind of data processing scheme would you use to find these periodic sources?

References

[1] Butrica, A. J., *To See the Unseen: A History of Planetary Radar Astronomy*, Diane Publishing Co., 1997 (also available on the Web).
[2] Goldsmith, P. ed. *Instrumentation and Techniques for Radio Astronomy*, IEEE Press,1988. (This book is prefaced with 12-page historical overview.)
[3] Kraus, J. D., *Radio Astronomy*, 2nd edn, Cygnus-Quasar Books, 1986.
[4] Ostro, S. J., Planetary radar astronomy, *Rev Mod Phys*, Vol. **65**, No. 4, pp. 1235–1279, October 1993.
[5] http://www.skatelescope.org/PDF/SKABrochure_2007.pdf

27 Radio spectrometry

Spectrometry or spectral analysis is the statistical characterization of random (*stochastic*) signals such as the IF voltage in a radio astronomy receiver, as described in Chapter 26. The spectrometers discussed in this chapter are all *multiplex* spectrometers, meaning that they measure N points on the spectrum simultaneously. This is to be distinguished from swept-frequency spectrum analyzers, which measure spectral points sequentially. Multiplex spectrometers are used when long integration times are needed to pull a signal out of the noise, as in radio astronomy. They are also used for low-frequency spectrum analysis, where narrow channel bandwidths require long measurement times to process a sufficient number of independent samples. Most often the signal is Gaussian; if a fine-grained histogram of samples of the signal's amplitude is scaled to make the area below the curve equal to unity, the average curve will be the Gaussian probability density function, $f(V) = (2\pi\sigma^2)^{-1/2} \exp(-V^2/2\sigma^2)$. You can verify (Problem 27.1) that σ is the rms value of V, i.e., σ^2 is the power, $\langle V^2 \rangle$. But total power does not completely characterize the signal. A complete description is contained in the *power spectral density function*, $S(\omega)$, the distribution of power vs. frequency. A set of bandpass filters and power meters, i.e., a set of radio-meters, serves to measure points on the spectral density function (usually called the PSD or simply the power spectrum). However, the simplest mathematical definition of the power spectrum uses the signal's *autocorrelation function*, $R(\tau)$, a function of time delay. The value of the autocorrelation function for a given time delay, τ, is defined as the average value of the "lagged product"

$$R(\tau) = \langle V(t) \cdot V(t+\tau) \rangle. \tag{27.1}$$

For a signal whose characteristics are unchanging (a *stationary process*), $R(\tau) = R(-\tau)$. We will see that the Fourier transform of the autocorrelation function is the power spectral density so, directly or indirectly, we describe the signal in terms of sine-wave basis functions, i.e., we find the magnitudes (but here not the phases) of a set of sine waves whose superposition has the same power spectrum as the original signal. Why are sine waves preferred to other sets of basis functions? Often the process under study is a spectral line which shows up clearly in a few adjacent Fourier coefficients (frequency bins) or a Doppler shift which just displaces the spectrum. The frequency spectrum is clearly a natural way to represent such signals. Of course we deal with sine waves all the time as the characteristic functions of linear systems; a sinusoidal signal of a given frequency remains a sinusoid with the same frequency after passing through any arbitrary chain or network of linear elements such as amplifiers, filters, and transmission lines. A variety of instruments have been developed for spectrometry. In approximate historical order, these include analog filterbanks, autocorrelators, chirp-z spectrometers, Fourier transform spectrometers, and acousto-optical spectrometers.

27.1 Filters and filterbanks

We have already seen the analog filterbank in Chapter 26, as part of a radiometer bank. Usually the signal is converted down to a convenient intermediate frequency, e.g., tens of MHz, and the filters are the various LC bandpass filters discussed earlier or crystal filters or digital filters. A traditional disadvantage of an analog filterbank spectrometer has been that it is not "elastic," i.e., the filters have a fixed width; if narrower or wider filters are needed, another filterbank must be built. With high-speed digital-to-analog conversion and fast memory, this restriction can be lifted by using an interesting spectral expansion technique. Sequences of data are read into the memory at whatever slower rate is needed and then played back at a higher rate into a single wideband filterbank.

27.2 Autocorrelation spectrometry

Autocorrelation spectrometers estimate the power spectrum by taking pairs of samples with a given time separation, multiplying them, and averaging these "lagged products." (In Fourier transform spectral analysis the samples are multiplied by sine waves but the resulting coefficients are squared before averaging. In both cases powers, rather than voltages, are averaged.) When the time separation is zero, the average lagged product is just the total power. But if we measure other lagged products as well (where the time separation is not zero), we can indirectly find the power spectrum because the autocorrelation function (ACF) and the power spectrum form a Fourier transform pair:

$$S(\omega) = \int_{-\infty}^{\infty} R(\tau)e^{-j\omega\tau}d\tau \qquad (27.2)$$

$$R(\tau) = \frac{1}{2\pi}\int_{-\infty}^{\infty} S(\omega)e^{j\omega\tau}d\omega. \qquad (27.3)$$

Figure 27.1. Single-channel radiometer.

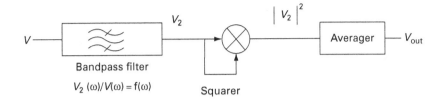

This relation, the Wiener–Khinchin theorem, is often used to define the power spectrum. To see that $S(\omega)$ as defined by Equation (27.2) agrees with our intuitive ideas of the power spectrum, consider the single-channel radiometer shown in Figure 27.1. The (in situ) voltage transfer function of the bandpass filter is defined in the frequency domain by $H(\omega) = V_2(\omega)/V(\omega)$. The power response function or filter shape is therefore given by $|H(\omega)|^2$, so we expect that the time average of $(V_2)^2$ will be given by

$$\left\langle |V_2(t)|^2 \right\rangle = \int_{-\infty}^{\infty} S(\omega)|H(\omega)|^2 (2\pi)^{-1} d\omega \qquad (27.4)$$

where $S(\omega)$ is the spectral power density function. Note that $(2\pi)^{-1}d\omega = df$; the power spectral density has units of watts/Hz.

Let us now show that the output of this radiometer is indeed given by Equation (27.4), and that $S(\omega)$ is given by Equation (27.2). We can express V_2 in the time domain as

$$V_2(t) = \int_{-\infty}^{t} V(t')h(t - t')dt', \qquad (27.5)$$

where $h(\tau)$, the impulse response function, is the inverse Fourier transform of $H(\omega)$.[1] The output of the squarer is then given by

$$(V_2(t))^2 = \int_{-\infty}^{t} \int_{-\infty}^{t} V(t')V(t'')h(t - t')h(t - t'')dt'dt''. \qquad (27.6)$$

Since we assume $V_{\text{in}}(t)$ is a stationary process, the value of t is arbitrary. Let us pick $t = 0$, which gives

$$(V_2(t))^2 = \int_{-\infty}^{0} \int_{-\infty}^{0} V(t')V(t'')h(-t')h(-t'')dt'dt''. \qquad (27.7)$$

---

[1] The impulse function, $h(t)$, is the filter output voltage at t, in response to a delta function input at $t = 0$. Note that $h(t)$ is real and that $h(t) = 0$ for $t < 0$.

From causality, $h(t) = 0$ for $|t| < 0$, so the upper limits of integration can be changed from 0 to infinity. Doing this, making a change of variable, $t'' = t' + \tau$, and taking the time average, we find

$$\left\langle (V_2(t))^2 \right\rangle = \int_{-\infty}^{\infty} \int_{-\infty}^{\infty} \langle V(t')V(t' + \tau) \rangle h(-t')\, h(-t' - \tau) dt' d\tau. \quad (27.8)$$

Using the definition of the ACF, $R(\tau) = \langle V(t)V(t+\tau) \rangle$, we find

$$\left\langle (V_2(t))^2 \right\rangle = \int_{-\infty}^{\infty} R(\tau) \left[\int_{-\infty}^{\infty} h(-t')h(-t' - \tau) dt' \right] d\tau. \quad (27.9)$$

Applying the convolution theorem,[2] the Fourier transform of the term in square brackets is $|H(\omega)|^2$. Therefore, the bracketed term is the inverse Fourier transform of $|H(\omega)|^2$ or

$$\left[\int_{-\infty}^{\infty} h(-t')h(-t' - \tau) dt' \right] = \frac{1}{2\pi} \int_{-\infty}^{\infty} e^{j\omega\tau} |H(\omega)|^2 d\tau. \quad (27.10)$$

Substituting this expression into Equation (27.9) produces

$$\left\langle (V_2(t))^2 \right\rangle = \int_{-\infty}^{\infty} R(\tau) \left[\frac{1}{2\pi} \int_{-\infty}^{\infty} e^{j\omega\tau} |H(\omega)|^2 d\omega \right] d\tau. \quad (27.11)$$

Finally, interchanging the order of integration gives us

$$\left\langle (V_2(t))^2 \right\rangle = \int_{-\infty}^{\infty} |H(\omega)|^2 \left[\int_{-\infty}^{\infty} e^{j\omega\tau} R(\tau) d\tau \right] (2\pi)^{-1} d\omega. \quad (27.12)$$

After all this effort, we can now compare Equation (27.12) with Equation (27.4) and confirm that the power spectrum defined by Equation (27.2) agrees with what we would expect to measure with a bandpass filter followed by a squarer and an averager. (Don't worry about the sign discrepancy in the $e^{j\omega t}$ term in Equation (27.12); since $V_2(t)$ and $R(\tau)$ are real, we can replace this term by $e^{-j\omega t}$.)

27.2.1 Hardware autocorrelators

A hardware autocorrelator is a special-purpose parallel signal processor that calculates an averaged ACF, usually in real time. A single Fourier transform operation in a computer then turns the ACF into the power spectrum with the same number of frequency bins (points on the spectrum) as the number of points

---

[2] If $H(\omega)$ is the Fourier transform of $h(\tau)$, then $|H(\omega)|^2$ is the Fourier transform of $\int h(t)h(t+\tau) dt$.

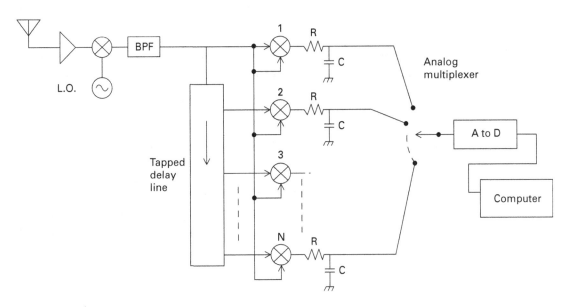

Figure 27.2. Analog autocorrelator.

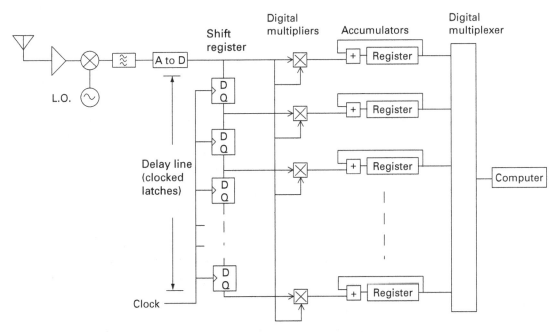

Figure 27.3. Digital autocorrelator.

on the ACF. Autocorrelators use a simple expandable architecture. Figure 27.2 shows an analog version of the autocorrelator.

Figure 27.3 shows a digital autocorrelator. The analog delay line is replaced with a digital shift register. The analog multipliers and averagers are replaced by digital multipliers and accumulators.

The autocorrelators of Figures 27.2 and 27.3 produce values only for discrete points on the ACF. But, as long as the input signal is band-limited (usually by a bandpass "anti-aliasing" filter), the spectrum can be represented as a Fourier *series*, so sampled ACF points are sufficient to compute the continuous power spectrum. In practice, however, we cannot measure an infinite number of points on the ACF. The spectral estimate obtained by transforming ACF points that extend only to τ_{max}, i.e., by transforming an ACF "windowed" with a rectangular function that is unity for $\tau_{max} < \tau < \tau_{max}$ and zero outside this range, produces a function which is the true spectrum convolved by the function $\sin(\omega \tau_{max})/(\omega \tau_{max})$. A sharp line in the spectrum therefore appears as a broadened line, surrounded by sidelobes. It is common to apply a smoother windowing function such as $\cos(\frac{1}{2}\pi \tau/\tau_{max})$ to reduce the sidelobes even though the main load will be broadened – a trade-off between spectral resolution and leakage between channels.

27.2.2 One-bit autocorrelation

One-bit autocorrelation is a technique that greatly reduces the complexity of the digital circuitry. The input analog voltage is fed to a comparator whose output (one bit) indicates the sign, i.e., the polarity, of the voltage. The continuous range of analog voltages is compressed down to just two values: plus one and minus one. (In the digital hardware, a digital "1" indicates a value of $+1$ and a digital "0" indicates a value of -1.) The correlator's delay line shift register needs to be only one bit wide and the one-bit $\times$ one-bit multipliers can be exclusive NOR gates, since the output of an exclusive NOR gate is $+ 1$ when the inputs are the same and -1 when they are different. The integrators that follow the multipliers are simple counters. Although the output of a one-bit correlator is a distorted version of the actual correlation function, the distortion can be entirely corrected. Because the input signal has Gaussian amplitude statistics, the output of the one-bit correlator turns out to be:

$$\rho_{1-bit}(\tau) = (2/\pi) \sin^{-1}[\rho(\tau)] \tag{27.13}$$

where ρ is the normalized autocorrelation function: $\rho(\tau) = R(\tau)/R(0) = R(\tau)/\sigma^2$. This functional relation was derived during World War II by J. H. van Vleck while studying the spectrum of clipped noise. (Jammers can have much better power efficiency if jamming with clipped noise works as well as jamming with noise having Gaussian amplitude statistics.)

The averaged values from a one-bit autocorrelator can, therefore, be inverted to get the true correlation function.

$$\rho(\tau) = \sin[(\pi/2)\rho_{1-bit}(\tau)]. \tag{27.14}$$

A straightforward proof of the van Vleck relation (Problem 27.4) proceeds from the Gaussian bivariate probability density,

$$f(x_1, x_2) = \frac{1}{2\pi\sigma_1\sigma_2(1-\rho^2)^{1/2}} \exp\left(\frac{-1}{2(1-\rho^2)}\left(\frac{x_1^2}{\sigma_1^2} - 2\rho\frac{x_1 x_2}{\sigma_1\sigma_2} + \frac{x_2^2}{\sigma_2^2}\right)\right).$$
(27.15)

Here x_1 and x_2 are the voltages $V(t)$ and $V(t+\tau)$, so $\sigma_1 = \sigma_2$. One gives up some radiometric sensitivity for the simplification of one-bit processing; the integration time has to be increased by a factor of $\pi^2/4$ compared to a correlator using multi-bit quantization.

Several generations of digital correlators have been used in radio astronomy. Custom CMOS correlator chips have thousands of lag channels and clock rates above 100 MHZ. A variety of serial and parallel multiplexing techniques are used to combine these chips to increase the number of lag channels and to increase input data rates to several hundred MHZ.

27.3 Fourier transform spectrometry

Fourier transform methods are more efficient than autocorrelation spectrometry simply because the FFT (Fast Fourier Transform algorithm) is an efficient way to compute the DFT (Discrete Fourier Transform). Processing a sequence of N data points requires on the order of $N \log(N)$ operations for an FFT vs. N^2 operations for either autocorrelation or a brute force DFT.

27.3.1 DFT spectrometer

The Fourier transform is, by itself, a spectrometer. The DFT of a sequence of voltages, V_n, is given by

$$u_k = \sum_{n=0}^{N-1} V_n e^{-j2\pi nk/N}.$$
(27.16)

We can write this DFT in terms of time and frequency as

$$u_{\omega_k} = \sum_{n=0}^{N-1} V_{t_n} e^{-j\omega_k t_n},$$
(27.17)

where $t_n = n/f_{\text{sample}}$ and $\omega_k = 2\pi k f_{\text{sample}}/N$. Note that the transform is equivalent to a bank of N pairs of frequency converters with sine and cosine L.O.s. The k-th pair multiplies the input signal by $\cos(\omega_k t)$ and $\sin(\omega_k t)$, thus shifting the input band to place ω_k at zero frequency. For every converter, N consecutive multiplier output values are summed so one output value is produced every N/f_{sample} seconds. This summation is equivalent to lowpass filtering and decimation (reducing the sample rate). Successive values of u_k are, therefore, equivalent to (complex) voltages from the signal at and around ω_k, after being shifted to zero frequency and then lowpass filtered. When the input voltage is a random

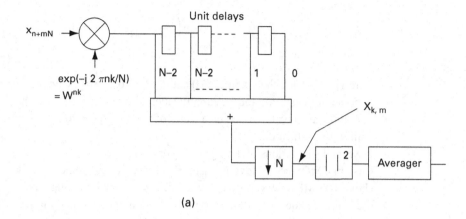

(a)

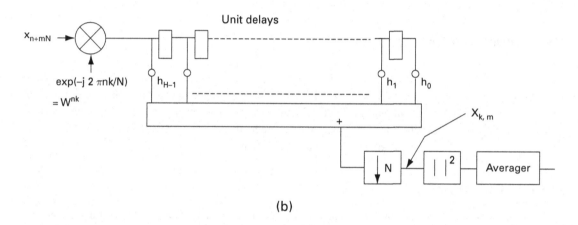

(b)

Figure 27.4. (a) One channel of
a straight DFT spectrometer
(periodogram averager);
(b) filter channel using a longer
sum and with arbitrary weights
on the inputs to the summer.

process, u_k is also a random process. Its mean is zero. The squared magnitude of
u_k represents power, and the set of points, $\{|u_k|^2\}$, is known as a *periodogram*.
To estimate the power spectrum of a random process, many periodograms must
be averaged together. Figure 27.4(a) shows one channel of a straight periodogram spectrometer.

A shortcoming of the DFT spectrometer (periodogram averager) is that the
equivalent filter shape of each channel is not rectangular but, instead, has a
$\sin(x)/x$ form, due to the uniform boxcar lowpass filter weights inherent in the
DFT. It therefore suffers "leakage" from adjacent and even not-so-adjacent
channels. We saw the same problem with the autocorrelation spectrometer. There
are obvious remedies for this problem. The application of a weighting function
to the input data blocks to give relatively low weight to the data points near
the beginning and end of the blocks is equivalent to tapering the weights of the
low-pass filter and therefore reduces the amplitude of the filter sidelobes. Such
weighting, however, broadens the main lobe and increases the variance of the
spectral estimates, since the spectrum is estimated mostly from the data points
around the center of the data blocks. This loss of sensitivity can be remedied by

computing periodograms from data blocks that overlap each other and including these periodograms in the average. (Even with uniform data weighting, there is a loss of sensitivity when periodograms are computed only from nonoverlapping blocks of data.)

27.3.2 Combined FIR filter/DFT spectrometer

An "FIR/Fourier transform" spectrometer design that fixes the problems of the DFT spectrometer begins with the filter channel shown in Figure 27.4(b). Like the DFT filter, this filter channel will produce one output value for every N input samples. But now, the mixer is followed by a lowpass FIR (finite impulse response) filter which forms a running weighted sum of H successive mixer output values by summing the numbers, weighted by factors, h_i, at the taps of a digital delay line.[3] This would be the same as the DFT filter of the periodogram if all the weights, h_i, were equal to unity and the number of taps, H, were equal to N. Here, however, the number of taps will be an integral multiple of N, $H = RN$. This can be quite a large number, e.g., $H = 4 \times 1024 = 4096$, and the FIR filter can therefore produce a nearly ideal rectangular channel response. As with the straight DFT spectrometer, the bandwidth reduction from lowpass filtering lets us decimate the output of the summer, keeping only one sum per N-point block of input data. Again, the magnitudes of the output of the summer must be squared and averaged to form an estimate of the power spectrum. Note that, because the filter window, H, is longer than the block length, N, the blocks processed by the filter are overlapped; while one output value is produced for every N input samples, the output values are produced from H input samples, where $H > N$. We will designate the input voltage samples as x_n. They enter in blocks of N samples. The output of the summer is $X_{k,m}$, where m is the block number and k, which runs from zero to N, designates the point on the spectrum. Defining W as $W = \exp(-2\pi \, \mathrm{j}\,/N)$, inspection of Figure 27.4 allows us to write

$$X_{k,m} = \sum_{n=0}^{H-1} h_n W^{-k(mN+n)} x_{mN+n}. \tag{27.18}$$

Now, since $W^{-kmM} = 1$, we can rewrite this equation as

$$X_{k,m} = \sum_{n=0}^{H-1} h_n W^{-kn} x_{mN+n}. \tag{27.19}$$

Let us next express n as $rN + \rho$ where the index r goes from 0 to $R-1$, where $R = H/M$, and the index ρ goes from 0 to $M-1$. Using this notation, we have

---

[3] The finite (duration) impulse response is obvious; when an input sample makes its way down the line and falls off at the end, it no longer contributes the output. A FIR filter can be given a desired frequency response function by selecting the tap weights to form the desired impulse function, i.e., the Fourier transform of the desired frequency response.

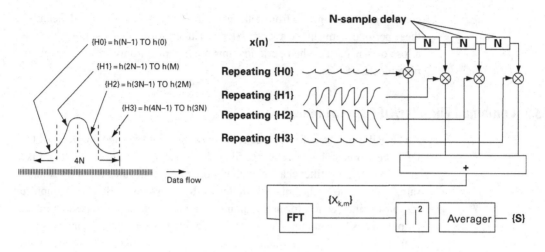

Figure 27.5. Weighted overlap-add filterbank.

$$X_{k,m} = \sum_{\rho=0}^{N-1} \left(\sum_{r=0}^{H/N-1} h_{rN+\rho} x_{mN+n} \right) W^{-k\rho}. \tag{27.20}$$

Note that Equation (27.20) is just an N-point DFT of the term in brackets, and that this term is independent of k, the frequency index. Consider this term for the case where $H = 4N$. The bracketed term is given by

$$(\ ) = h_\rho X_{Nm+\rho} + h_{N+\rho} X_{N(m+1)+\rho} + h_{2N+\rho} X_{N(m+2)+\rho} + h_{3N+\rho} X_{N(m+3)+\rho}. \tag{27.21}$$

In the first term on the right-hand side, successive samples are multiplied by h_0 through h_{N-1}. In the second term, delayed samples are multiplied by h_N through h_{2n-1}, and so on. An architecture that forms this bracketed term is shown in Figure 27.5. Four multipliers are used, together with three delay lines, each of length N. The set of H FIR coefficients is split into four sections, each of which feeds one of the four multipliers in a continuous cycle. An adder sums the products flowing from the multipliers. This preprocessor section feeds a standard FFT block. This architecture is called a *weighted overlap-add* (WOLA) filterbank. Another way of producing the bracketed term yields the *polyphase filter bank* [3].

27.4 *I* and *Q* mixing

In the discussions above, it was implicit that the spectrum of the band-limited signal voltage, $V(t)$, extends from zero to some cutoff frequency, B. This baseband signal can be produced from an IF signal that has been bandpass filtered and then down-converted to put the lower edge of the passband at zero frequency. For digital spectrum analysis, the sampling frequency must be at least $2B$ to avoid aliasing (the stroboscope effect). In the case of the DFT spectrometer and the improved WOLA version, the N channels extend in frequency from

0 to the sampling rate, i.e., from 0 to $2B$. Therefore, only the first $N/2$ output channels can be used, since those channels correspond to the frequency range 0 to B. (The second $N/2$ channels will contain the same information; channel $N-k$ will have the same data as channel k.) This effective waste of half the DFT channels can be avoided with a technique known as *complex sampling*. In this technique, the IF signal of bandwidth B is down-converted into baseband "I" and "Q" (in-phase and quadrature) channels by using two mixers having L.O. signals $\cos(\omega_c t)$ and $\sin(\omega_c t)$, where ω_c is the center of the IF band. We can express the original IF signal in terms of the mixer outputs, $I(t)$ and $Q(t)$, as

$$V_{\mathrm{IF}}(t) = I(t)\cos(\omega_c t) + Q(t)\sin(\omega_c t). \qquad (27.22)$$

To find the spectral value at the point $\omega_c + \Delta$, we must multiply (mix) the IF voltage by $e^{-j(\omega_c + \Delta)t}$. Carrying out this multiplication and ignoring terms of frequency $2\omega_c$, we find

$$V_{\mathrm{IF}}(t)e^{-j(\omega_c+\Delta)t} = [I(t) - jQ(t)]e^{-j\Delta t}. \qquad (27.23)$$

The term on the right-hand side is equivalent to mixing a complex-valued signal, $I(t) - jQ(t)$, so that the spectral point at $\omega_c + \rho$ is shifted to zero frequency. The beauty of this is that both $I(t)$ and $Q(t)$ have bandwidth $B/2$, since $B/2$ is the greatest separation between ω_c and any point in the IF band. Therefore we can feed $I(t) - jQ(t)$ as a sequence of complex numbers to the straight DFT spectrometer or the improved WOLA version and the N output values will span the entire bandwidth B. The spectrum from $\Delta = 0$ to $B/2$ will appear in the first $N/2$ channels, while the spectrum from $\Delta = -B/2$ to 0 will appear in the second $N/2$ channels. When I and Q mixing is used in autocorrelation spectrometry, both autocorrelation functions and cross-correlation functions are computed from the I and Q signals.

27.5 Acousto-optical spectrometry

The acousto-optical spectrometer (AOS) uses an optical crystal which becomes a diffraction grating by virtue of mechanical waves (sound waves at radio frequencies) propagating through it. These waves produce corrugations in the refractive index along the length of the crystal. The waves are launched into this *Bragg cell* by an electrical-to-acoustic transducer (usually a piezoelectric crystal), driven by the IF signal to be analyzed. Figure 27.6 shows how the crystal is illuminated by a laser and how a multi-element CCD array detects (and averages) the diffraction pattern.

The linear CCD array, like those used in supermarket checkout scanners, accumulates charge along its length at a rate given by the incident light intensity. For read-out, the charge packets are clocked down the length of the CCD in a bucket brigade fashion. The voltage at the end of the array, which is proportional to the charge at that position, is digitized and made available to the data-taking computer.

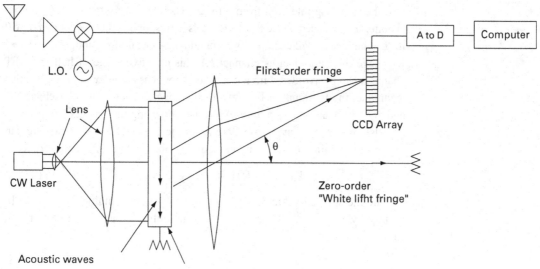

Figure within shows labels: A to D, Computer, Flirst-order fringe, CCD Array, L.O., Lens, θ, Zero-order "White lifht fringe", CW Laser, Acoustic waves, Bragg cell = "Programmable diffraction grating"

Figure 27.6. Acousto-optical spectrometer (AOS).

A figure of merit for any spectrometer is its number of frequency bins, i.e., its analysis range divided by its resolution. For the AOS, the effective number of channels is given by the time–bandwidth product of the Bragg cell. To see this we note that the angular size (in radians) of the light incident on the CCD is given by $\lambda_{\text{laser}}/L_{\text{cell}}$, the ratio of the laser wavelength to the length of the cell, just as the beam size for an aperture antenna with diameter D is given by λ/D. The first-order diffraction condition (so that all the rays arrive in phase at the CCD) is given by $\lambda_{\text{grating}}\sin\theta = 1 \times \lambda_{\text{laser}}$ where λ_{grating} is the corrugation wavelength in the cell. Now λ_{grating} can be written as v_{cell}/f where v_{cell} is the sound velocity and f is the input RF frequency. Approximating $\sin\theta$ by θ, we find the total angular range over the total frequency range given by $\Delta\theta/\Delta f = \lambda_{\text{laser}}/v_{\text{cell}}$. The number of channels is then given by the range divided by resolution,

$$N = \frac{\lambda_{\text{laser}}\Delta f L_{\text{cell}}}{v_{\text{cell}}\lambda_{\text{laser}}} = \frac{L_{\text{cell}}}{v_{\text{cell}}}\Delta f = T\Delta f, \qquad (27.24)$$

where T is the propagation time through the length of the cell. The quantity $T\,\Delta f$ is known as the *time–bandwidth product* of the cell. Materials used for Bragg cells include quartz, lithium niobate, and glass. A spectrometer for radio astronomy might use cells with a 200-MHz bandwidth and a time–bandwidth product of one thousand.

27.6 Chirp-z spectrometry

The chirp-z spectrometer is shown in Figure 27.7. Also known as the microscan or compressive receiver, it uses a dispersive filter and a swept L.O. While it appears to be a swept spectrum analyzer, it is actually a multiplex analyzer. The

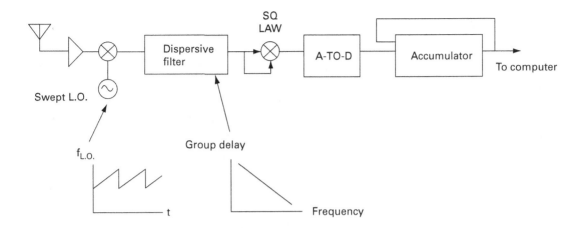

Figure 27.7. Chirp-z spectrometer.

filter has a group delay characteristic that is linear with respect to frequency. In Figure 27.7, the delay is greatest for frequencies at the low end of the filter band. The local oscillator frequency is given a linear sweep or *chirp*, repeated in a sawtooth fashion with period T. The ramp rate of the L.O. (MHz/sec) is made to be the reciprocal of the group delay slope (seconds per megahertz) of the filter.

Consider a continuous wave (cw) input signal. Since the L.O. is swept, the mixer output will also sweep in frequency. In the case of a high-side mixer (L.O. frequency above the input signal frequency), the mixer output sweeps up in frequency at the same rate as the L.O. sweep. As the L.O. sweeps upward, the first signals entering the filter are at the filter's low-frequency end. These signals, as they travel through the filter, are delayed the most. Frequencies entering the filter farther up the sweep are delayed less. But all frequencies arrive simultaneously at the filter output, producing a single sharp output pulse. If the input signal consists of two cw tones, there will be two output pulses per sweep. The position of the pulses in time will indicate the frequencies of the cw input signals. Of course this linear system can handle any combination, i.e., spectrum, of input signals. The output data rate from the squarer is equal to the input data rate so additional circuitry (the digital accumulator in Figure 27.7) is required for signal averaging.

The principal application of chirp-z spectrometers has been in electronic warfare – as surveillance receivers to sense radar or communication signals over a wide spectrum and with a high probability of intercept. To find the effective number of channels in this spectrometer we can first consider the output pulse produced by a single cw input signal. This output pulse is produced by a coherent superposition of frequencies over the entire bandwidth, B, of the filter. The width of this pulse is therefore given by $1/B$. Dividing the total output time T by the pulse width, the effective number of channels is given by TB, the time–bandwidth product, just as with the Bragg cell spectrometer.

Dispersive filters used for chirp-z spectrometry are usually surface acoustic wave devices. Piezoelectric transducers convert the input signal into mechanical surface waves and then back again to an electrical output signal. In one design, diffraction gratings on the surface of the crystal are arranged to make the input-to-output path length longer for low-frequency waves than for higher frequency waves. Surface-wave dispersive filters can have bandwidths of hundreds of MHZ and time–bandwidth products of several hundred. Dispersive filters are also made from charge-coupled devices (CCDs). Generically these filters are allpass networks and the first dispersive filters were cascades of many second-order allpass sections. (The number of sections required is equal to the time–bandwidth product.) Note: the filters are not just allpass networks; they are usually given some amplitude "taper," i.e., the amplitude response is made to roll off at the band edges, in order to eliminate the "sidelobes" of the instrumental function that result from uniform amplitude response. The same is done with the AOS; the transmission of the Bragg cell is reduced near the ends.

Problems

Problem 27.1. For the Gaussian probability density, $f(V) = (2\pi\sigma^2)^{-1/2} \exp(-V^2/2\sigma^2)$, verify that

$$\int_{-\infty}^{\infty} f(V)\mathrm{d}V = 1 \quad \text{and} \quad \int_{-\infty}^{\infty} f(V)V^2\mathrm{d}V = \sigma^2.$$

Problem 27.2. (a) The noise-like signals observed in radio astronomy can be modeled as a comb of delta functions in frequency space – a picket fence of sine waves. Use your computer to generate 50 sine waves spanning the angular frequency range 1.00–1.10. Let these frequencies be somewhat random, e.g., $\omega_n = 2\pi(1+0.002n +0.05 \text{ rnd}(1))$ where rnd(1) is a random variable in the range zero to one. Give each sine wave a random starting phase, $\phi_n = 10$ rnd(1). Plot the sum of the sine waves as a function of time for $t = 0$ to $t = 20$ using a time interval of 0.1. This narrowband spectrum should look quite sinusoidal but with a slowly varying phase and amplitude.

(b) Repeat this exercise but with the angular frequency range extending from 1 to 2, i.e., $\omega_n = 2\pi(1 + 0.05n + 0.05 \text{ rnd}(1))$. This wideband spectrum should look like random noise.

(c) In either case, (a) or (b), the resulting voltage (sum) behaves like a random variable with Gaussian statistics. Plot a histogram of voltage samples to verify this.

Problem 27.3. Suppose a random signal is produced by applying an ideal lowpass filter to white noise. The resulting power spectrum is flat from dc to the filter cutoff frequency ω_c. Show that the normalized autocorrelation function is given by $\rho(\tau) = \sin(\omega_c\tau)/(\omega_c\tau)$.

Problem 27.4. Derive the VanVleck relation, Equation (27.13), that gives the one-bit autocorrelation function in terms of the (normalized) autocorrelation function. Use the bivariate Gaussian probability density function, Equation (27.15), where x_1 and x_2 represent the voltages $V(t)$ and $V(t+\tau)$, and $\sigma_1 = \sigma_2$. Since the product sign(x_1) sign(x_2) is equal to +1 in the first and third quadrants and −1 in the second and fourth quadrants, the expectation of sign(x_1) sign(x_2) is the integral of $f(x_1,x_2)$ over x_1 and x_2, weighted by $+1$ or -1 according to the quadrant. Hint: change to polar coordinates and keep a table of integrals handy.

Problem 27.5.   If the dispersive filter used in the chirp-z spectrometer or pulse compression radar is given an impulse (delta function in time), what sort of output waveform does it produce?

References

[1] Blackman R. B. and Tukey, J. W., *The Measurement of Power Spectra*, New York: Dover, 1959.

[2] Childers, D. B. ed., *Modern Spectrum Analysis*, New York: IEEE Press, 1978.

[3] Crochiere, R. E. and Rabiner, L. R., *Multirate Digital Signal Processing*, Englewood Cliffs: Prentice-Hall, 1983.

[4] Jack, M. A., Grant P. M. and Collins, J. H., The theory, design, and applications of surface acoustic wave Fourier-transform processors. *Proc. IEEE*, vol. 68, pp. 450–468, 1980.

[5] Kesler, S. B. ed. *Modern Spectrum Analysis, II*, New York: IEEE Press, 1986.

[6] Thomas, J. B., *An Introduction to Statistical Communication Theory*, New York: John Wiley, 1969.

[7] Thompson, A. R., Moran, S. M. and Swenson, Jr. G. W., *Interferometry and Synthesis in Radio Astronomy*, Malabar, Florida: Krieger Publishing Company, 1991.

28 S-parameter circuit analysis

The S-parameter (scattering parameter) analysis technique for linear circuits has become the de facto standard in RF engineering, especially for microwave work, where it had its origins in WWII radar development. Physicists-turned-engineers, who were accustomed to analyzing the scattering (collisions) of atomic particles, used that term to describe the reflection and transmission of electromagnetic waves in electrical circuits. The resulting version of circuit theory, in terms of waves rather than currents and voltages, is convenient when working with circuits whose interconnections are transmission lines, especially when those lines are waveguides, where the definitions of voltage and current are somewhat arbitrary. The Smith chart, introduced in 1939, provided a familiar connection to the S-parameter method. In the 1960s, Hewlett Packard introduced the first network analyzers for direct S-parameter measurements.

28.1 S-parameter definitions

Keep in mind that the S parameters are just another set of transfer parameters, like the Y and Z parameters, to describe a linear circuit or circuit element at a given frequency. A two-port network is described equally well by its four complex Y parameters, its four complex Z parameters, or its four complex S parameters. Knowing one set of parameters, we can calculate any other set.

28.1.1 S parameter of a one-port network

A resistor, the simplest circuit element, can be described by a single parameter, its resistance, R, the ratio of voltage to current. It can be equally well described by its conductance, G, the ratio of current to voltage. The conversion between these parameter choices is particularly simple: $G = 1/R$. When a resistor is mounted in a coaxial connector as a dummy load, we regard it as a one-port RF device and often describe it by its reflection coefficient, $\Gamma = (R-Z_0)/(R+Z_0)$, where Z_0, the reference impedance, is often 50 ohms. We saw in Chapter 10 that Γ is the ratio of the reflected wave amplitude to the incident wave amplitude when a wave arrives at the resistor by way of a transmission line of characteristic impedance Z_0. This ratio can be the ratio of either the voltages or currents of the reflected and incident waves, since these ratios are equal. The reflection coefficient is denoted as S_{11}, and is the one and only S parameter for a one-port device. The subscripts refer to a wave leaving port 1 as a result of a wave entering port 1. Using the familiar formula for reflection coefficient, the S parameter of an arbitrary one-port is given by $S_{11} = (Z - Z_0)/(Z + Z_0)$, where Z is the impedance upon which the wave is incident and Z_0 is a reference impedance.

28.1.2 *S* parameters for a general *n*-port network

Let us now define the S parameters of a general n-port device, examples of which include attenuators, filters and amplifiers (two-port devices), two-way power splitters (three-port devices), and directional couplers (four-port devices). For any *properly terminated n*-port device, S_{ij} is the ratio of the (complex) amplitude of the wave exiting port i to the (complex) amplitude of a wave incident on port j. Note that, when $i = j$, the S parameter is a *reflection* coefficient and, when $i \neq j$, the S parameter is a *transmission* coefficient. An n-port device will have n^2 S parameters. Figure 28.1, shows a two-port network reviewing how the input and output currents are written in terms of the amplitudes of forward and reverse waves at each port.

The S parameters for a two-port network are defined as follows:

$$S_{11} = \frac{V_{1\text{rev}}}{V_{1\text{fwd}}}\Bigg|_{V_{2\text{fwd}} = 0} \qquad S_{12} = \frac{V_{1\text{rev}}}{V_{2\text{fwd}}}\Bigg|_{V_{1\text{fwd}} = 0}$$

$$S_{21} = \frac{V_{2\text{rev}}}{V_{1\text{fwd}}}\Bigg|_{V_{2\text{fwd}} = 0} \qquad S_{22} = \frac{V_{2\text{rev}}}{V_{2\text{fwd}}}\Bigg|_{V_{1\text{fwd}} = 0}. \tag{28.1}$$

Figure 28.1. Forward and reverse waves at the ports of a network.

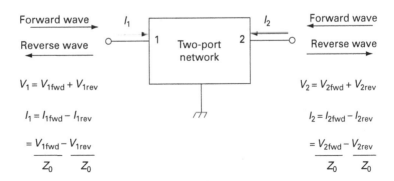

The condition that the device be properly terminated means that when the S-parameters are measured or calculated, each port must be connected to a load impedance equal to the value assigned to that port. Port impedance assignments are usually obvious. If an amplifier is designed to be driven by a 50-ohm source and to feed a 50-ohm load, the input and output ports will both be designated as 50-ohm ports. The one-port dummy load discussed above, if it contains a 50-ohm resistor, would surely be specified by the supplier as having $\Gamma = 0$ (or some very small value). Note, however, that the same 50-ohm device would have a reflection coefficient of $(50-100)/(50+100) = -1/3$, at a port with an assigned impedance of 100 ohms. When port impedances have been assigned and the corresponding S parameters are measured or calculated, the device is totally characterized; if necessary, one can use this set of S parameters to calculate a new set of S parameters, corresponding to different port impedances.

28.1.3 *S* parameters for an example network

Let us calculate the S parameters for the network shown in Figure 28.2, which contains two passive elements Z_1 and Z_2, which can be resistances or reactances or both, i.e., $Z_1 = R_1 + jX_1$ and $Z_2 = R_2 + jX_2$. Interconnection wires are considered to be infinitesimally short. We will assume 50-ohm port assignments.

To calculate S_{11}, we apply the proper termination, 50 ohms, to port 2 as shown in the figure. S_{11} is then the reflection coefficient at port 1, i.e., $S_{11} = (Z_{in} - Z_0)/(Z_{in} + Z_0)$, where Z_{in} is the impedance looking into port 1. By inspection we see that $Z_{in} = Z_1 + Z_2 \,||\, 50 = Z_1 + (50\,Z_2)\,/(\,Z_2 + 50)$. Substituting this expression into the expression for Γ_{in}, we find

$$S_{11} = \frac{Z_1 - 50 + 50Z_2/(Z_2 + 50)}{Z_1 + 50 + 50Z_2/(Z_2 + 50)}. \tag{28.2}$$

Figure 28.2. Example network whose S parameters are calculated.

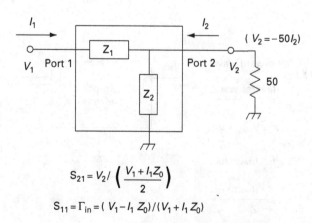

$$S_{21} = V_2 / \left(\frac{V_1 + I_1 Z_0}{2} \right)$$

$$S_{11} = \Gamma_{in} = (\,V_1 - I_1\,Z_0\,)/(V_1 + I_1\,Z_0)$$

We calculate S_{22} with the same method, but now terminating port 1 with 50 ohms. Then S_{22} is the reflection coefficient looking into port 2, and we find

$$S_{22} = \frac{Z_2(Z_1 + 50) - 50(Z_1 + Z_2 + 50)}{Z_2(Z_1 + 50) + 50(Z_1 + Z_2 + 50)}. \qquad (28.3)$$

To calculate S_{21} (out 2, in 1) we put a 50-ohm load on port 2 and apply an arbitrary voltage V_1 to Port 1. We can then calculate the amplitude of the wave leaving port 2. Because port 2 is properly terminated, there will be no wave reflected back into it. Therefore, the amplitude of the voltage appearing at port 2, V_2, is the amplitude of the wave leaving that port. However, since S_{11} is not zero, V_1 is the sum of an incident wave and a reflected wave. Recalling the transmission line theory of Chapter 10, $V_1 = V_{\text{fwd}} + V_{\text{rev}}$ and $I_1 = I_{\text{fwd}} - I_{\text{rev}} = V_{\text{fwd}}/Z_0 - V_{\text{rev}}/Z_0$. Combining these two equations gives us the amplitude of the forward wave: $V_{\text{fwd}} = \frac{1}{2}(V_1 + Z_0 I_1)$. Summing up, a procedure to calculate S_{21} is to apply a voltage V_1 to Port 1, solve for V_2 and I_1, and then calculate $S_{21} = V_2/[\frac{1}{2}(V_1 + Z_0 I_1)]$. Carrying out this procedure, we find, after a few steps of algebra,

$$S_{21} = \frac{2(50 Z_2)}{50 Z_2 + (Z_1 + 50)(Z_2 + 50)}. \qquad (28.4)$$

If we do the same thing from the other side, to calculate S_{12}, we find that $S_{12} = S_{21}$, which, as we will see later, is not a coincidence.

If, instead of directly calculating S_{11} from the reflection coefficient formula, we had applied a voltage V_1 to port 1, as above, we would have used $S_{11} = (V_1 - I_1 Z_0)/(V_1 + I_1 Z_0)$. Note, then, that we can write general expressions for both S_{21} and S_{11} in terms of conditional ratios of linear combinations of voltage and current, without reference to transmission lines or reflections (or the internals of the two-port), just as we define the Y or Z parameters:

$$S_{11} = \left. \frac{V_1 - Z_0 I_1}{V_1 + Z_0 I_1} \right|_{V_2 = -Z_0 I_2} \qquad (28.5)$$

$$S_{21} = \left. \frac{V_2}{0.5(V_1 + Z_0 I_1)} \right|_{V_2 = -Z_0 I_2}. \qquad (28.6)$$

To get the expressions for S_{22} and S_{12}, simply change ones to twos and twos to ones in Equations (28.5) and (28.6).

Before leaving this example, let us look at an alternative way to calculate S_{12} or S_{21}. Rather than connecting an arbitrary voltage generator directly to port 1, we use a 2-V generator and make the connection through a 50-ohm resistor. If the resulting source (the 2-V generator and the 50-ohm series resistor) were connected to a properly terminated 50-ohm cable (or just to another 50 ohm resistor), it would produce a forward wave with an amplitude of 1 V. And if it is connected to a mismatched load, it will still produce a 1-V forward wave. A mismatched load will produce a reflection, but since the source has a 50-ohm

Figure 28.3. Alternate circuit for calculation of S_{21}.

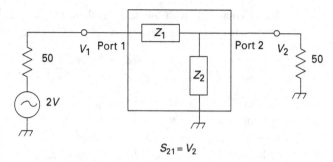

$$S_{21} = V_2$$

impedance, the reflected wave is totally absorbed and the forward wave maintains its 1-V amplitude. Thus, to calculate S_{21} we simply excite the network with this source and calculate V_2, since $S_{21} \equiv V_{2\text{fwd}} / V_{1\text{fwd}} = V_2 / V_{1\text{fwd}} = V_2$. This is shown in Figure 28.3.

28.1.4 Normalization

When the ports of a device are not all assigned the same impedance, it is common to normalize S-parameter magnitudes, i.e., the magnitude of S_{ij} is taken as the square root of the ratio of power emerging from port i to power entering port j when all other ports are terminated in their assigned impedances.

28.2 Circuit analysis using *S* parameters

Writing the relations between the independent variables (ingoing or "forward" waves) and dependent variables (outgoing or "reverse" waves) for a two-port network in matrix-like form, we have

$$V_{1\text{rev}} = S_{11} V_{1\text{fwd}} + S_{12} V_{2\text{fwd}} \tag{28.7}$$

$$V_{2\text{rev}} = S_{21} V_{1\text{fwd}} + S_{22} V_{2\text{fwd}}. \tag{28.8}$$

Let us use these two equations to solve a basic network problem: finding the reflection coefficient at port 1 of a two-port network when port 2 is connected to an arbitrary load, as shown in Figure 28.4.

As we have noted, the load can be specified by its reflection coefficient, Γ_L. That immediately requires that $V_{2\text{fwd}} = \Gamma_L V_{2\text{rev}}$, i.e., a fraction Γ_L of the wave leaving port 2 is reflected back into port 2. We will set $V_{1\text{fwd}}$ to unity so that the reflection coefficient at port 1 will be $V_{1\text{rev}}$. Equations (28.7) and (28.8) then become

$$V_{1\text{rev}} = S_{11} \cdot 1 + S_{12} \Gamma_L V_{2\text{rev}} \tag{28.9}$$

$$V_{2\text{rev}} = S_{21} \cdot 1 + S_{22} \Gamma_L V_{2\text{rev}}. \tag{28.10}$$

Figure 28.4. A two-port network with an arbitrary load.

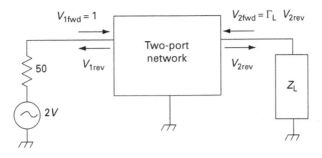

Solving Equation (28.10) for $V_{2\text{rev}}$ and substituting into Equation (28.9), we find

$$\Gamma_{\text{in}} = V_{1\text{rev}} = S_{11} + \frac{S_{12}\Gamma_L S_{21}}{1 - S_{22}\Gamma_L}. \tag{28.11}$$

If we simply turn things around, we see that the reflection coefficient seen at port 2 is obtained by interchanging 1's and 2's and *S*'s and *L*'s in this expression, i.e.,

$$\Gamma_{\text{out}} = V_{2\text{rev}} = S_{22} + \frac{S_{21}\Gamma_S S_{12}}{1 - S_{11}\Gamma_S}. \tag{28.12}$$

28.2.1 Signal flow graphs

A convenient aid to S-parameter circuit analysis is the signal flow graph – a directed graph that represents a set of linear equations. A signal flow graph for the circuit discussed above is shown in Figure 28.5(a).

The nodes (dots) on the graph correspond to the magnitudes of the forward and reverse waves. The branches (lines) are labeled with the transfer coefficients (*S* parameters) that determine the node values. Equations (28.7) and (28.8) are contained in this graph. Consider for example the node at the upper right, $V_{2\text{rev}}$. Only two branches are directed toward this node, one from $V_{1\text{fwd}}$ and one from $V_{2\text{fwd}}$. Summing the contributions that arrive through these branches we write $V_{2\text{rev}} = S_{21}\,V_{1\text{fwd}} + S_{22}\,V_{2\text{fwd}}$, which is just Equation (28.8). On this graph, the load is represented by a branch which indicates that $V_{2\text{fwd}} = \Gamma_L V_{2\text{rev}}$. The source, which could be the 2-V generator and series 50-ohm resistor, produces a forward wave with a magnitude of one, which is represented on the graph by assigning the value unity to the left-most node, $V_{1\text{fwd}}$. Let us now see how we can use the graph to arrive at our previous result for the reflection coefficient, $V_{1\text{rev}}$. Inspecting the graph, we see that if we first find the value of $V_{2\text{rev}}$, we will immediately be able to write

$$V_{1\text{rev}} = V_{2\text{rev}}\Gamma_L S_{12} + 1 \cdot S_{11} \tag{28.13}$$

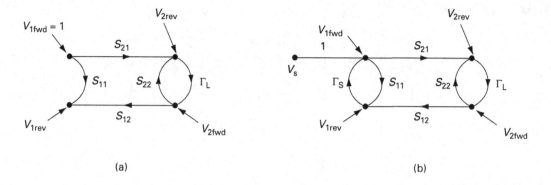

Figure 28.5. (a) Signal flow graph for the circuit of Figure 28.4; (b) same, but for a source whose impedance is not Z_0.

by following the two paths into the node $V_{1\text{rev}}$. Focusing on the node $V_{2\text{rev}}$, see that two branches are directed into it, so we write $V_{2\text{rev}} = 1 \cdot S_{21} + V_{2\text{rev}} \Gamma_L S_{22}$, from which $V_{2\text{rev}} = 1 \cdot S_{21}/(1-\Gamma_L)$. Substituting this into Equation (28.13) produces $V_{1\text{rev}} = S_{11} + \Gamma_L S_{12} S_{21}/(1-\Gamma_L)$, as before.

28.2.2 Two-port transmission and reflection with arbitrary source and load impedances

Let us now look at the same circuit, but with an imperfect source, i.e., a source having a reflection coefficient $\Gamma_S \neq 0$. The signal flow graph, Figure 28.5(b), acquires two extra branches. One is Γ_S, which steers some of the reflection from the two-port around, contributing to $V_{1\text{fwd}}$. The other branch comes from a node, V_S, which is the forward wave the source would produce at a matched input port. Our goal is to solve for $V_{2\text{rev}}$, the wave transmitted to the load. Examining the graph, we write simultaneous equations for $V_{2\text{rev}}$ and $V_{1\text{fwd}}$:

$$V_{2\text{rev}} = V_{1\text{fwd}}S_{21} + V_{2\text{rev}}\Gamma_L S_{22} \tag{28.14}$$

$$V_{1\text{fwd}} = V_S + V_{1\text{fwd}}S_{11}\Gamma_S + V_{2\text{rev}}\Gamma_L S_{12}\Gamma_S. \tag{28.15}$$

Eliminating $V_{1\text{fwd}}$, we have our desired result:

$$V_{2\text{rev}} = \frac{V_S S_{21}}{(1 - \Gamma_L S_{22})(1 - \Gamma_S S_{11}) - S_{12}S_{21}\Gamma_L\Gamma_S}. \tag{28.16}$$

Note: if the identification of the equations for $V_{2\text{rev}}$ and $V_{1\text{fwd}}$ is too great a reach, the reader can readily identify equations for all four variables:

$$V_{2\text{rev}} = V_{1\text{fwd}}S_{21} + V_{2\text{fwd}}S_{22}$$
$$V_{2\text{fwd}} = V_{2\text{rev}}\Gamma_L$$
$$V_{1\text{rev}} = V_{2\text{fwd}}S_{12} + V_{1\text{fwd}}S_{11}$$
$$V_{1\text{fwd}} = V_S + V_{1\text{rev}}\Gamma_S$$

and then eliminate $V_{2\text{fwd}}$ and $V_{1\text{rev}}$ to obtain Equations (28.14) and (28.15).

When the two-port is an amplifier, we often want to find the gain. There are several definitions of gain. The *operating power gain* is just the ratio of the net power delivered to the load to the net power delivered by the source. The *transducer gain* is the ratio of the net power delivered to the load to the power available from the source, and the *available power gain* is the ratio of the power available from the amplifier divided by the power available from the source. Let us calculate the operating power gain. The net power delivered to the load is the power of the wave into the load, $|V_{2rev}|^2$, minus the power of the wave leaving the load, $|\Gamma_L V_{2rev}|^2$, i.e., $P_{load} = |V_{2rev}|^2 (1 - |\Gamma_L|^2)$. The net power delivered by the source is the power of the wave from the source minus the power reflected back into the source, i.e., $P_{source} = |V_{1rev}|^2(1 - |\Gamma_{in}|^2)$, where Γ_{in} is the reflection coefficient already calculated (Equation 28.11). To find V_{1fwd}, we can substitute the expression for V_{2rev} back into Equation (28.13). Stirring this all together, we find

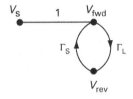

$$G_{operating} = \frac{|S_{21}|^2(1 - |\Gamma_L|^2)}{|1 - \Gamma_L S_{22}|^2(1 - |\Gamma_{in}|^2)}. \qquad (28.17)$$

To calculate the other gain expressions, we must find expressions for the available powers. Figure 28.6 shows a load connected to a generator. Again, this is a generator that would supply V_S volts to a resistive load of Z_0 ohms.

From this signal flow graph we see that $V_{fwd} = V_S + V_{fwd}\Gamma_L\Gamma_S$ or $V_{fwd} = V_S/(1 - \Gamma_L\Gamma_S)$. The net power to the load is $|V_{fwd}|^2/Z_0 - |V_{rev}|^2/Z_0 = |V_{fwd}|^2 (1 - |\Gamma_L|^2)$. We know that maximum power is transferred to the load when $Z_L = Z_S^*$, from which $\Gamma_L = \Gamma_S^*$. Putting this together, the maximum available power from the source becomes $|V_S|^2/[Z_0 (1 - |\Gamma_S|^2)]$. In the same way, the maximum available power from the network is $|V_{2rev}|^2/[Z_0 (1 - |\Gamma_{out}|^2)]$. Finally, we can also find the voltage amplification factor $A_V = V_2/V_1$, since we know that $V_1 = V_{1fwd} - V_{1rev}$ and $V_2 = V_{2fwd} - V_{2rev}$.

Figure 28.6. Signal flow graph for a source driving a load. The source would supply a forward wave of amplitude V_S to a load for which $\Gamma_L = 0$.

28.3 Stability of an active two-port (amplifier)

If the input impedance measured at any port of a device has a negative real part – a negative resistance – the device can supply power. Obviously this can only occur with active devices, such as amplifiers, which are connected to an external source of power. An impedance with a negative resistance value produces a reflection coefficient, Γ, whose magnitude is greater than unity (Problem 28.2). Thus, if $|S_{ii}| > 1$, the device has the potential for oscillation. It will oscillate unless the port is terminated with an impedance whose real part is larger than the negative resistance of the port. To see if an amplifier is potentially unstable, we check whether there are any values of load impedance that will make $|\Gamma_{in}| > 1$ and also whether there are any values of source impedance that will make $|\Gamma_{out}| > 1$. Consider first the load impedances. We use the expression for Γ_{in}

Figure 28.7. Output stability
circle in the Γ_L plane.

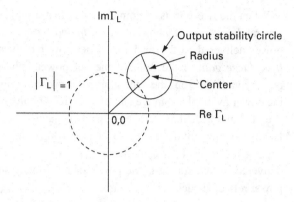

given in Equation (28.11) and find all the values of Γ_L that will make $|\Gamma_{in}| = 1$. These values of Γ_L will form a closed curve, always a circle,[1] that divides the Γ_L plane into two regions. The values of Γ_L in one region result in $|\Gamma_{in}| < 1$, while the values of Γ_L in the other region result in $|\Gamma_{in}| > 1$ (potential instability). You can use Equation (28.11) to show that the circle in the Γ_L plane, called the *output stability circle*, will be centered at the point

$$\Gamma_L(\text{center point}) = \left| \frac{S_{22}^* - \Delta^* S_{21}}{|S_{22}|^2 - |\Delta|^2} \right|, \qquad (28.18)$$

where $\Delta = S_{11}S_{22} - S_{12}S_{21}$, and will have a radius given by

$$\text{rad}_L = \left| \frac{S_{12}S_{21}}{|S_{22}|^2 - |\Delta|^2} \right|. \qquad (28.19)$$

Figure 28.7 shows an example of an output stability circle. Suppose we have found that the "good" values of Γ_L are in the region outside the output stability circle (for example, by verifying that $|\Gamma_{in}| < 1$ for $\Gamma_L = 0$). For this example, the output stability circle partially overlaps the $|\Gamma_L| = 1$ circle. Values of Γ_L in the overlap region are "bad" values; they will give the amplifier a negative input resistance, even though these Γ_L values correspond to passive loads. The "bad" values outside the overlap region all have $|\Gamma_L| > 1$ so they represent loads with negative resistance. These values could be important if the load is not passive, for example, the input of an amplifier which might have $|\Gamma_{in}| > 1$.

---

[1] The locus of points $|\Gamma_{in}| = 1$ is a circle in the Γ_{in} plane. The expression for Γ_{in} (Equation 28.11) can be written in the form $\Gamma_{in}(z) = (a + b\,\Gamma_L)/(c + d\,\Gamma_L)$, where a, b, c, and d are constants. This form, a *bilinear transformation*, maps circles to circles. Thus the values of Γ_L that map to a circle in the Γ_{in} plane must form a circle in the Γ_L plane. In the same way, the constant resistance and constant reactance circles on the Smith chart (Γ plane) are mapped from the Z-plane, where they are segments of infinite-radius circles (lines).

As you can imagine, *input stability circles* are found in the same way. Interchange twos and ones in Equations (28.18) and (28.19) to get the radius and center point for the input stability circle. If both the output and input stability circles lie totally outside the $\Gamma < 1$ circle, an amplifier is unconditionally stable, i.e., no combination of passive terminations will cause it to oscillate.

28.4 Cascaded two-ports

Two-port devices are often connected in series, as in cascades of amplifiers and filters or in amplifiers, as cascades of components. Obviously a cascade of two two-port devices must be equivalent to a single two-port device, and the S parameters of the equivalent device must be functions of the S parameters of the individual devices. In Figure 28.8, a cascade is fed by a reflectionless source providing a wave of amplitude unity. We place a reflectionless load at the far end. Therefore, the wave reflected back into the source will be S_{11}, while the wave incident on the load will be S_{21}.

Because of the matched load and matched source, this circuit is quite simple to analyze. Let x represent the value of the signal at the top center, the forward wave into device B. By inspection of the graph we can immediately write an equation for x: $x = S^A_{21} + x \, S^B_{11} S^A_{22}$ or $x = S^A_{21}/(1 - S^B_{11} S^A_{22})$. Therefore $S_{11} = S^A_{11} + x S^B_{11} S^A_{12}$ or

$$S_{11} = S^A_{11} + S^A_{21} S^B_{11} S^A_{12}/(1 - S^B_{11} S^A_{22}). \qquad (28.20)$$

Note that we had earlier found the reflection coefficient of a two-port terminated by an arbitrary load. In this case the arbitrary load is just S^B_{11}. Again by inspection, S_{21} is just $x S^B_{21}$ or

$$S_{21} = S^B_{21} S^A_{21}/(1 - S^B_{11} S^A_{22}). \qquad (28.21)$$

Interchanging S^A's with S^B's and ones with twos gives us the remaining two S parameters:

$$S_{22} = S^B_{22} + S^B_{12} S^A_{22} S^B_{21}/(1 - S^A_{22} S^B_{11}) \qquad (28.22)$$

$$S_{12} = S^A_{12} S^B_{12}/(1 - S^A_{22} S^B_{11}). \qquad (28.23)$$

Figure 28.8. Circuit for calculation of overall *S* parameters for a cascade of two two-port devices having *S*-parameter sets {S^A} and {S^B}.

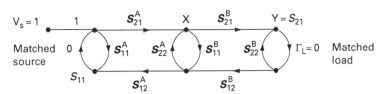

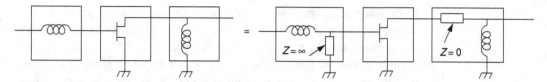

Figure 28.9. An amplifier drawn as a cascade of two-port networks.

In an amplifier, transistors are obvious two-port devices. Shunt and series impedances can also be treated as two ports, as shown in Figure 28.9, which shows that they are just special cases of the network of Figure 28.2.

28.5 Reciprocity

In circuit theory, a common statement of the reciprocity theorem is that if a voltage source V is inserted into any branch of a circuit, branch n, say, and it causes a current of I amps to flow in branch m, then the same current, I, would flow in branch n if the voltage source had been inserted in branch m. This is a conditional theorem; the circuit must contain only passive linear elements (no tubes or transistors or other dependent sources). Moreover, the circuit must not contain nonreciprocal elements, i.e., *circulators* or *isolators* which are microwave components that contain ferrites – magnetic materials in which the permeability is a tensor ($\boldsymbol{H}$ and $\boldsymbol{B}$ are not in the same direction). We used this theorem in Chapter 20 to find the relation between an antenna's receiving performance (effective capture area) and its transmitting performance (directional gain). In the context of S-parameter analysis, the theorem requires that $S_{ij} = S_{ji}$, i.e., the S matrix must be symmetric, even when the circuit itself is not symmetric, as in Figure 28.2. (Reciprocity takes the same form with the Z and Y parameters; $Z_{ij} = Z_{ji}$ and $Y_{ij} = Y_{ji}$.)

28.6 Lossless networks

A lossless network contains no resistive elements. For such networks, conservation of energy requires that the average total power going into the network be equal to the average total power leaving the network. (In S-parameter analysis we are dealing with ac, so the averaging time, strictly speaking, should be an integral number of cycles.) Let us see how the constraint of energy conservation puts constraints on the S parameters.

Suppose we connect a generator of output impedance Z_0 to port i of an N-port network while all the other ports are properly terminated. Let this generator supply a forward wave whose amplitude is unity. No power will flow into the network from the other ports, since they all have nonreflective terminations. Assuming normalized impedances, the total power flowing into the network is therefore 1. The total power flowing out will be the sum of $|S_{ii}|^2$, the power reflected back into the generator plus $\sum |S_{ji}|^2$, summed over the other ports. Putting this into a single sum and equating the powers, we have

$$\sum_{j=1}^{N} |S_{ji}|^2 = 1. \tag{28.24}$$

This says that the sum of the squared magnitudes of any column of the S matrix must be unity, if the network is lossless. And, since $S_{ij} = S_{ji}$, the same is true for any row. Next, consider the situation where power is fed simultaneously into *two* ports. Let us apply a forward wave of amplitude V_i to port i and a forward wave of amplitude V_k to port k. Again, we use reflectionless generators ($Z_{\text{out}} = Z_0$) and terminate the other $N-2$ ports in Z_0. The total input power will be $|V_i|^2 + |V_k|^2$. This must equal the total output power, which is the sum of the squares of the magnitudes of the waves leaving the ports, i.e.,

$$|V_i|^2 + |V_k|^2 = \sum_{j=1}^{N} \left| (V_i S_{ji} + V_k S_{jk}) \right|^2. \tag{28.25}$$

Expanding the right-hand side we find

$$|V_i|^2 + |V_k|^2 = |V_i|^2 \sum_{j=1}^{N} |S_{ji}|^2 + |V_k|^2 \sum_{j=1}^{N} |S_{jk}|^2 + 2\text{Re}\left(V_i V_k^* \sum_{j=1}^{N} S_{ji} S_{jk}^* \right). \tag{28.26}$$

From Equation (28.24), we see that the first two terms on the right are equal to the first two terms on the left, leaving us with

$$0 = 2\text{Re}\left(V_i V_k^* \sum_{j=1}^{N} S_{ji} S_{jk}^* \right). \tag{28.27}$$

Since V_i and V_k are arbitrary, their product could be entirely real or entirely imaginary which requires that, for $i \neq k$,

$$\sum_{j=1}^{N} S_{ji} S_{jk}^* = 0. \tag{28.28}$$

Thus, for a lossless network, the "dot product" of one column of the S matrix times the complex conjugate of any different column, must be zero. The same is true for rows since $S_{ij} = S_{ji}$.

Let us apply Equations (28.24) and (28.28) to a lossless two-port network. A general two-port is described by four complex S parameters, i.e., eight real numbers. If the network is reciprocal, $S_{ij} = S_{ji}$, reducing the eight real numbers to six. If the network is lossless, Equation (28.24) gives us $|S_{21}|^2 + |S_{11}|^2 = 1^2$. Thus the magnitude of the transmitted wave is determined completely by the magnitude of the reflected wave, $|S_{21}|^2 = 1 - |S_{11}|^2$, i.e., we could determine the bandpass shape of a lossless filter by just measuring its input reflection coefficient (with the output correctly terminated). Equation (28.24) also gives us $|S_{12}|^2 + |S_{22}|^2 = 1$. With these two equations, the number of real parameters drops to four. Note that, from these two equations, it also follows that $|S_{22}|^2 = |S_{11}|^2$. Next we apply Equation (28.28), giving us

$$S_{11}S_{12}^* + S_{21}S_{22}^* = 0. \qquad (28.29)$$

This appears to be two equations (real and imaginary parts), but actually amounts to only one equation [1]. To see this we express the S parameters in polar form as $S_{11} = |S_{11}|e^{j\theta_1}$, $S_{22} = |S_{11}|e^{j\theta_2}$, and $S_{21} = (1-|S_{11}|^2)^{1/2} e^{j\phi}$. Putting these expressions, which contain four parameters, into Equation (28.28) will give us $[e^{j(\theta_1-\phi)} + e^{j(\phi-\theta_2)}] = 0$, from which $\theta_2 = 2\phi - \theta_1 + (2n+1)\pi$ where n is an integer. Any odd multiple of π will be equivalent, so we can set $\theta_2 = 2\phi - \theta_1 + \pi$. Thus we see that a lossless two-port network can be completely specified by only *three* real numbers, $|S_{11}|$, θ_1 and ϕ.

If we have a vector reflectometer (a one-port network analyzer) and measure S_{11} and S_{22} for a lossless two-port, we will have values for $|S_{11}|$, θ_1, $|S_{22}|$ and θ_2. Writing ϕ in terms of θ_1 and θ_2 gives $\phi = [\theta_2 + \theta_1 - (2n+1)\pi]/2$. If n is increased by one, ϕ changes by π. Therefore, while measurements of either S_{11} or S_{22} let us calculate the magnitude of S_{12}, measurements of both S_{11} and S_{22} allow us also to calculate the phase of S_{12}, but with a 180° ambiguity. (To see that this must be the case, suppose we attach a half-wave cable to one side of a two-port network. Then S_{11} and S_{22} will remain unchanged, but S_{12} will change by 180°.)

Since it can be described by three real numbers, a lossless two-port can also be modeled (for one frequency) as a network containing three parameters. As an example of such a network, suppose we start with the capacitor, which, if paralleled with a 50-ohm resistor, would produce an impedance Z, such that $|(1-Z)/(1+Z)| = |S_{11}|$. Let this capacitor be a shunt element to ground. To its left, install a 50-ohm cable with the length needed to get the correct phase for S_{11}. To its right, install a 50-ohm cable whose length produces the specified phase for S_{22}. One capacitor value and two lengths make up the three parameters.

Finally, since a lossless two-port must contain only reactors (no resistors), Z or Y-parameter analysis will show that the Z or Y parameters must be completely imaginary. Add in reciprocity, and four imaginary numbers become three imaginary numbers, showing us again that a lossless two-port can be specified by three parameters.

Three-port lossless networks

We saw in Chapter 15 that hybrids are four-port networks that are completely matched, that is, the impedance looking into any port will be equal to Z_0 if the other three ports are terminated in Z_0. Let us show by contradiction that a lossless reciprocal three-port network cannot have this property. Suppose, then, that we have a matched three-port, i.e.,

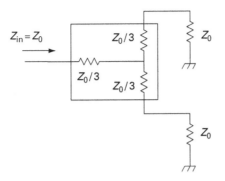

$$S = \begin{bmatrix} 0 & S_{12} & S_{13} \\ S_{12} & 0 & S_{23} \\ S_{13} & S_{23} & 0 \end{bmatrix}, \qquad (28.30)$$

where we have also assumed reciprocity, i.e., $S_{21}=S_{12}$, $S_{23}=S_{32}$, and $S_{13}=S_{31}$.
Applying Equation (28.28) to rows one and two, we find $S_{13}\,S_{23}=0$. One of
these parameters must be zero, say, S_{13}. The S matrix then becomes

$$S = \begin{bmatrix} 0 & S_{12} & 0 \\ S_{12} & 0 & S_{23} \\ 0 & S_{23} & 0 \end{bmatrix}. \qquad (28.31)$$

Finally, applying Equation (28.24) to row one yields $S_{12}=0$ and applying it to
row three yields $S_{23}=0$.

Thus, if a lossless reciprocal three-port network is matched, all of its elements
are zero. Power going in any port is totally absorbed – the opposite of lossless.
Three resistors, each of value $Z_0/3$ in "T" connection, as shown in Figure 28.10,
form a matched but lossy three-port network that is often used as a power
divider.

A circulator is a matched nonreciprocal device; it contains ferrite material.
You can show (Problem 28.4) that any lossless matched three-port must be a
circulator. Power entering port 1 exits from port 2; power entering port 2 exits
from port 3; and power entering port 3 exits from port 1.

Directional couplers

In Chapter 15 we looked at hybrids, which are 3-dB directional couplers. From
Equations (28.24) and (28.28) it can be shown that any lossless matched ($S_{ii}=0$
for all i) reciprocal four-port network must be a directional coupler and that,
conversely, and directional coupler will be matched. Figure 28.11 is a signal
flow graph of the general directional coupler. Opposite ports are isolated from
each other, i.e., $S_{14}=0$ and $S_{23}=0$. The real number c is the coupling coefficient.
If $c=0.1$ for example, the power coupling is $c^2=0.01$ and the device is a 20-dB
directional coupler. For hybrids, $c^2=1/2$.

Figure 28.11. Directional
coupler signal flow graph.

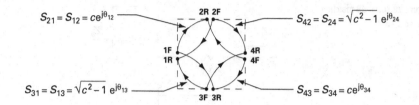

From Equation (28.28) you can show that

$$e^{j(\theta_{12}-\theta_{24})} + e^{j(\theta_{13}-\theta_{34})} = 0, \qquad (28.32)$$

which imposes the condition

$$\theta_{12} - \theta_{24} = \theta_{13} - \theta_{34} \pm \pi. \qquad (28.33)$$

Problems

Problem 28.1.   The net power absorbed by a load is $P = (|V_{\mathrm{fwd}}|^2 / Z_0) - 1/2\,(|V_{\mathrm{rev}}|^2 / Z_0)$, i.e., the net power delivered to the load. Show that this is equal to the power expressed by the current voltage product at the load, $P = \mathrm{Re}(VI^*)$.

Problem 28.2.   Show that if the real part of Z is negative, the reflection coefficient, $(Z - Z_0)/(Z+Z_0)$, will have a magnitude greater than unity.

Problem 28.3.   Derive an expression for S_{11} of a two-port in terms of its Y parameters. Hint: use Equation (28.5), together with the definition of the Y parameters: $I_1 = V_1 Y_{11} + V_2\,Y_{12}$ and $I_2 = V_1 Y_{21} + V_2 Y_{22}$. Remember that port 2 is terminated in Z_0.

Problem 28.4.   Three 50-ohm transmission lines are connected as shown in the figure to form a three-port network. The electrical lengths of these lines are θ_1, θ_2, and θ_3. Find the S parameters of this network. Hint: first find the S parameters when lengths are zero and then determine the appropriate phase factors.

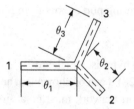

Problem 28.5.   Draw a circuit for a three-port network whose S parameters are all zero.

Problem 28.6.   Show that a matched three-port device must be a circulator. Hint: write down the S matrix with zeros on the diagonal (the matching condition) but do not assume $S_{ij} = S_{ji}$ for the off-diagonal elements. Use Equations (28.24) and (28.28).

Problem 28.7. A linear three-port network has ports 2 and 3 terminated with Γ_2 and Γ_3, respectively. Find the reflection coefficient, Γ, looking into port 1. Hint: the signal flow graph is shown below. A forward wave of magnitude 1 is assigned to port 1. Select the two intermediate points, y and z, and find their values by first writing an equation for each one by inspection of the graph. Then, by inspection of the graph, $\Gamma = S_{11} + y\,\Gamma_3\,S_{13} + z\,\Gamma_2\,S_{12}$.

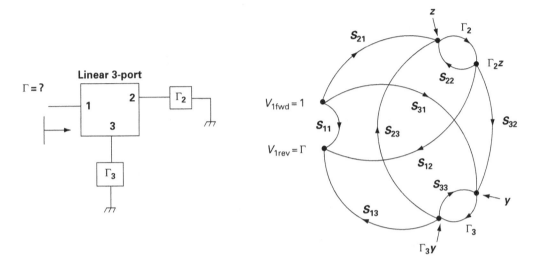

References

[1] Collin, R. E., *Foundations for Microwave Engineering*, New York: McGraw-Hill, 1966.

[2] Gonzalez, G., *Microwave Transistor Amplifiers – Analysis and Design*, Englewood Cliffs: Prentice-Hall, 1984.

[3] Montgomery, C. G., Dicke, R. H. and Purcell, E. M., Principles of Microwave Circuits, Volume 8 of the MIT Radiation Laboratory Series, New York: McGraw Hill 1948. (Reprinted by various publishers.)

[4] Pozar, D. M., *Microwave Engineering*, 2nd edn, New York: Wiley, 1998.

29 Power supplies

Power supplies provide the constant dc voltage(s) used to power amplifiers, oscillators, and logic circuitry. As switching frequencies increase, RF engineers need to understand switching power supply techniques, which are often incorporated in the designs of high-efficiency RF amplifiers, and power supply designers need to learn RF techniques. Conventional power supplies, originally known as "battery eliminators," operate from the ac power line and are usually transformer/rectifier/filter combinations. Dc-to-dc converters are called switching power supplies. Both conventional and switching power supplies are discussed in this chapter.

29.1 Full-wave rectifier

Figure 29.1. Choke-input power supplies: (a) full-wave; (b) full-wave bridge.

Probably the most straightforward circuit combines a full-wave rectifier with a choke-input filter. A "choke" is an inductor which ideally has infinite inductance and no dc resistance. Two such power supplies are shown in Figure 29.1.

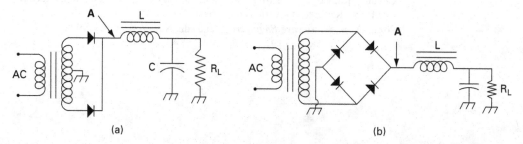

The full-wave rectifier delivers power on both the positive and negative halves of the input ac waveform. The bridge version does not require a center-tapped transformer and needs only half as much copper in the secondary winding, since it uses the entire secondary winding on both halves of the cycle. The two-rectifier circuit, on the other hand, has only half as much rectifier voltage drop and, for that reason, is usually chosen for low-voltage supplies (five volts

Figure 29.2. Voltage waveform at the filter input. The average value of this waveform is the dc voltage at the output of the filter.

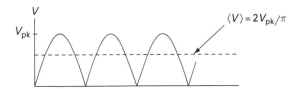

or less) in order to maintain high efficiency. For either circuit, the voltage at point "A" should be as shown in Figure 29.2.

29.1.1 Inherent regulation of the choke-input power supply

Here regulation means voltage stability with respect to changes in the load, R_L. (A separate consideration is regulation with respect to variations in the input ac voltage.) It is important to see that a choke-input power supply provides constant output voltage as long as the current through the choke is never zero. Here is why: If the current through the choke is not zero, then either one rectifier (or rectifier pair) or the other is forward biased and has nearly zero voltage drop. The left side of the choke is therefore always connected to the transformer secondary and the voltage will be as shown in Figure 29.2, complete half-cycle cusps from zero to V_{pk}, the peak secondary voltage. This waveform, in as much as it derives from an almost ideal voltage source (the power line), is practically independent of load current. Its average value is $V_{pk}(2/\pi)$. Since there can be no dc voltage across an (ideal) inductor, the dc component on R_L is the same, V_{pk} $(2/\pi)$, and the regulation is perfect except for the small ohmic losses in the rectifiers, transformer windings and choke, and any leakage inductance in the transformer.

Let us see what is necessary to maintain a nonzero current in the choke. The L-section LC filter suppresses ripple, so the ac voltage at the right-hand side of the inductor is negligible compared to the ac voltage at the left-hand side. At the left-hand side, most of the ac voltage is the 120-Hz component of the rectifier output. Since the peak rectifier voltage is V_{pk}, the peak voltage of the 120-Hz Fourier component is $4V_{pk}/3\pi = 0.42V_{pk}$. The peak ac current through the choke will be $4V_{pk}/(3\pi\omega L)$. To keep the current from turning off the rectifiers and bottoming out, the dc current must be greater than this peak ac current. Therefore we have $2V_{pk}/(\pi R_L) > 4V_{pk}/(3\pi\omega L)$ or $L > 2R_L/3\omega$. This minimum inductance is known as the critical inductance,

$$L_{\text{critical}} = 2R_L/3\omega. \qquad (29.1)$$

In some applications the load may vary such that the current drops to very low values, even zero. A minimum current can be guaranteed by paralleling a "bleeder" resistor across the load. Unfortunately, the resistor will consume

Figure 29.3. Power supply with a resonant filter.

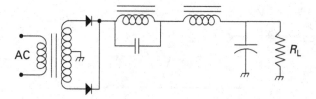

some power. High-voltage power supplies often include a bleeder for safety; it discharges the filter capacitor when the supply is turned off. Another method to maintain constant current is to use a "swinging" choke which has high inductance when needed (i.e., at low current) but presents lower, though still sufficient, inductance at high current, where the core goes into partial saturation.

Ripple

The L and C form a simple L-section lowpass filter. In a full-wave power supply the peak value of the "ripple," the 120-Hz sine wave at the input of the filter (left-hand side of the choke) is $0.42V_{\text{pk}}$. The capacitor will normally have a reactance much smaller than the load resistance, R_L, so we can find the output ripple voltage by considering just the L and C as a voltage divider:

$$V_{\text{ripple(peak)}} = \frac{0.42V_{\text{pk}}X_C}{X_L - X_C}. \tag{29.2}$$

Sometimes a resonant LC circuit is used, as in Figure 29.3, as a bandpass filter to suppress the primary ripple frequency even more.

29.2 Half-wave rectifier

The simplest rectifier circuit, shown in Figure 29.4, uses a single diode. With just a load resistor (no filtering), the output waveform is zero during the negative half-cycles of the 60-Hz line voltage.

Obviously this arrangement does not make good use of the transformer secondary. Moreover, if we add the series L, shunt C filter to get a smooth dc output, the voltage waveform at a point "A" will float upward by an amount depending on the load and there will be no inherent voltage regulation. The solution is to add a "freewheeling" diode, as shown in Figure 29.5.

When the voltage at the top of the secondary becomes negative and "disconnects" the top diode, the freewheeling diode conducts and continues to supply current (from ground) to the inductor. The voltage at the left-hand side of the choke is therefore always the positive secondary voltage or zero volts, just as in Figure 29.4(b), and the circuit can have the inherent regulation of the full-wave choke-input power supply.

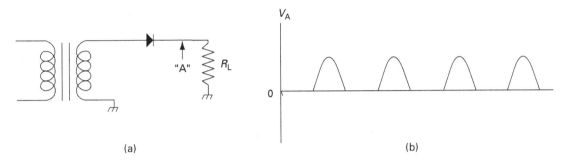

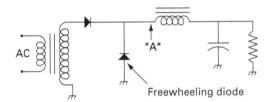

Figure 29.4. Simple half-wave rectifier circuit.

Figure 29.5. A half-wave circuit can also have choke-input regulation if a freewheeling diode is included.

The simplest power supplies use an *RC* filter, substituting a resistor for the inductor. Power is wasted in the resistor and regulation is poor; the average voltage at the rectifier output depends on the load. Of course, for many low-power devices that draw constant current or can tolerate voltage variations, these disadvantages are slight, and this is the kind of circuit you will find in "wall wart" dc power adapters.

29.3 Electronically regulated power supplies

If a basic power supply, such as any of those described above, is followed by a series "pass transistor," a feedback circuit can control the pass transistor's resistance to maintain an extremely constant output voltage. A typical circuit is shown in Figure 29.6.

Note that the pass transistor is used as a series "dropping resistor." An alternate regulator circuit uses a series resistor followed by a parallel transistor as an active bleeder resistor to make a shunt-regulated power supply. These regulator circuits, really just class-A dc amplifiers with negative feedback, operate by dissipating power. Laboratory power supplies use this kind of circuitry to achieve precise regulation and to allow the user to set the desired voltage, as well as a protective current limit. If the load tries to exceed the current limit, the circuitry switches from a voltage regulation mode to a current regulation mode.

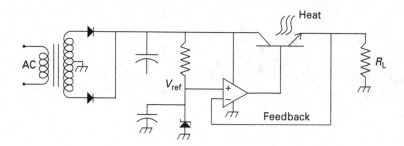

Figure 29.6. Electronically regulated power supply.

29.4 Three-phase rectifiers

Big power supplies, like big motors, benefit from three-phase power which allows more efficient power transmission and lighter transformers. Let us examine a few common circuits. Figure 29.7 shows a three-phase half-wave supply.

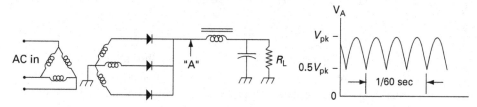

Figure 29.7. Three-phase half-wave power supply.

This is already an improvement over the full-wave single phase supply. The voltage at the input to the choke has a higher ripple frequency (180 rather than 120 Hz) and much less ripple, since the voltage at point "A" is always greater than $V_{pk}/2$. Choke-input filters can be used without the need for freewheeling diode(s). A three-phase full-wave rectifier is shown in Figure 29.8.

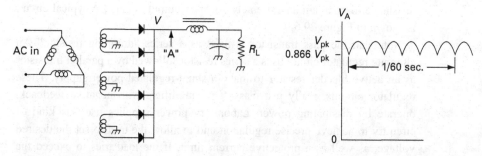

Figure 29.8. Three-phase full-wave power supply.

Figure 29.9. Three-phase full-
wave bridge power supply.

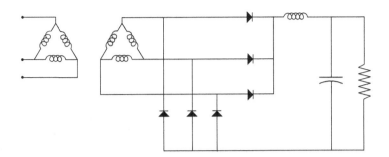

This circuit uses center-tapped secondary windings. The voltage at the input
to the choke now has very low ripple and the fundamental ripple frequency is
360 Hz (which allows even smaller values of L and C). Note that this is really a
six-phase circuit; each rectifier connects to one of the six phases. Note again that
the voltage from a three-phase or six-phase rectifier set never goes to zero.
There is power available from a multiphase ac line at every instant. By comparison, an ordinary single-phase ac line suffers 120 power interruptions every
second. A dc power supply operating off a single-phase line must include at
least one energy storage element (C or L) to ride through these "power failures."
On the other hand, an electronic regulator circuit like the one shown in
Figure 29.6 could be powered by the three-phase rectifier of Figure 29.8 to
make a dc supply having no energy storage elements. Such a supply would
dissipate some of its input power in removing the ripples. When delivering
maximum voltage, $0.866\ V_{\mathrm{pk}}$, the efficiency would be 91%. A three-phase full-
wave bridge circuit is shown in Figure 29.9.

This bridge rectifier has the same relation to the previous circuit as the four-
diode full-wave bridge has to the two-diode ordinary full-wave circuit. Again
there are six equivalent phases. You may recognize this circuit as the one used in
automobile alternators.

29.5 Switching converters

Switching converters are dc-to-dc power converter circuits based on switching
elements, often one transistor and one diode, and an inductor (sometimes the
magnetizing inductance of a transformer). They are often used as local "board-
level" power converters in systems with a single power bus. For example, a bus
voltage of 20 volts dc can be converted wherever needed to power 5-volt logic
circuitry, to furnish various positive and negative voltages for analog circuitry,
and to drive high-voltage display systems. A switching converter can be
preceded by a rectifier and filter capacitor to derive its input power directly
from the ac power line. This combination, a *switching power supply*, is found,
for example, in every line-powered computer. By varying the duty cycle of the

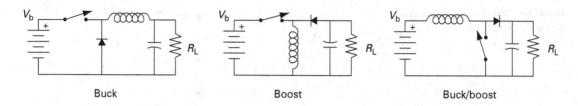

Buck   Boost   Buck/boost

Figure 29.10. Three common dc-to-dc converters.

switching element, the dc output can be precisely regulated. If the duty cycle is deliberately modulated, the switching converter becomes a high-efficiency "class-D" power amplifier used, for example, as an AM modulator for a class-C RF amplifier or as a high-efficiency audio amplifier in a battery-operated MP3 player or hearing aid. The virtues of switching converters are their high efficiency and light weight. Three common converter circuits are shown in Figure 29.10. All include a switch, an inductor, a diode, and a filter capacitor. (The dc input supply is shown as a battery.)

In all these circuits an inductor provides the means by which energy is transferred from the supply side to the load side.

29.5.1 Buck circuit – continuous mode

The buck circuit is the same as a choke-input power supply fed by a half-wave rectifier with a freewheeling diode, except that here the filter sees rectangular voltage pulses rather than the positive halves of a sine wave. When the switch is open, the diode provides the left side of the inductor with a current path from ground. If the choke inductance is large enough to maintain nonzero positive current in the inductor, the average voltage at the input of the choke will be the supply voltage, V_b, times the duty factor of the switch (the fraction of the time the switch is on). The output voltage is equal to this average voltage, so it will remain constant with respect to load changes. (Perfect inherent regulation with respect to load changes requires ideal components, e.g., no dc resistance in the choke.) The duty factor may, of course, be changed to control the output voltage. Feedback control of the duty factor can be used to obtain regulation with respect variations of the input voltage. Feedback control will also provide very precise regulatizon with respect to load changes but, due to the inherent regulation, it will not have to make more than slight adjustments.

The current through the choke ramps up while the switch is closed and ramps down when it is open. Since $V = L dI/dt$, the slope of the current up-ramp is a constant, equal to the supply voltage minus the output voltage divided by the inductance. The slope of the down-ramp is also constant, equal to the output voltage divided by the inductance. This operation is shown in Figure 29.11.

When $V_{out} < V_b/2$ (i.e., the duty factor is < 1/2) the charging slope is steeper than the discharging slope and vice versa.

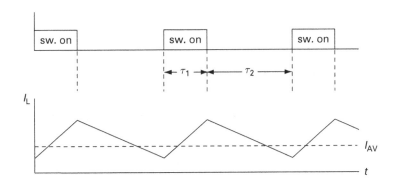

Figure 29.11. Buck supply operation cycle (continuous mode).

Buck circuit – discontinuous mode

If the inductor value is too low, the down-ramp will drain the inductor dry (zero current). The freewheeling diode will stop conducting and the left-hand side of the choke will float. The inherent regulation will be lost and output voltage will now be a function of the load. Figure 29.12 shows the operation cycle in this discontinuous mode. We will assume the capacitor is large enough to allow us to treat the output voltage as a constant.

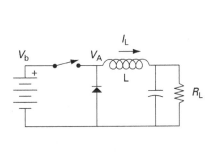

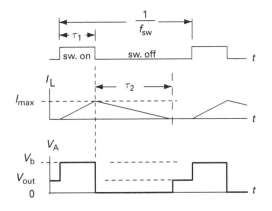

Figure 29.12. Buck supply operation cycle (discontinuous mode).

The length of the up-ramp is τ_1, the *on* time of the switch. In the discontinuous mode, the length of the down-ramp is the time needed for the inductor current to reach zero. To find the output voltage, we first note that the average current through the choke must be equal to the current in the load, V_{out}/R. From Figure 29.12, the average current through the choke is the area under the triangle divided by the length of the cycle, $1/f_{sw}$ where f_{sw} is the switching frequency. The duration of the down-ramp, τ_2, is given by $L\,I_{max}/V_{out}$. Finally, the up-ramp shows that I_{max} is equal to $(V_b - V_{out})\tau_1/L$. Putting all this together yields a quadratic equation for the output voltage, V_{out}.

$$\left(\frac{V_{\text{out}}}{V_{\text{b}}}\right)^2 + \frac{f_{\text{sw}}\tau_1^2 R}{2L}\left(\frac{V_{\text{out}}}{V_{\text{b}}}\right) - \frac{f_{\text{sw}}\tau_1^2 R}{2L} = 0. \tag{29.3}$$

But while this lets us find V_{out} in the discontinuous mode, we would prefer to avoid this mode altogether. To find the necessary critical inductance, we note that the discontinuous mode becomes continuous when there is no space between the triangular current pulses in Figure 12.12. The average current is then just half the height of the triangle, $I_{\text{max}}/2$ or $\frac{1}{2}\tau_1(V_{\text{b}} - V_{\text{out}})/L$. Since this average current has to be equal to V_{out}/R, we have

$$L_{\text{critical}} = \frac{\tau_1 R}{2}\left(\frac{V_{\text{b}}}{V_{\text{out}}} - 1\right). \tag{29.4}$$

29.5.2 Buck/boost circuit – continuous mode

The buck/boost circuit provides negative voltage, whose magnitude can be greater than the supply voltage. In this circuit, energy is pumped into the inductor while the switch is closed and is then transferred to the load when the switch is open. The inductor current vs. time, shown in Figure 29.11, also applies for the buck/boost circuit in the continuous mode. While the switch is closed, the inductor current ramps up linearly. The diode is back-biased during this time so the load side is disconnected. When the switch opens, the inductor current remains continuous, supplied now by the load via the diode. Since current is drawn from the load, the output voltage is negative. For the up-ramp we have $\Delta I = V_{\text{b}}\tau_1/L$. For the down-ramp we have $\Delta I = -V_{\text{out}}\tau_1/L$. The up and down ramps have equal current excursions, ΔI, so we have

$$V_{\text{out}} = -V_{\text{b}}\frac{\tau_1}{\tau_2}. \tag{29.5}$$

Note that, in the continuous mode, the output voltage is also inherently regulated, i.e., independent of the load.

Buck/boost circuit – discontinuous mode

Figure 29.13 shows the voltage and current waveforms for this circuit in the discontinuous mode. The analysis is not much different from that of the buck circuit. The maximum inductor current is given by $I_{\text{max}} = \tau_1 V_{\text{b}}/L$ and also by $I_{\text{max}} = -\tau_2 V_{\text{out}}/L$ so $\tau_2 = \tau_1 V_{\text{b}}/|V_{\text{out}}|$. Again, the load current is given by $|V_{\text{out}}|/R$ and must equal the average current furnished by the inductor. The latter is the area under the τ_2 triangle divided by the cycle length, $1/f_{\text{sw}}$, so we have

$$\frac{|V_{\text{out}}|}{R} = \frac{1}{2}\tau_2 I_{\text{max}} f_{\text{sw}} = \frac{1}{2}\frac{V_{\text{b}}\tau_1}{|V_{\text{out}}|}\frac{V_{\text{b}}\tau_1}{L}f_{\text{sw}}. \tag{29.6}$$

Solving for V_{out} gives

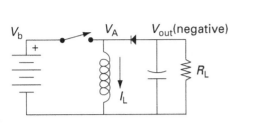

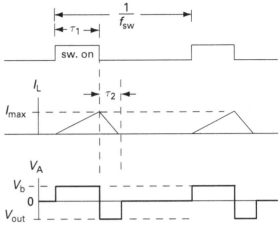

Figure 29.13. Buck/boost operation (discontinuous mode).

$$V_{\text{out}} = V_{\text{b}}\tau_1 \sqrt{\frac{f_{\text{sw}}R}{2L}}. \qquad (29.7)$$

As before, the output voltage in this mode is a function of the load, so inherent regulation is lost. The critical inductance needed to stay out of this discontinuous mode is the subject of Problem 29.6(a).

29.5.3 Boost circuit – continuous mode

The boost circuit, like the buck/boost circuit, uses the inductor for energy storage. For the boost circuit, Figure 20.11 again represents the current in the inductor for the continuous mode. The inductor's charge and discharge become $\Delta I/\tau_1 = V_{\text{b}}/L$ and $\Delta I/\tau_1 = (V_{\text{out}}-V_{\text{b}})/L$. The boosted voltage is therefore

$$V_{\text{out}} = V_{\text{b}}\left(1 + \frac{\tau_1}{\tau_2}\right). \qquad (29.8)$$

Once again the continuous mode provides inherent regulation.

Boost circuit – discontinuous mode

Finally, we have the case of the discontinuous mode for the boost circuit. Figure 29.14 shows I_{L} and V_{A} vs. time. Again the current ramps up from zero when the switch is closed and ramps back down to zero when it is opened. As before, we assume the filter capacitor, C, is large so that V_{out} is constant. In this case the maximum inductor current is given by $I_{\text{max}}= \tau_1 V_{\text{b}}/L$ and also by $I_{\text{max}}= \tau_2(V_{\text{out}}-V_{\text{b}})/L$ so $\tau_2= \tau_1 V_{\text{b}}/(V_{\text{out}}-V_{\text{b}})$. Again, the current into the load is given by V_{out}/R and must equal the average current furnished by the inductor. The latter is the area under the τ_2 triangle divided by the cycle length, $1/f_{\text{sw}}$, so we have

$$\frac{V_{\text{out}}}{R} = \frac{1}{2}\tau_2 I_{\text{max}}f_{\text{sw}} = \frac{1}{2}\frac{V_{\text{b}}\tau_1}{V_{\text{out}} - V_{\text{b}}}\frac{V_{\text{b}}\tau_1}{L}f_{\text{sw}}. \qquad (29.9)$$

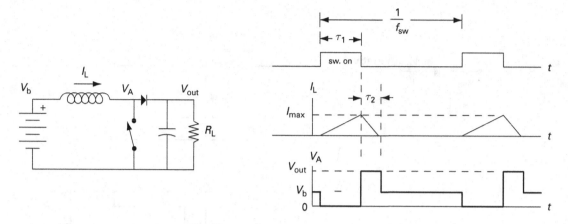

Figure 29.14. Boost supply operation cycle (discontinuous mode).

Rearranging gives us a quadratic equation for V_{out},

$$V_{out}(V_{out} - V_b) = V_b^2 \tau_1^2 \frac{f_{sw}R}{2L}.$$ (29.10)

29.5.4 Ćuk converter

In all three converter designs discussed above, an inductor is used as an energy storage element, to pass power from the input side to the output side. The Ćuk converter circuit, shown in Figure 29.15, uses a capacitor for this purpose.

The input and output inductors filter out ac currents (ripple). Their inductances can be large and, hence, they will store little pulsating energy. Unlike the previous converters, this converter will have negligible ripple currents at both input and output sides. Note also that the capacitor provides dc isolation between the input and output circuits, a useful feature for some applications. Like the buck/boost converter, the output voltage is negative and, for the continuous mode, is given by $V_{out} = -V_{in} \, \tau_1/\tau_2$. This can be seen as follows: Assume that the current in each inductor is essentially just a dc current, i.e., that the inductances are large enough to make any ripple currents very small. When the switch is off (open), current I_1 (from L_1) and current I_2 (from L_2) converge and flow through the diode to ground. During this time, V_A will, therefore, be zero, except for a small voltage drop across the diode. During this *off* period, of duration τ_2, the capacitor will receive an amount of charge $Q_{charge} = (I_1 + I_2)\tau_2$. When the switch is closed, V_A will go negative. The diode therefore stops conducting and I_2 flows into the capacitor. During this *on* period, of duration τ_1, the capacitor will therefore be (dis)charged by an amount $Q_{discharge} = I_2\tau_1$. In steady state, the capacitor must have same charge at the end of every cycle, so we can equate $Q_{discharge}$ and Q_{charge} to find

$$(I_1 + I_2)\tau_2 = I_2\tau_1.$$ (29.11)

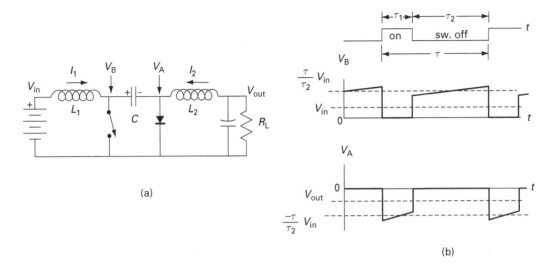

Figure 29.15. Ćuk converter uses a capacitor to transfer energy from input to output.

Next, consider that the input power from the dc supply must equal the power delivered to the load, since there are nominally no lossy components in the circuit. Setting $I_{in}V_{in} = I_{out}V_{out}$, we have

$$I_1 V_{in} = -I_2 V_{out}. \tag{29.12}$$

Combining these two equations and using the fact that $I_2 = -V_{out}/R$, we confirm that

$$V_{out} = -V_{in}(\tau_2/\tau_1). \tag{29.13}$$

The waveforms are shown in Figure 29.15(b), for the continuous mode, which will be the operating mode unless the inductor values are too small. To verify these waveforms, remember that during the *on* period, $V_A = -V_{CAP}$ and during the *off* period, $V_B = V_{CAP}$. Moreover, there can be no average dc across the inductors, so $\langle V_A \rangle \tau_1/\tau = V_{out}$ and $\langle V_B \rangle \tau_2/\tau = V_{in}$, where $\tau = \tau_1 + \tau_2$.

29.5.5 Transformer-coupled converters

When the dc input is derived from a rectifier connected directly to the ac power line, the three basic converters are modified by incorporating an isolation transformer. The transformer provides isolation from the power line, required for safety whenever external connections will be made to the circuitry powered by the converter. Operating at a high frequency, this transformer can be much smaller and lighter than a 60-Hz transformer of equivalent power. Transformer versions of the buck/boost and boost converters are known as *flyback* converters and can provide voltage of either polarity. Flyback converters require no inductor apart from the transformer. The inductance needed for energy storage is provided by the magnetizing inductance of the transformer. Positive and negative output flyback circuits are shown in Figure 29.16.

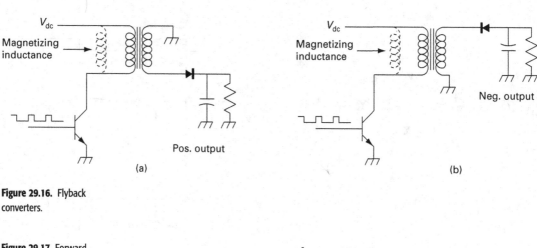

Figure 29.16. Flyback converters.

Figure 29.17 Forward converter.

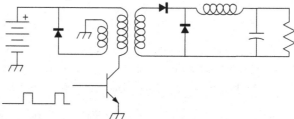

When the switch is closed, the diode disconnects the secondary and a ramp current flows in the primary, storing magnetic energy. When the switch is opened the transformer discharges, just as in the simple buck/boost circuit, except that now the discharge is through the secondary winding. (The magnetic energy does not care which winding carries the discharge current.)

The transformer version of the buck converter is the *forward converter* shown in Figure 29.17. Here the original choke remains; the transformer is an additional component.

While the switch is pulsed on, the transformer secondary supplies power to the load. Here the transformer is used as a real transformer rather than as an energy storage device. The turns ratio of the transformer, as well as the duty factor of the switch and the primary supply voltage, determine the output voltage. There is a problem, however, with opening the switch since there will be current flowing in the magnetizing inductance. (Remember that the magnetizing inductance is an equivalent shunt inductance on the primary side equal to the inductance of the primary winding or, just as well, a shunt inductance on the secondary side equal to the inductance of the secondary winding.) Even if the transformer has a high magnetizing inductance, there will be some stored energy when the switch is opened. A third winding can be provided, as shown above, to discharge this energy back into the primary supply, so that it is not wasted. This winding is sometimes called a *reset winding* since it resets the transformer to zero energy, readying it for the next pulse. In the forward converter, the transformer secondary delivers a stiff voltage pulse, i.e., a pulse from an equivalent low-impedance voltage generator. This circuit therefore needs a choke and a freewheeling diode to provide inherent voltage regulation.

Problems

Problem 29.1. The half-wave power supply shown below supplies 25 volts dc at 0.2 amps to the load (shown as a resistor). Assume the components are ideal and that the capacitor is large enough so that the ac voltage at the output (ripple) is very small.

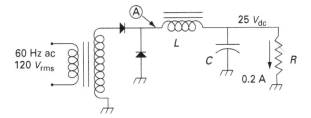

(a) Assuming the inductance of the choke is large enough to maintain positive current, sketch the voltage waveform at point "A."
(b) The value of the average voltage at "A" must be 25 volts since there is no dc drop across the choke. Find the rms voltage, $V_0/\sqrt{2}$, across the transformer secondary. Answer: 55.5 V rms.
(c) Calculate the peak value of the dominant ac component (the 60-Hz component) of the voltage at "A," i.e., find its Fourier amplitude. If you have trouble with this, estimate the magnitude of this component from your sketch. Answer: 39.3 V peak.
(d) Use the result from (c) to find the critical inductance of the choke, i.e., the inductance just large enough to keep the current in the choke always positive. Answer: 0.52 H.
(e) What value should the capacitor have to make the 60-Hz ripple at the output be less than 100 mV rms? Answer: 3700 μF.

Problem 29.2. Consider the common half-wave capacitor-input power supply shown below.

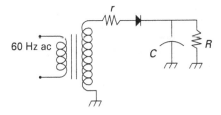

The resistors r and R represent, respectively, the ohmic resistance of the diode and the load resistance. Assume that the transformer is perfect and that C is large enough to make any ac voltage at the output negligible. Suppose that the peak voltage on the transformer secondary is 10 volts, and that the values of r and R are 1 ohm and 20 ohms. Calculate the output voltage, V_{dc}. Hint: the average current through the diode is equal to the current in the load, $V_{dc}/20$. Current flows in the diode only when the voltage at the top of the transformer secondary is greater than V_{dc}; the instantaneous value of the current is $(10 \cos \theta - V_{dc})/1$. Integrate this current over one cycle to find its average and set this expression equal to $V_{dc}/20$. You willl need a calculator to solve for the angle at which the current stops flowing. Answer: 7.51 V.

Problem 29.3. A not-so-common choke-input half-wave power supply is shown below. This circuit, without the free-wheeling diode, will not provide the inherent regulation of the circuit of Problem 19.1. If the load is constant, however, regulation is not important and with this circuit the voltage can be lowered (with no power loss) by increasing the inductance of the choke. Assume as before that the value of the capacitor is large enough to make any ac voltage at the output negligible.

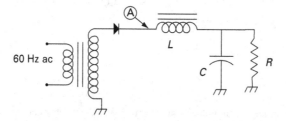

(a) Draw a sketch showing the voltage at point "A" and the current in the inductor. Hint: remember that $V = LdI/dt$. It may help you to consider the cases where R is very large and where R is very small.

(b) If the 60-Hz voltage at the transformer secondary is $10 \sin(\omega t)$ and the output voltage is 5 volts, find duty cycle of the diode, i.e., the angle at which the diode goes into conduction (where the diode becomes forward biased) and the angle at which it goes out of conduction (where the choke goes dry, i.e., choke current goes to zero).

(c) If the output current is 1 ampere, find the value of L.

Problem 29.4. Show that when an uncharged capacitor is brought to potential V by connecting it through a resistor to a battery of voltage V, the energy supplied by battery is twice the energy deposited in the capacitor (CV^2 versus $CV^2/2$), i.e., the transfer is only 50% efficient even if the value of the resistor is made very small.

Problem 29.5. Draw the voltage waveform at point "A" for continuous mode operation of the buck, buck/boost, and boost converters (see Figures 29.12, 29.13, and 29.14).

Problem 29.6. Find the critical inductance needed to stay out of the discontinuous mode (a) for the buck/boost converter of Figure 29.13, and (b) for the boost converter of Figure 29.14.

Problem 29.7. The diagram below shows a circuit for a class-S amplifier (switching amplifier). This could be an audio amplifier, in which case the load resistor represents a loudspeaker. The transistors operate as switches, i.e., fully on or fully off. The voltage on the left-hand side of the inductor is always either zero or V_{cc}. The average value of this voltage appears at the right-hand side of the inductor and also on the load (except that C_2 blocks dc).

(a) When the signal is zero, the duty cycle must be 50%. For this case, draw the waveforms of the currents in the inductor, Q_1, D_1, Q_2, and D_2.

(b) For this zero signal case, explain why there is no net energy flow from the power supply.

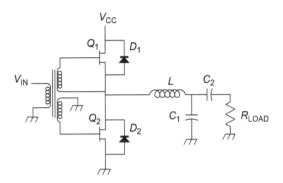

Problem 29.8. The circuit shown below, a blocking oscillator, is essentially a self-excited switching converter. The transformer has three windings. Assume they are perfectly coupled and have equal inductance, L. The current from the dc supply is a positive ramp during the first half of the cycle while the transistor is on. If the magnetic core does not saturate, this ramp ends when the transistor goes out of saturation, i.e., when $I_c = \beta I_b$. The transistor quickly turns off and the stored energy in the transformer is returned to the power supply via the diode.

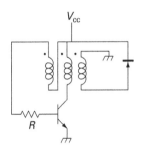

(a) Show that the frequency is given by $f = R/(4\beta L)$. (Assume the base-to-emitter junction is a short circuit.) Sketch the waveforms of the collector voltage and of the collector and diode currents.

(b) Show that, if the magnetic core saturates before the transistor goes out of saturation, this circuit is a VCO or voltage-to-frequency converter.

References

[1] Kassakian, J. G., Schlect, M. F., and Verghese, G. C. *Principles of Power Electronics*, Addison Wesley, 1991.

[2] Mohan, N., Undeland, T. M. and Robbins, W. P. *Power Electronics*, 2nd edn, New York: John Wiley, 1995.

30 RF test equipment

In RF work, as in low-frequency work, test instruments generally are designed either to analyze signals, or to determine parameters of components and systems. Signal analysis ranges from simple voltage, current, and power measurements to measurements of frequency, relative phase, frequency stability, modulation characteristics, RF field strength, and spectrum analysis. Parameter measurements range from simple resistance and reactance measurements through measurements of transfer characteristics of linear multiport devices (filter shapes, amplifier frequency response, etc.), and distortion (dynamic range, intermodulation). Parameter measurements require a signal generator to provide the stimulus whose response is measured. It is now common for measurement instruments to include built-in signal generators, but free-standing signal generators are widely used in specialized test setups.

30.1 Power measurements

Power measurements, as pointed out in Chapter 18, are best made with a square-law detector. If the waveform is unknown or noisy (i.e., has a random component), a square-law detector *must* be used, as power, by definition, is proportional to the average square of voltage. If the signal waveform is known, e.g., a sine wave, the power can also be calculated, for example, from a measurement of $\langle |V(t)| \rangle$. Commercial RF power meters normally present a 50-ohm impedance to source being measured, allowing a reflectionless connection to a 50-ohm cable.

Thermistor (temperature-dependent resistor) power meters are true square-law instruments and have a dynamic range typically from $1\,\mu\text{W}$ ($-30\,\text{dBm}$) to $100\,\text{mW}$ ($20\,\text{dBm}$). The power to be measured heats a thermistor which is one leg of a Wheatstone bridge circuit.[1] A feedback circuit supplies dc or

---

[1] With one end of the thermistor at ground, the RF signal is coupled to the other end through a dc blocking capacitor. A series RF choke isolates the input signal from the rest of the bridge circuit.

low-frequency ac power to heat the a second *reference* thermistor in an adjacent leg in the bridge. When the heater power is equal to the input RF power, the two thermistors come to equal resistance values and the bridge is balanced. A feedback control circuit keeps the bridge balanced. The voltage × current product of the reference thermistor is calculated and this power is displayed. These instruments use small "bead"-type thermistors with very small heat capacities, so that when the power changes, the reading will come to equilibrium within less than one second. More refined circuits provide better compensation for changes in ambient temperature. Similar instruments use thermocouples instead of thermistors.

Powers down to about 10^{-3} μW (-60 dBm) can be measured with carefully calibrated *diode-based power meters*. An upper limit results from the need to keep the signal low enough that the fourth and higher even power terms in the diode I vs. V curve will be negligible (see Chapter 18). A typical upper limit is 100 mW (20 dBm). Thus, the nominal dynamic range of a diode-based power meter is three orders of magnitude greater than that of the thermistor-based power meter. However, when used to measure signals with high crest factors (signals such as noise or multiple carriers, whose peak powers greatly exceed their average power), the upper limit of the diode meters must be derated in order to preserve measurement accuracy.

Power meters measure the power a source would deliver to a standard impedance, usually 50+j0. Directional power meters measure the power traveling in each direction along a transmission line. This is best done with two power meters connected to a directional coupler with a low coupling factor. Negligible power is diverted from the main line and, in principle, the device introduces no reflections. Directional power meters are often installed permanently in the transmission line connecting an antenna to a transmitter as they show both that the transmitter is putting out its rated power and that the antenna is operating properly. Many times the reflected power detector is used to turn the transmitter off automatically or reduce power when a disturbance to the antenna has caused abnormal reflected power.

Flow calorimeters are used for high-power measurements and sometimes for low-power mm-wave signals. These contain a liquid-cooled load resistor ("dummy load"). The fluid temperature is measured on both sides of the load to get the temperature difference (a thermocouple pair can be used). The flow rate must also be measured. Note that all the power measurements discussed above are "true rms" measurements; they do not require that the unknown voltage be sinusoidal.

30.2 Voltage measurements

Probing any circuit with a voltmeter always inevitably "loads down" the circuit to some degree, so it is necessary that the impedance of the voltmeter be much

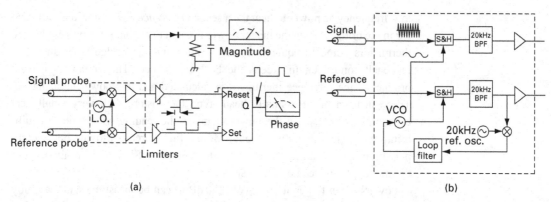

(a)   (b)

Figure 30.1. (a) Superheterodyne vector voltmeter; (b) phase-locked low-frequency oscillator drives sampling mixers.

greater than the equivalent Thévenin source impedance of the node being probed. For low-frequency work, it is sufficient that the meter probe present high shunt resistance. For RF work, the meter must present high shunt reactance as well.

An early innovative instrument was the *vector voltmeter* that displayed phase as well as amplitude. Since phase is relative, the instrument had an auxiliary "reference" input. A simplified circuit is shown in Figure 30.1(a). The signal and reference signal are both converted down to a low IF frequency, 20 kHz, and amplified. By mixing with identical L.O. signals, the relative phases are left unchanged. One channel is fed to a diode peak detector, to measure the amplitude. Both channels are fed to limiting amplifiers. The outputs of the limiters are used to produce a pulse whose duty factor is proportional to the phase difference or to operate a start/stop counter.

Figure 30.1(b) shows how the IF frequency was kept constant over the 1 MHz to 1 GHz range of the instrument. Sample and hold (S&H) circuits were used as mixers: an integral multiple of the sampling frequency differed by 20 kHz from the input frequency, causing the outputs of the S&H units to form a staircase version of a 20 kHz sine wave.[2] A phase lock loop controlled the sampling to keep the IF signal at 20 kHz. This scheme was also used in the receiver sections of vector network analyzers (VNAs), discussed below, instruments which have superseded the vector voltmeter.

30.3 Spectrum analysis

Spectrum analyzers, once called "panoramic receivers", are superheterodyne receivers with swept local oscillators. A nonlinear IF amplifier provides a logarithmic output voltage, i.e., the detector output is proportional to the log of the input signal power. This output is displayed on the Y-axis of a display

---

[2] With the sampling frequency low compared to the input frequency, the output of the sampler is an "aliased" version of the input, like the stroboscopic effect in movies or television that makes fast turning wheels slow down or appear to rotate backward.

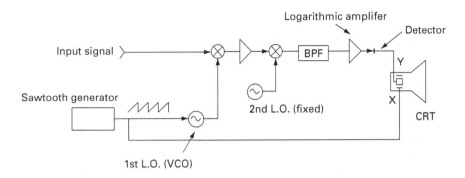

Figure 30.2. Spectrum analyzer.

while the *X*-axis is swept in sync with the local oscillator. The display is therefore a graph of signal strength vs. frequency. RF spectrum analyzers usually have a wide range such as 10 MHz to 20 GHz. A block diagram for such an instrument is shown in Figure 30.2.

A swept-frequency spectrum analyzer using a CRT has objectionable flicker in situations where the sweep time must be greater than about 1/20 second. The sweeping filter must spend a time at least equal to the reciprocal of its bandwidth on each frequency to get the desired resolution. The total sweep time must therefore be greater than the effective number of points divided by the resolution bandwidth which is the same as the frequency span divided by the square of the resolution bandwidth. For example, if it is necessary to analyze a span of 20 kHz with 200 effective points, the resolution bandwidth would be 20 000/ 200 = 100 Hz and the sweep time would have to be two seconds. Digital storage is a solution to the flicker problem when the unknown spectrum is stable (not noise-like) and when the total sweep time (time to acquire the spectrum) is not objectionable.

Multiplex spectrum analyzers, as opposed to swept-frequency spectrum analyzers, evaluate all the spectral points simultaneously. The probability of intercept is 100%, i.e., a signal that pops up momentarily will not be lost because the instrument is off scanning a different part of the band. Moreover, at low frequencies, parallel processors vastly increase the measurement speed. DSP chips calculate fast Fourier transforms in low-frequency spectrum analyzers and can be built into swept spectrum analyzers to extend the low-frequency range. FFT processors are even included as a feature in many oscilloscopes.

30.4 Impedance measurements

The simplest instrument for impedance measurement was the so-called *Q meter*, shown in Figure 30.3, an instrument for testing inductors. It provided the values of both the reactive part and the resistive part of the impedance. The circuit, shown in Figure 30.3, is very simple, and can be set up quickly in the lab in lieu

Figure 30.3. Q meter.

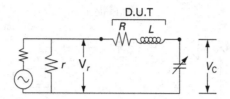

of a suitable instrument. A signal generator provides a fairly constant sine wave voltage V_r across r, a low-value resistor. The variable capacitor is tuned to maximize V_C, the voltage across the capacitor. This maximum occurs when $X_C = X_L$ (series resonance). The operator can read the capacitance dial and, knowing the generator frequency, calculate the unknown inductance. At series resonance, the current through the device-under-test (D.U.T.) is just V_r/r so $V_c = X_C I = (X_C/r)V_r = Q_{DUT} V_r$ and, therefore, $Q = V_C/V_r$.

The operation of the Q meter assumes that the Q of the variable capacitor is very much higher than the Q of the inductor to be tested. This is almost always true in RF work, and for this reason, there was no similar instrument to measure capacitors.

30.4.1 Impedance bridges

At low frequencies (where lumped components are practical) impedance can be measured directly with a straightforward bridge such as shown in Figure 30.4.

Figure 30.4. Impedance bridge.

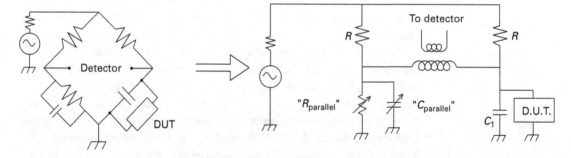

If the device under test (D.U.T.) is purely resistive, the variable capacitor, C, will be set to the same value as the fixed capacitor C_1. It is convenient to use a variable capacitor with a midpoint capacitance equal to C_1. The front panel "$C_{PARALLEL}$" dial is offset so that this point is marked zero. If the load has an inductive component, the bridge can be balanced by setting C to less than C_1. The dial then indicates a negative capacitance which the operator interprets as a parallel inductance. The value of the variable resistor is indicated on its dial as "$R_{PARALLEL}$." This kind of bridge was made famous by the Boonton Electronics Corporation's "RX meter." The RX meter used an amplified detector, actually a superheterodyne circuit.

Amateur radio operators use a version of this bridge, the "noise bridge," in which the signal generator is replaced with a wideband noise generator, usually a Zener diode followed by a wideband amplifier, and the null detector is replaced with a tuned receiver, e.g., a communications receiver.

30.4.2 Reflectometers

A directional coupler, e.g., a hybrid, allows the measurement of the reflection coefficient magnitude for one-port devices and of gain, as well as input and output reflections of two-port devices. In Figure 30.5, any power reflected by the D.U.T. is divided between the generator port, where it is absorbed by the matched generator and the cross-port, where it can be measured.

Figure 30.5. Reflectometer.

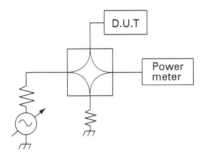

To calibrate the instrument, the D.U.T. is temporarily replaced by a perfect reflector (usually a short) and the power meter gain is adjusted for a reading of unity (0 dB). This kind of circuit, when used with a swept-frequency RF source (sweep generator) and power display, makes a *scalar network analyzer*, as shown in Figure 30.6, where a switch is provided for two-port devices, such as amplifiers, to measure either reflected power or transmitted power.

Figure 30.6. Scalar network analyzer.

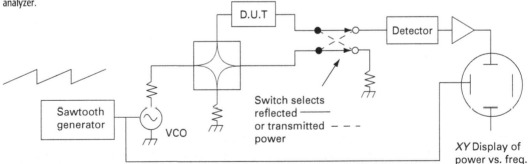

A *vector reflectometer* measures the phase of the reflection coefficient as well as its magnitude. This can be done by combining a vector voltmeter with the above circuit. But it is easy to build a circuit that produces a polar display, i.e., a Smith chart on a CRT screen. Such a circuit is shown in Figure 30.7.

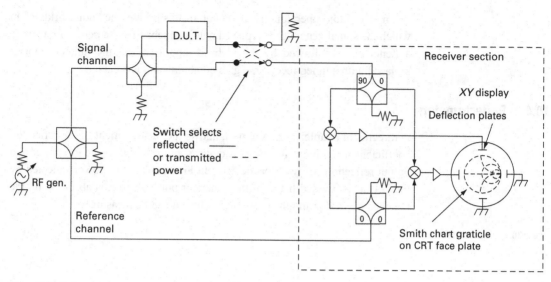

Figure 30.7. Vector reflectometer.

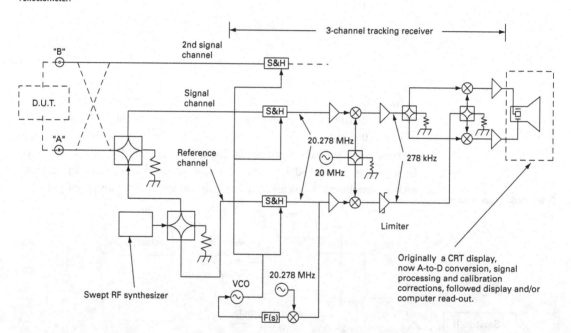

Figure 30.8. Vector network analyzer with a three-channel tracking receiver.

A swept-frequency vector reflectometer is called a *vector network analyzer* (VNA) and is the ultimate RF test instrument. It uses a superheterodyne receiver section. Figure 30.8 shows how one commercial network analyzer uses two RF conversions before the final conversion to baseband. This design uses sample-and-hold mixers with the same phase lock L.O. circuit used in the vector voltmeter (Figure 30.1b). Other VNAs use straightforward fundamental mixing. Only one complete signal channel is shown, but usually two are provided, together with a dual-trace display. Often a reversing switch is included to interchange ports "A" and "B" so that all four S parameters of a two-port device can be measured without having to turn the device around. All newer vector network analyzers digitize the signals after the final conversion and apply calibration corrections before presenting the data on the display, sending it to a computer, or storing it for subsequent retrieval.

30.4.3 VNA calibration

When using a VNA, coaxial cables are normally used to connect the device under test to the analyzer. Reflection and transmission coefficients will therefore be rotated on the display screen, the Γ plane, by an amount proportional to the length of the connecting cables. Moreover, these coefficients will appear too small if the cables have nonnegligible loss. Unless told otherwise, the analyzer knows nothing of these cables, and will assume the D.U.T. is right up against its front panel. Knowing the propagation constant, loss factor, and length of the cables, the operator could calculate the necessary corrections. But modern analyzers have built-in data processors to make these corrections automatically, once the operator has gone through a calibration procedure. At its simplest, this procedure simply lets the analyzer determine electrical length and loss of the cables.

30.4.3.1 One-port calibration procedures

If only reflection coefficients are to be measured, the analyzer could prompt the operator with a message such as "Substitute a short circuit for the device under test. Then press the CAL button and wait for the CALIBRATION COMPLETED prompt." When the operator presses the CAL button, the analyzer steps through the selected set of frequencies and reads the phase and amplitude of the reflected or transmitted signal at each frequency. It stores these numbers and uses them to correct each data point when the operator replaces the short with the D.U.T. Most network analyzers provide this calibration as an option. However, this simple procedure is often not adequate. Consider the situation in which the D.U.T. does not have a 50-ohm port. An adapter must be used to make the transition from the end of the 50-ohm cable to the device port. Examples of adapters are coax-to-waveguide transitions and microprobes, used to probe RF integrated circuit chips. Adapters are seldom perfect; they nearly always introduce a reflection of their own. In addition, an adapter will introduce a phase shift, since its length is not zero. The 50-ohm test cable may not be perfect; it might have worn or otherwise imperfect connectors or it may have been kinked, leaving it with an internal point where some power is reflected. The instrument can correct errors caused by all these defects, as well as by the line loss and delay, if it has first been calibrated with a procedure that uses a sequence of *three* calibration standards rather than just one. Often the three standards are chosen to be a short, an open, and a load. In Figure 30.9, the D.U.T. is a microwave horn antenna and

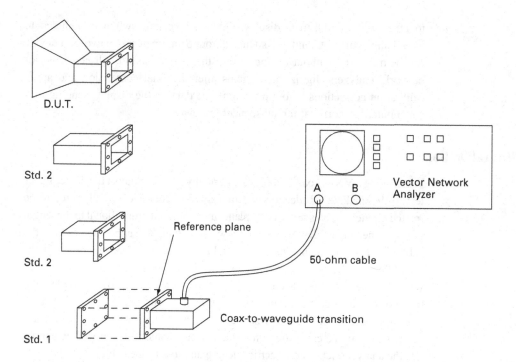

Figure 30.9. VNA setup to measure reflection coefficient vs. frequency of a microwave horn antenna.

the test setup includes a coax-to-waveguide transition. In this example, the three standards comprising the calibration kit are three waveguide shorts of different lengths. Let us look in detail to see why an accurate calibration requires three standards.

Network analyzer calibration for S_{11} measurements using three standards

In Figure 30.10, a signal flow diagram, the two-port network represents everything between the analyzer and the device to be tested, usually a cable plus an adapter. Either a standard or the D.U.T. is put at the right-hand side of this hypothetical two-port network. The S parameters of this network will be designated A_{11}, A_{12}, A_{22}, and A_{21} to distinguish them from the S parameters of the D.U.T., which are S_{11}, S_{12}, S_{22}, and S_{21} or just S_{11}, if the D.U.T. is a one-port device.

For the setup of Figure 30.9, for example, the two-port network is the combination of the cable and the coax-to-waveguide transition. Again, neither the cable nor the transition needs to be perfect, nor do they need to be previously characterized. The analyzer outputs a wave of unit amplitude. The wave Γ_{OBS} is the signal reflected back into the analyzer. The external device, either one of the three standards or the device under test, has reflection coefficient Γ_{EXT}. If the external device is the D.U.T., then Γ_{EXT} will be equal to S_{11} if the D.U.T. is a one-port device or if it is a multiport device with the other ports terminated in matched loads. Let us first find Γ_{OBS} as a function of Γ_{EXT}.

Figure 30.10. Signal flow diagram for reflection measurement setup.

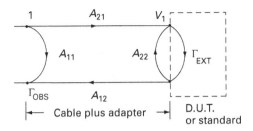

By inspection of the diagram, we see that the forward wave into the external device, V_1, can be expressed as

$$V_1 = A_{21} + V_1 \Gamma_{\text{EXT}} A_{22} \quad \text{or} \quad V_1 = \frac{A_{21}}{1 - \Gamma_{\text{EXT}} A_{22}}. \tag{30.1}$$

It follows that

$$\Gamma_{\text{OBS}} = A_{11} + \frac{\Gamma_{\text{EXT}} A_{21} A_{12}}{1 - \Gamma_{\text{EXT}} A_{22}}. \tag{30.2}$$

Solving for Γ_{EXT}, we find

$$\Gamma_{\text{EXT}} = \frac{\Gamma_{\text{OBS}} - A_{11}}{A_{21} A_{12} - A_{22} A_{11} + \Gamma_{\text{OBS}} A_{22}} \quad \text{or} \quad \Gamma_{\text{EXT}} = \frac{\Gamma_{\text{OBS}} - A}{B + \Gamma_{\text{OBS}} C} \tag{30.3}$$

where A, B and C are three constants ($A = A_{11}$, $B = A_{21} A_{12} - A_{22} A_{11}$, and $C = A_{22}$). Measuring Γ_{OBS} for each of the three known values of Γ_{EXT} provided in the CAL kit provides three simultaneous linear equations that can be solved for the values of A, B, and C. This is done automatically by the analyzer when one performs a full S_{11} calibration. Moreover, it is done at each discrete frequency contained between the start and stop frequencies. Once the calibration procedure has been run, the device under test is attached and $\Gamma_{\text{EXT}} = \Gamma_{\text{DUT}}$ is calculated from Γ_{OBS} using Equation (30.3).

In the discussion above, we assumed the two-port network of Figure 30.9 included just the test cable and any adapter. But this network can represent everything between a hypothetically perfect network analyzer and the D.U.T; the three-standard calibration procedure corrects not only for the cable and adapter, but also for deficiencies of the analyzer itself: imperfect directivity and port reflection. You can see that none of these corrections are possible with the scalar network analyzer of Figure 30.6.

Full calibration for two-port measurements requires that we consider the second cable and any adapter as another two-port network. Both networks must be characterized using three standards. In addition, a "through" standard is required. In a coaxial cable set-up, the "through" might be a coaxial male-to-male "bullet." In the waveguide setup, no actual "through" device is needed; the coax-to-waveguide adapters connected by cables to the analyzer's "A" and "B" ports are simply placed face to face.

30.5 Noise figure meter

The automatic noise figure meter, Figure 30.11, consists of a tuneable receiver, power meter, and calibrated noise source.

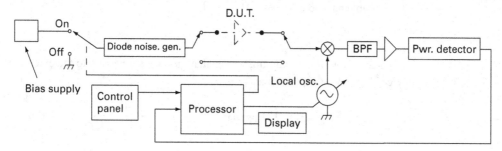

Figure 30.11. Automatic noise figure meter.

In principle, the noise figure of a D.U.T., often an amplifier, can be determined without the use of an external noise source. The noise spectral density at the D.U.T. output can be measured using a power meter, a filter whose bandpass shape is known, and, usually, an auxiliary high-gain amplifier whose gain and noise figure are known. The D.U.T. gain can be measured if a calibrated sine wave source is also available. But when a calibrated noise source is used, the shape of the filter and the gain and noise figure of the auxiliary amplifier need not be known. We define T_{sys} as the equivalent system noise temperature at the input of the D.U.T. when the noise source is turned off and equivalent to a 50-ohm resistor at the ambient temperature, T_0. When turned on, the calibrated noise generator has an equivalent noise temperature T_{hot}. The instrument turns the noise generator on and off and calculates the power ratio $Y = P_{on}/P_{off}$. As explained in Chapter 17, the system temperature is given by $T_{sys} = (T_{hot} - YT_0)/(Y-1)$. If the D.U.T. has high gain, its equivalent noise temperature, T_{DUT}, is essentially equal to T_{sys}. But if the D.U.T. has low gain, correction must be made for the noise in the rest of the system, i.e., the rest of the noise meter. Note that, by operating the switch that bypasses the D.U.T., the noise meter can measure its own noise temperature and then calculate the noise figure of the D.U.T.

Problems

Problem 30.1. In testing a high-power transmitter with a water-cooled dummy load it is found that the water leaving the dummy load is hotter by 10 degrees Celsius than the water entering the load. If the flow rate is 10 liters/second, what is the output power of the transmitter? (1 calorie = 4.18 joules.)

Problem 30.2. Sketch a digital circuit to provide a numerical phase readout for the vector voltmeter of Figure 30.1. Hint: use a start/stop counter in place of the S/R flipflop. Include provision to count multiple turns, i.e., give the readout a range of 360 degrees $\times n$.

Problem 30.3.   Design a three-short vector network analyzer calibration kit for WR430 waveguide (width = 4.30 inches). Assume that the midband frequency is 2.38 GHz. Determine the lengths of the three shorts so that, at midband, their reflection coefficients are as different as they can be, in order to reduce systematic errors.

Problem 30.4.   After building the calibration kit of Problem 30.3, how would you measure the reflection coefficient of the waveguide-to-coax transition?

References

Hewlett Packard Application Note 1287–1 "Understanding the fundamental principles of vector network analyzers."

Hewlett Packard Application Note 1287–2 "Exploring the architecture of vector network analyzers."

Hewlett Packard Application Note 1287–3 "Applying error corrections to vector network analyzer measurements."

Index

Printed in the United States
by Baker & Taylor Publisher Services